MODERN GEOMETRIES

Non-Euclidean, Projective, and Discrete

SECOND EDITION

MICHAEL HENLE

Oberlin College
Oberlin OH

PRENTICE HALL
UPPER SADDLE RIVER,
NEW JERSEY 07458

Library of Congress Cataloging in Publication Data

Henle, Michael.
    Modern geometries: non-Euclidean, projective, and discrete/Michael Henle - 2nd
ed.
    p. cm.
Includes bibliographical references and index.
ISBN 0-13-032313-6
1. Geometry, Modern I. Title.
QA473.H46 2001
516'.04 - dc21   00-065274

Acquisition Editor: **GEORGE LOBELL**
Editor in Chief: **SALLY YAGAN**
Vice President/Director of Production
    and Manufacturing: **DAVID W. RICCARDI**
Executive Managing Editor: **KATHLEEN SCHIAPARELLI**
Senior Managing Editor: **LINDA MIHATOV BEHRENS**
Production Editor: **BARBARA MACK**
Manufacturing Buyer: **ALAN FISCHER**
Manufacturing Manager: **TRUDY PISCIOTTI**
Marketing Manager: **ANGELA BATTLE**
Marketing Assistant: **VINCE JANSEN**
Director of Marketing: **JOHN TWEEDDALE**
Editorial Assistant: **GALE EPPS**
Art Director: **JAYNE CONTE**
Cover Design: **KIWI DESIGN**
Cover Photo: *Marathon Runners* by David Svenson

 © 2001, 1997 by Prentice-Hall, Inc.
Upper Saddle River, New Jersey 07458

Printed in the United States of America

10 9 8 7 6 5 4 3 2 1

ISBN 0-13-032313-6

Prentice-Hall International (UK) Limited, London
Prentice-Hall of Australia Pty. Limited, Sydney
Prentice-Hall Canada Inc., Toronto
Prentice-Hall Hispanoamericana, S.A., Mexico
Prentice-Hall of India Private Limited, New Delhi
Prentice-Hall of Japan, Inc., Tokyo
Pearson Education Asia Pte. Ltd.
Editora Prentice-Hall do Brasil, Ltda., Rio de Janeiro

PREFACE

What Is Modern Geometry?

For most of recorded history, Euclidean geometry has dominated geometric thinking. Today, however, while Euclidean geometry is still central to much engineering and applied science, other geometries also play a major role in mathematics, computer science, biology, chemistry and physics. This book surveys these geometries, including non-Euclidean metric geometries (hyperbolic geometry and elliptic geometry) and nonmetric geometries (for example, projective geometry). The study of such geometries complements and deepens the knowledge of the world contained in Euclidean geometry.

Modern geometry is a fascinating and important subject. Above all, it is pure mathematics filled with startling results of great beauty and mystery. It also lies at the foundation of modern physics and astronomy, since non-Euclidean geometries appear to be the geometry of physical reality in several different ways: on the surface of the earth, as well as in the universe as a whole at very small and very large scales. Finally, modern geometry plays an important role in the intellectual history of Western civilization. Its development, in the nineteenth century, radically altered our conception of physical and geometric space, creating a revolution in philosophical, scientific, artistic and mathematical thought comparable to the Copernican revolution, which changed forever the relationship between science, mathematics, and the real world. The main purpose of this book is to describe the mathematics behind this revolution.

Analytic versus Synthetic Geometry

Despite its crucial influence on modern scientific thought, non-Euclidean is not well known. In part, this is due to the general neglect of geometry in the mathematics curriculum. However, another factor is the axiomatic format in which geometry is usually presented. Axiomatic geometry (also called *synthetic* geometry) is a legacy of the Greeks and suited admirably their view of geometry. However, it has no particular relevance to the modern viewpoint and indeed tends to obscure connections among the real world, geometry, and other parts of mathematics. In an age in which applications are paramount, there is no reason why geometry alone among mathematical subjects should be singled out for a quasi-archaic treatment that conceals its practical value. An axiomatic approach is particularly inappropriate with the current generation of mathematics students, since they have had only the briefest exposure to it in high school geometry classes. It makes much more sense to base a geometry course on analytic methods, with which students are much more familiar.

The purpose of this book is to provide a brief, but solid, introduction to modern geometry using *analytic* methods. The central idea is to relate geometry to familiar ideas from analytic geometry, staying firmly in the Cartesian plane and building on skills already known and extensively practiced there. The principal geometric concept used is that of congruence or geometric transformation. Thus, the *Erlanger Programm*, fundamental to modern geometry, is introduced and used explicitly throughout. The hope is that the resulting treatment will be accessible to all who are interested in geometry.

At the same time, synthetic methods should not be neglected. Hence, we present (in Part VI) axiom systems for Euclidean and absolute geometry. Presenting alternative systems demonstrates how axiomatics can clarify logical relationships among geometric concepts and among different geometries.

Use of the Book

This book is intended for an undergraduate geometry course at the sophomore level or higher. Parts I, II, and VII are the heart of the book. The remaining parts are almost independent of each other. They include material on solid geometry, projective geometry, discrete geometry, and axiom systems. (See the Dependency Chart on page ix.) These parts can be added to a syllabus, depending on the time available, and the interests of the instructor and students. In this way, a wide variety of different courses can be constructed.

Prerequisites

A previous acquaintance with (high school) Euclidean geometry and analytic geometry is needed to read this book. Specific Euclidean topics used include the sum of the angles of a triangle, the congruence of triangles, and the theory of parallels. From analytic geometry, the reader needs to be familiar with Cartesian and polar coordinates, and to be able to graph straight lines and circles from their equations (including parametric equations). Some familiarity with vector operations (addition and scalar multiplication) is also useful, and some linear algebra is used (but only in later chapters on projective geometry and discrete geometry).

ACKNOWLEDGMENTS

This work owes a lot to previous expositors of the material it presents. In particular, the author wishes to acknowledge the inspiration of works by Lars Ahlfors [A1], Sansone and Gerretsen [A3-4], and Fishback [D2].

The following sources are acknowledged for graciously allowing the use in this book of copyrighted material:

Quotations

Wiley and Sons, Inc., for permission to quote from *Introduction to Geometry* by H. S. M. Coxeter. Copyright 1969 by John Wiley and Sons, Inc.

Oxford University Press for permission to quote from *Mathematical Thought from Ancient to Modern Times* by Morris Kline.

Bantam Books, a division of Bantam Doubleday Dell Publishing Group, Inc., for permission to quote from *The Brothers Karamazov* by Fyodor Dostoevsky (translated by Andrew H. MacAndrew). Translation copyright 1970 by Bantam, a division of Bantam Doubleday Dell Publishing Group, Inc.

Applause Theater Book Publishers for permission to quote from *The Madman and the Nun and Other Plays* by Stanislaw Witkiewicz, translated by Daniel Gerould and C. S. Durer. Copyright 1987 by Applause Theater Book Publishers.

Faber and Faber, Inc., for permission to quote from *Arcadia* by Tom Stoppard.

Dover Publications, Inc., for permission to quote from *Geometry* by Felix Klein and *Science and Hypothesis* by Henri Poincaré.

Springer-Verlag for permission to quote from *A History of Non-Euclidean Geometry: Evolution of the Concept of Space* by B. A. Rosenfeld.

Marcel Dekker, Inc., for permission to quote from *The Shape of Space* by Jeffrey Weeks.

Illustrations

The Wallraf-Richartz-Museum for permission to reproduce *Madonna in the Rose Garden* by Stephan Lochner.

Bildarchiv Preusscher Kulturbesitz for permission to reproduce *Demonstration of Perspective Drawing of a Lute* by Albrecht Dürer, from the Kupferstichkabinett, Staatliches Museum zu Berlin, West Berlin.

The Cleveland Museum of Art for permission to reproduce *Piazza San Marco, Venice* by Bernardo Bellotto.

The Metropolitan Museum of Art, New York, for permission to reproduce *Corpus Hypercubicus* by Salvador Dali.

The Museum of Modern Art, New York, for permission to reproduce *Girl before a Mirror* by Pablo Picasso.

ARS for permission to reproduce *Family Promenade and Seascape with a Heavenly Body* by Paul Klee.

Cordon Art, Baarn, for permission to reproduce *Circle Limit IV* by M. C. Escher.

Wolfram Research, Inc., for permission to reproduce a hyperbolic dodecahedron generated by Mathematica.

Reviewers

Numerous useful suggestions were made by Susan Colley, Roger Cooke, Jan E. H. Johanson, Elizabeth Magarian, Joe Malkevitch, J. Donald Monk, Jeff Nicholls, Barbara Nimershiem, Steen Pedersen, Kevin M. Pilgrim, Cecil C. Rousseau, Russell J. Rowlett and Anthony Thompson.

CONTENTS

DEPENDENCY CHART

Large rectangles represent parts of the book; small rectangles represent chapters. Arrows outline logical sequences in which the chapters can be read.

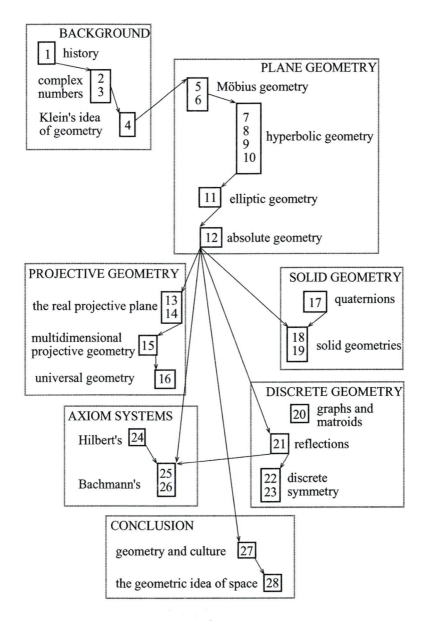

INTRODUCTION

Over the course of the last 200 years, doubts concerning the physical truth of Euclidean geometry have emerged. This book is about those doubts.

For example, we begin by asking: Does the sum of the angles of a triangle really equal 180 ?

What is the meaning of this outrageous question?

In one sense, of course, the answer is obviously: NO. If a triangle is drawn on a piece of paper and its angles are measured, the sum will not be 180 exactly, because no pen is fine enough to draw, nor any protractor accurate enough to measure, the angles of a perfect triangle. Ever since the Greeks began to study geometry systematically over 2,000 years ago, people have realized that physical shapes only approximate the properties of *ideal* geometric objects. Euclidean geometry, like all mathematical theories, concerns ideal objects. For *physical* objects, such theories are only approximately true.

However, our intention is to doubt more fundamentally the physical truth of Euclidean geometry.

It is possible to argue that the sum of the angles of physical triangles, measured with greater and greater accuracy, approximates not 180 , but some other quantity. In other words, it may be that Euclidean geometry is not the right geometry to apply to the physical universe.

This book is about alternative geometries, **non-Euclidean geometries**, which just possibly may model our physical universe more accurately than Euclidean geometry does. Non-Euclidean geometries embody alternative ideas of space or more poetically, alternative universes. In these universes, some Euclidean conceptions are misconceptions, and many Euclidean theorems are false.

One example of a non-Euclidean geometry is **spherical geometry**, the natural geometry of the surface of a sphere. On a sphere, the role of what are called straight lines in Euclidean geometry is played by arcs of great circles, circles whose center is also the center of the sphere. Great circles are the straightest circles that can be drawn on a sphere, since they have the greatest possible radius on that sphere (equal to the radius of the sphere). Great circles are, in fact, the straightest curves that can be drawn on a sphere and supply paths of shortest length between any two points on the sphere.

1

In spherical geometry, the sum of the angles of a triangle is greater than 180 .   There are even triangles with three right angles. (See Figure 1.)

Figure 1. A spherical triangle with three right angles

At times, some people have thought the earth was flat. This was an error, as we see, for instance, by examining photographs of the earth sent back from space. What may also be the case is that not only the two-dimensional *surface* upon which we walk, but also the three-dimensional *space* through which we move, is curved. It took longer to recognize the curvature of space than the curvature of the earth because it is not possible to step outside space.

Spherical geometry exists not only in the two-dimensional version just described, but also in three (and higher) dimensional versions. These geometries are called **elliptic geometries**. But the most important non-Euclidean geometries are **hyperbolic geometries**. Like spherical geometry, hyperbolic geometry is curved; however, it curves the "other way". (See Chapter 12.) The curvature of hyperbolic geometries (as well as of elliptic geometries) may be so gentle as to make it difficult for the residents of a hyperbolic (or elliptic) universe to recognize that their space is curved.

Our own universe may be elliptical or hyperbolic in curvature. Or even Euclidean (to be perfectly honest).

That the universe might be non-Euclidean in its geometry created as great a revolution in scientific and mathematical thought, when it was conceived, as the realization that the earth is round. This non-Euclidean revolution extended over the period from the end of the eighteenth century through the beginning of the twentieth. The ideas described in this book led, among other things, to Einstein's special and general theories of relativity.

This non-Euclidean revolution is still in progress, for it is unknown today what is the best geometry with which to model our universe.

PART I

BACKGROUND

GEOMETRY AND ART: PLATE I

from *The Grammar of Ornament* by Owen Jones [G3]

Geometry plays a major role in all decorative arts. Repeating designs like these appear on fabrics, tilings, tapestries, rugs, wallpaper, and many other places. Although thousands of years old, these Egyptian patterns are still in use and influential in design today.

I have resolved to publish a work on the theory of parallels as soon as I have put the material in order. . . . The goal is not yet reached, but I have made such wonderful discoveries that I have been almost overwhelmed by them. . . . I have created a new universe from nothing.

–János Bolyai (1823), in a letter to his father (quoted in Coxeter [B2])

There are two ways to present geometry.

Synthetic geometry begins with a set of axioms or postulates. From these axioms, results are deduced by pure reasoning. This approach, pioneered by the Greeks, emphasizes the abstract side of geometry, its otherworldliness, and its separation from any application that might be made of its results. Such a presentation brings out the fundamentally artificial or "synthetic" quality of geometry (and, indeed, of all mathematics).

Analytic geometry begins with a model of the geometry being studied: in a plane, on a sphere, or in some other mathematical structure. Results are deduced by computation within the mathematical realm surrounding the model. This approach emphasizes geometry's connection with other parts of mathematics, and the real world. Such a presentation is called "analytic" because it usually depends on coordinates (the same coordinates used in the analytic geometry taught in high school).

Since one theme of this book is the interpretation of geometric theorems in the real world, analytic geometry will be emphasized. (Synthetic geometry appears in Part VI.)

In Part I, we present the historical and analytic background needed to understand non-Euclidean geometries.

Chapter 1 outlines, very roughly, the history of geometry. However, in the main, our story is best *not* told historically, since modern developments (meaning especially developments since 1872!) have remade geometry into almost a different subject from what it was for most of its history. Instead, we emphasize the transformational point of view pioneered by Felix Klein in his *Erlanger Programm*.

To understand the *Erlanger Programm*, it is necessary to study transformations of the Euclidean plane, and, since the complex

5

numbers are the most convenient algebraic language to express these transformations, we will study them also.

Chapter 2 presents the necessary background on complex numbers which, in addition to their role in geometry, are a powerful tool in many branches of mathematics. Chapter 3 applies complex functions to the study of geometric transformations. Finally, Chapter 4 introduces the *Erlanger Programm*; a unifying vision of all geometry.

1 SOME HISTORY

The Beginning

Geometry began as practical science and engineering. The Egyptians, among other early civilizations, pioneered in the application of geometry to problems in surveying (the Nile river valley after the floods receded each spring) and construction (the pyramids). The word "geometry" reveals its origin in applied mathematics: "geo" + "metry" = "earth measure" (as transliterated from the Greek).

Egyptian geometry encompassed no theoretical knowledge, only empirically derived rules or procedures for accomplishing various tasks. The systematic foundation of mathematical theory was accomplished by the Greeks. The (Ionic) Greek, Thales, is supposed to have visited Egypt (~600 BC) and brought geometry back to Greece. He began the Greek tradition of abstract thinking in geometry. This continued with Pythagoras and his disciples, including a certain Hippocrates (no relation to the founder of Greek medicine) who was the first to arrange theorems in a chain, each derived from preceding ones (~400's BC). Hippocrates was also the first to use the method of indirect proof.

Still later, another Greek, Euclid, wrote down the elements of geometry (~300 BC) in the form in which they have been studied right up to the present. In Euclid's version, all geometric theorems are deduced from just ten assumptions divided among five axioms and five postulates. Euclid's axioms (also called "common notions") are algebraic statements such as "equals added to equals are equal". The postulates are geometrical in nature and thus embody the essence of Euclidean geometry.

Euclid's postulates are

> POSTULATE 1: *A straight line may be drawn from a point to any other point.* (In other words, two points determine a straight line. Euclid meant to say a *unique* straight line. This is one of numerous ways in which Euclid's language falls short of modern standards of precision.)

POSTULATE 2: *A finite straight line may be produced to any length*. (This postulate is usually interpreted to mean that a straight line is infinite, but it is ambiguous, a useful feature, as we shall see later.)

POSTULATE 3: *A circle may be described with any center and any radius*.

POSTULATE 4: *All right angles are equal*. (Euclid means "congruent.")

POSTULATE 5: *If a straight line meet two other straight lines so as to make the two interior angles on one side less than two right angles, the other straight lines meet on that side of the first line*. (This is the famous parallel postulate. It is equivalent to the more familiar: *There exists exactly one parallel line through a point not on a given line*. The latter is known as Playfair's postulate; however, Euclid avoids mentioning parallel lines until proposition 27.)

Euclid's *Elements* is probably the most studied book in the world, religious works, such as the Bible and the Koran, possibly excepted. This is in spite of the fact that the original is lost. Remarkably, our knowledge of Euclid depends on commentaries, notes, remarks, and so forth, handed down by contemporaries and successors, from which *The Elements* has had to be reconstructed. One of the most important sources is a Greek manuscript made in the tenth century AD, that is a copy of an edition of Euclid from some period before the fourth century AD This manuscript was unknown generally until the seventeenth century, when it was found in the Vatican library. It is now available over the Internet.

The Parallel Postulate

Right from the beginning there was concern with the parallel postulate. Unlike the other postulates, it was not regarded as self-evident. Coxeter [B2] says that this is evidence of Euclid's genius, that he was able to find a single postulate upon which so much depends.

The question of the self-evidence of the fifth postulate was of the utmost importance to geometers, from the Greek period all the way through the nineteenth century. The reason for their obsession with this "self-evidence" lies in the practical roots of geometry. Historically, geometry has above all been an applied subject, intended to describe reality. For Euclid and the mathematicians who followed him, geometry was important because it applied to the physical world.

But how could they be sure that an *abstract* geometry, derived from postulates, would genuinely describe the *real* world? Only because the postulates self-evidently were true about that world. From this point of view, the usefulness of Euclidean geometry depends entirely on the self-evidence of its fundamental assumptions: the five postulates of Euclid.

However, the matter went beyond usefulness. Once geometry was established as the premier mathematical subject, on the presumption that it described the real world, it became a standard against which the truth of other branches of mathematics, as well as science, were measured. Mathematicians came to believe that only to the extent that mathematics in general referred to reality (meaning geometric reality) was mathematics certain. Geometry and this vision of mathematical "truth" became a sacred part of the worldview of Western civilization, so much so that when Isaac Newton finally set down the principles of his calculus (~1687), he deliberately adopted the language and axiomatic style of Euclidean geometry. Similarly, it was only when Carl Friedrich Gauss and others showed that the complex numbers were really just another way to describe the Euclidean plane (~1799) and therefore, complex numbers were geometric quantities rather than comparatively mysterious algebraic quantities, that they could be treated seriously by other mathematicians and scientists. Even nonmathematical subjects were sometimes presented in a mathematical style, so powerfully attractive was the certainty attached to Euclidean geometry. The *Ethics* of Spinoza and Hobbes's *Leviathan* are examples.

Discovery of non-Euclidean Geometry

Euclid himself appears to be uncomfortable with the fifth postulate, avoiding its use as long as possible, thereby becoming (in Coxeter's phrase) the first non-Euclidean geometer.

Some such assumption is logically necessary, as we shall see, in order to develop Euclidean geometry. Nonetheless, for centuries mathematicians struggled to prove the parallel postulate from the first four. They succeeded only in replacing it with statements that are equivalent to it (assuming the first four postulates). Examples are

> (a) Two parallel lines are equidistant. (Posidonius, 100 BC)
> (b) If a line intersects one of two parallel lines, then it
>   intersects the other. (Proclus, fifth century AD)
> (c) Given a triangle, a similar triangle can be constructed of
>   any size. (Wallis, seventeenth century)
> (d) The sum of the angles of a triangle is 180 . (Legendre,
>   eighteenth century)

Naturally, all who worked on the parallel postulate believed in its truth. Indeed, they firmly believed that Euclidean geometry was the only possible geometry.

Among all the fruitless attempts to prove the parallel postulate, the saddest story is that of Saccheri (1667–1733). He attempted to prove the parallel postulate by the indirect method. Saccheri started with a quadrilateral *pqrs* in which the sides *pr* and *qs* are assumed equal, and there are right angles at *p* and *q*. This is now known as a **Saccheri quadrilateral.** (See Figure 1.1.)

Saccheri first proved that $\angle r = \angle s$. Euclid's postulate, it turns out, is equivalent to $\angle r = 90°$. Assuming that $\angle r$ was obtuse, Saccheri obtained a contradiction. Assuming that $\angle r$ was acute, Saccheri obtained numerous peculiar theorems, but no outright contradiction. Nonetheless, he concluded his work, entitled *Euclides ab Omni Naevo Vindicatus*, by rejecting the hypothesis of an acute angle, saying, that the consequences were "repugnant to the nature of a straight line!"

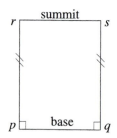

Figure 1.1 A Saccheri quadrilateral in all its misery

Actually, Saccheri had discovered many theorems of hyperbolic geometry, one of several geometries in which Euclid's fifth postulate is false and, accordingly, called **non-Euclidean geometries**. In hyperbolic geometry: (a) there are an *infinite* number of parallel lines through a given point not on a line, (b) the lines mentioned in the parallel postulate do *not* have to meet, (c) two coplanar lines are *never* equidistant, (d) a line may intersect one of two parallel lines, but *not* necessarily the other, (e) two similar triangles are always *congruent*, and (f) the sum of the angles of a triangle is *less* than 180°! (Hyperbolic geometry is described in detail in Chapters 7 through 10.)

The realization that there could be a geometry other than Euclidean geometry gradually dawned on a number of mathematicians toward the close of the eighteenth century and at the beginning of the nineteenth. This discovery came in two waves. First Klügel (1763), Lambert (1786), Schweikart (1816), and Taurinus (1826) realized that the fifth postulate could not be proven and that another geometry existed. The last three developed Saccheri's results further. Then Gauss, Lobatchevsky (1829), and Bolyai (1832) further realized that non-Euclidean geometry could possibly describe physical space as accurately as Euclidean geometry.

All of these ideas were truly revolutionary. Gauss, by common consent one of the greatest mathematicians that ever lived, never published his thoughts for fear of ridicule.

In any event, the work of all these people, published or not, and constituting "the most revolutionary development in mathematics since Greek times" (Kline [E1]) was entirely ignored for more than 30 years, until 1855, when, after his death, Gauss' notes on the subject were discovered and published.

Further Developments

In the meantime, the 19th century was proving to be the most active period in the history of geometry since the Greeks, even without non-Euclidean geometry. Much work was done in projective geometry (a geometry without parallel lines). Affine geometry (the geometry of linear algebra) and multidimensional geometries were invented. Differential geometry (the geometry of curved lines and surfaces) was in a state of continual development throughout the century. Thus, later in the nineteenth century, the revolutionary ideas of Gauss and the others fell on more receptive ears.

Many interesting points can be drawn from the story of the discovery of non-Euclidean geometry. One fascinating aspect is the striking way that important contributions were made, sometimes nearly simultaneously, by German, Russian, Italian, and French individuals, all part of a single international community of mathematicians. Another observation is that mathematicians are, on the whole, as conservative as any other group of people and are typically reluctant to embrace revolutionary ideas. It is also interesting to observe the effect on mathematicians of the spirit of their times. In 1733, Saccheri, believing religiously in the truth of Euclidean geometry (he was a Jesuit priest), dismissed as contradictory all the theorems of hyperbolic geometry that he had proven, without sound, logical grounds for this rejection. One hundred years later, Gauss, Bolyai, and Lobatchevsky determined to accept the same results as theorems of an alternative geometry, although they had no more reason to suppose that a contradiction did *not* exist in their geometry than Saccheri had reason to assume it *did*. Finally, work on the parallel postulate occurred only in mathematical cultures that pursued an axiomatic approach to geometry: the Hellenistic Greeks, medieval Muslims, and modern Europeans, all of whom started with Euclid. Other mathematical cultures, such as the Japanese, Chinese, and Hindu, which were satisfied with nonaxiomatic approaches, did not consider the possibility of non-Euclidean geometries.

Following Gauss' death in 1855, the discovery of non-Euclidean geometry created a revolution in mathematics and physics. Some of the effects of this revolution are discussed later (Chapters 27 and 28), after we have studied the geometry itself. Meanwhile, two further nineteenth-century developments deserve mention, since they have profoundly influenced the modern view of geometry.

In 1854, Bernhard Riemann observed that spherical geometry constituted a second non-Euclidean geometry, one in which there are *no* parallel lines. In this geometry, also called elliptic geometry, the sum of the angles of a triangle exceeds 180 . As in hyperbolic geometry, similar triangles are congruent. In elliptic geometry, great circles, which give the shortest paths between points on a sphere, play the role of straight lines. Several postulates other than the fifth need to be modified: Lines are not infinite, and two points can determine more than one line. Both of these problems can be dealt with. (Elliptic geometry is described in detail in Chapter 11.) In any event, the importance of Riemann's discovery is that there were now *two* non-Euclidean geometries.

In the second half of the nineteenth century, much effort was expended trying to find a single theory encompassing all the diverse geometries then making their definitive appearance. The culmination of these efforts was a synthesis of all then-existing geometries by Felix Klein in 1872. His ideas, known as the *Erlanger Programm*, after the university at which he lectured, created a revolution of their own in mathematics and physics, and provide a framework that still accurately describes most geometry.

In Chapter 4, we describe Klein's ideas in detail. They are then used to introduce and organize geometries throughout this book. First, however, it is necessary to learn something of the complex numbers (Chapter 2) and their ability to express geometric transformations (Chapter 3).

2 COMPLEX NUMBERS

The complex numbers are an alternative way to study analytic geometry in the Euclidean plane.

As with all the tools of analytic geometry, the key to understanding complex numbers is to realize that they simultaneously are algebraic and geometric in nature. Naturally, the geometry of complex numbers is emphasized here. At the same time, it is their algebraic properties that supply the computational power needed to pursue analytic geometry. The dual algebraic-geometric nature of the complex numbers is, in fact, the essence of analytic geometry.

The complex numbers are formed by adjoining the number i, a square root of -1, to the real numbers. The initial development of the complex numbers was purely algebraic: They were introduced to solve equations that would otherwise have no solutions, for example, quadratics, such as $x^2 + 1 = 0$, and also equations of higher degree. Initially, complex numbers were regarded as mysterious quantities and treated with suspicion by most mathematicians. Early in the nineteenth century, Gauss introduced the idea of treating the complex numbers as points in the plane. By making them geometric as well as algebraic, Gauss made complex numbers respectable.

We define the complex numbers geometrically (see Figure 2.1).

Definition *A complex number is a point* $z = (x, y)$ *in the Cartesian plane in which the x-axis is measured in ordinary (real) units and the y-axis is measured in a different (imaginary) unit i. The real number x is called the real part of z* $(x = \text{Re}(z))$*, and the real number y is called the imaginary part* $(y = \text{Im}(z))$*. (Note: Both the real and imaginary parts of z are real numbers.)*

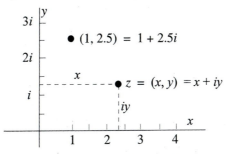

Figure 2.1 Complex numbers

Operations on Complex Numbers

As points in the plane, complex numbers inherit two operations familiar from multivariable calculus: vector addition and scalar multiplication. Each of these has an algebraic and a geometric aspect.

The (vector) sum of two complex numbers is found algebraically by adding separately real and imaginary parts of the numbers. The geometric rule corresponding to this is called the **parallelogram law**. (See Figure 2.2.)

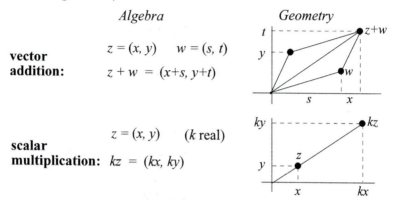

Algebra   *Geometry*

vector addition:

$z = (x, y) \quad w = (s, t)$

$z + w = (x+s, y+t)$

scalar multiplication:

$z = (x, y) \quad (k \text{ real})$

$kz = (kx, ky)$

Figure 2.2 Operations on complex numbers

Scalar multiplication simply multiplies both real and imaginary parts of a complex number by the same real number. Geometrically, the result is a complex number whose distance from the origin has been multiplied by this real number.

Using vector addition and scalar multiplication, we can write the complex number $z = (x, y)$ as

$$z = x(1, 0) + y(0, 1) = x + iy$$

since we consider the point $(1, 0)$ on the real axis as identical to the real number 1, and $(0, 1)$ is the complex unit i. The notation $x + iy$ is called the **Cartesian form** of z.

Questions[*]:

A–Find the Cartesian form of the complex numbers $(2 + 3i) + (1 - 6i)$ and $4(-2 + 5i) - (-7 - 2i)$.

B–Addition of complex numbers is commutative: $z + w = w + z$. Give two explanations: one algebraic and one geometric. (*Hint:* Use Figure 2.2.)

---

[*] It is best to read mathematics slowly. These questions (and others scattered through the text) are here to slow you down. Answer them to confirm your grasp of the subject.

Related Geometric Notions

Figure 2.3 introduces more complex number concepts. The most important is the modulus. Note that each concept has an algebraic notation and a geometric interpretation.

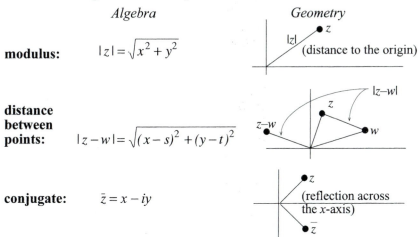

Algebra   *Geometry*

modulus: $|z| = \sqrt{x^2 + y^2}$ (distance to the origin)

distance between points: $|z - w| = \sqrt{(x-s)^2 + (y-t)^2}$

conjugate: $\bar{z} = x - iy$ (reflection across the x-axis)

Figure 2.3 More operations on complex numbers

The modulus has two important properties:

(a) Homogeneity: $|kz| = |k|\,|z|$ (where k is real)
(b) Triangle Inequality: $|z + w| \le |z| + |w|$

Proof of the triangle inequality: From the geometric interpretation of the vector sum (see Figure 2.2), $|z + w|$ is one side of a triangle, the other sides of which are $|z|$ and $|w|$. According to Euclidean geometry, one side of a triangle is always less than or equal to the sum of the other sides. This proves the triangle inequality. ∎

Note: This proof is deliberately based on *Euclidean* geometry. We systematically use results from Euclidean geometry (which we consider *known*) in order to make more rapid progress toward the proof of the *unknown* results of non-Euclidean geometry.

C–Let $z = 2 + i$. Plot z, $-z$, $\bar{z}$, and $-\bar{z}$ on the same pair of axes. What shape is outlined? Note its relation to the coordinate axes.
D–What is $|2 + i|$?

Complex Multiplication

Let $z = x + iy$ and $w = s + it$. The product of z and w is defined by

$$zw = (x + iy)(s + it) = (xs - yt) + i(ys + xt)$$

For example, $i^2 = -1$, $(1 + i)^2 = 2i$, and $(4 + 5i)(1 - i) = 9 + i$.

E–Check these examples!

Multiplication of complex numbers obeys all the usual rules of algebra (familiar as properties of the *real* numbers), namely, the commutative, associative, and distributive laws and the existence of inverses. The new feature is that -1 has a square root, since $i^2 = -1$. It turns out that every nonzero complex number has a square root (in fact, two). Although this can be proven algebraically, it is a simple consequence of the geometric interpretation of multiplication (explained shortly). The geometric interpretation of complex multiplication, in turn, depends on understanding the polar form of a complex number.

Polar Form

Polar coordinates (r, θ) provide another way to write a complex number z. The polar coordinate r, the distance of z from the origin, is, of course, the same as the modulus of z. The polar angle θ gives the direction of z from the origin. (See Figure 2.4.)

$$z = \quad x \quad + i \quad y$$
$$= r\cos(\theta) + i\, r\sin(\theta)$$
$$= r\,(\cos(\theta) + i\,\sin(\theta))$$

polar form: $\quad = |z|\qquad e^{i\theta}$

$$y = r\sin\theta$$
$$x = r\cos\theta$$

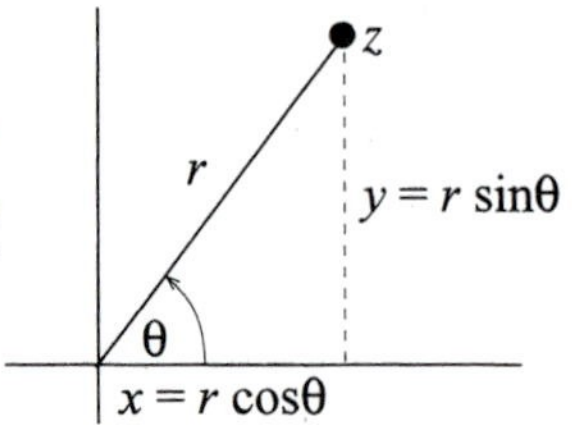

Figure 2.4 Polar form of a complex number

The polar form $z = |z|e^{i\theta}$ combines r and θ using the complex exponential:

Definition *The complex exponential function is defined by setting*
$$e^z = e^{x+iy} = e^x(\cos(y) + i\sin(y))$$
so that
$$e^{i\theta} = \cos(\theta) + i\sin(\theta) \qquad (Euler's\ formula)$$

The main reason for adopting this definition is that the fundamental law of exponents

$$e^{z+w} = e^z e^w$$

continues to hold for complex as well as real exponents. For example, using imaginary exponents, we have

$$e^{i\theta}e^{i\mu} = (\cos\theta + i\sin\theta)(\cos\mu + i\sin\mu)$$
$$= (\cos\theta\cos\mu - \sin\theta\sin\mu) + i(\sin\theta\cos\mu + \cos\theta\sin\mu)$$
$$= \qquad \cos(\theta+\mu) \qquad +i \qquad \sin(\theta+\mu)$$
$$= \qquad e^{i(\theta+\mu)}$$

Note how neatly complex exponentiation combines trigonometric identities and the laws of exponents. Euler's formula also fits with other known facts about exponentiation. (See Exercises 7 and 8.)

We now add a definition naming the θ coordinate:

Definition   *Let* $z = re^{i\theta}$ *be a complex number.   The* **argument** *of* z *[written* arg(z)] *is*

$$\arg(z) \;=\; \theta \quad (\text{mod } 2\pi)$$

The word "mod" means that two values of the argument are considered equal if they differ by 2π, or a multiple thereof. Geometrically this makes sense, because such values determine the same polar angle.

F–What is the polar form of these numbers: $1 + i$, $1 - i$?

The Geometry of Complex Multiplication

Polar form and the laws of exponents combine to make (some) geometric sense out of complex multiplication. (See Figure 2.5.)

If

$$z = |z|\,e^{i\theta} \text{ and } w = |w|\,e^{i\mu}$$

then

$$zw = |z||w|\,e^{i(\theta+\mu)}$$

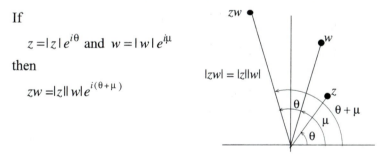

$$|zw| = |z||w|$$

Figure 2.5 Complex multiplication using polar form

In words, the figure says, "To multiply complex numbers (using polar coordinates) *multiply the radii* (the r coordinates, or *moduli*) and *add the polar angles* (the θ coordinates, or *arguments*)." In algebraic notation:

$$|zw| \;=\; |z||w|$$

and

$$\arg(zw) \;=\; \arg(z) + \arg(w) \quad (\text{mod } 2\pi)$$

Summary

The complex numbers are simply the points of the Cartesian plane. They can be combined algebraically using the operations of addition, scalar multiplication, and general multiplication. These operations have significant geometric meaning in terms of distances, angles, and movement about the plane.

EXERCISES

Complex Arithmetic

1. Find the Cartesian form of these complex numbers:
   (a) $4(1 + 2i) - 2(5 - i)$ (b) $(1 + i \div 2) + \div 2(p + i)$
   (c) $2(4 + i) - (1 + 6i)$ (d) $(1 - 2(i + 3(1 - 4i)))$

2. Find the Cartesian form of these complex numbers:
   (a) $(1 + i)(1 - i)$ (b) $(5 + 10i)(-2 + 3i)$
   (c) $(1 + i)^3$ (d) $(1 + 2i)(3 - 4i)(5 + 6i)$

3. Prove these properties of the conjugate and the modulus:
   (a) $|kz| = |k|\,|z|$ (k real)   (b) $|z|^2 = z\bar{z}$
   (c) $|wz| = |w|\,|z|$ (d) $\overline{zw} = \bar{z}\,\bar{w}$

   Hint: An algebraic proof of these results is simpler than a geometric proof.

   Example: Proof of (a): Let $z = x + iy$. Then

   $$|kz| = |kx + iky| = \sqrt{k^2 x^2 + k^2 y^2} = \sqrt{k^2}\sqrt{x^2 + y^2} = |k|\,|z|$$

4. Verify algebraically that complex multiplication satisfies the distributive law: $z(w + u) = zw + zu$.

The Complex Exponential

5. Evaluate these exponentials (that is, express them in Cartesian and/or polar form):

   $$e^{1+i\pi}, \qquad e^{i\pi/2}, \qquad e^{e^{\ln \pi + i\pi/2}}$$

6. Verify that the law of exponents holds with base e and arbitrary complex exponents.

7. Use the known power series

   $$e^x = 1 + x + \frac{x^2}{2} + \frac{x^3}{3!} + \frac{x^4}{4!} + \cdots$$

   $$\sin(x) = x - \frac{x^3}{3!} + \frac{x^5}{5!} - \frac{x^7}{7!} + \cdots$$

   and

   $$\cos(x) = 1 - \frac{x^2}{2!} + \frac{x^4}{4!} - \frac{x^6}{6!} + \cdots$$

to provide further support for Euler's definition of complex exponentiation.

8. Explain how the equation $e^{i0} = 1$ fits known facts about exponents and the trigonometric functions.

Polar Form

9. Express in polar form: $-1 + i\sqrt{3}$, $4i$, $5 - 5i\sqrt{3}$.

10. Express these fractions in Cartesian and/or polar form:

$$\frac{1}{i}, \quad \frac{1}{1+i}, \quad \frac{1+i}{i}, \quad \frac{4+i}{1-2i}, \quad \frac{i}{4}$$

*Hint: Multiply numerator and denominator by the conjugate of the denominator. This is called **rationalizing the denominator.***

Example:

$$\frac{2}{3+i} = \left(\frac{2}{3+i}\right)\frac{3-i}{3-i} = \frac{6-2i}{9-i^2} = \frac{6-2i}{10} = .6 - .2i$$

11. Use polar form to show that every complex number $z \neq 0$ has a multiplicative inverse $1/z$.

Note: This proves that division by nonzero complex numbers is always possible.

12. Use polar form to show that every complex number $z \neq 0$ has two square roots.

Hint: What happens (geometrically!) when a complex number is squared? Therefore, what is necessary to "unsquare" (that is to take the square root) of a number (geometrically)?

13. In Cartesian form, one square root of $x + iy$ when $y \geq 0$ is

$$\sqrt{x+iy} = \sqrt{\frac{\sqrt{x^2+y^2}+x}{2}} + i\sqrt{\frac{\sqrt{x^2+y^2}-x}{2}}$$

Verify this. What is the other root? What if $y < 0$?

14. Find all square roots of these complex numbers:
    (a) -1 (b) -4 (c) i (d) $2i$
    (e) $-i$ (f) $-1 - i\sqrt{3}$ (g) $2 + 2i$ (h) 4

15. Let a, b, and c be complex constants. Show that the quadratic equation

$$az^2 + bz + c = 0$$

has one or two roots.

Hint: Show that the usual proof of the quadratic formula, by completing the square, is justified with complex numbers as well as real numbers.

16. Let k be a complex constant. Show that

$$z\bar{z} - \bar{k}z - k\bar{z}$$

can be written $|z - k|^2 - |k|^2$. This is a *complex* form of completing the square.

17. Let p, q, and r be complex numbers. Explain why

$$\angle pqr = \arg\left(\frac{r-q}{p-q}\right)$$

Hint: Consider first the case $q = 0$. Use polar form.

Graphs

18. Sketch the set of complex numbers z satisfying the following conditions:

(a) $|z| = 1$
(b) $|z| = 2$
(c) $|z - 1| = 2$
(d) $|z + i| = 3$
(e) $|z/2 + 1/2| = 3$
(f) $|4z + 2i| = .5$
(g) $|3z - i| > 6$
(h) $|z/10 + 1 - i| < 5$
(i) $|z - 1| = |z|$
(j) $|z + i| = |z - i|$
(k) $\operatorname{Re}(z) = 1$
(l) $\operatorname{Im}(z) > -1$

Hint: These problems can be approached from several different directions. It is sometimes helpful to replace the complex variable z with two real variables: letting $z = x + iy$. The resulting Cartesian equation may be recognizable. Plotting points may be useful. However, first try simply to read the condition and think of what it means geometrically.

Example: $|z - 2| = 4$. The absolute value represents distance. Thus, this equation says that the distance from z to 2 equals 4. A circle is the set of points that are a given distance from a given point. Hence, the set of complex numbers satisfying the equation $|z - 2| = 4$ is the circle with center 2 and radius 4.

19. Sketch the set of complex numbers z satisfying the following conditions:

(a) $\operatorname{Re}\left(\dfrac{1}{z}\right) = 1$
(b) $\operatorname{Im}\left(\dfrac{1}{z}\right) = \dfrac{1}{2}$

(c) $\operatorname{Re}\left(\dfrac{1}{z-1}\right) = 1$
(d) $\operatorname{Im}\left(\dfrac{z}{z-i}\right) < \dfrac{1}{2}$

(e) $\left|\dfrac{1}{z-1}\right| = 1$
(f) $\left|\dfrac{z}{z+i}\right| = 4$

(g) $\left|\dfrac{iz}{z+2}\right| = \dfrac{1}{2}$
(h) $\left|\dfrac{z+1}{iz-2}\right| = 3$

Example: $|z/(z - i)| = 2$. Here algebraic manipulation is definitely useful. Since

$$\left|\frac{z}{z-i}\right| = \frac{|z|}{|z-i|}$$

the given equation can be put in the simpler form

$$|z| = 2|z-i| \qquad (+)$$

Now,

$$|z-i| = |x+iy-i| = |x+i(y-1)| = \sqrt{x^2 + (y-1)^2}$$

so (+) becomes

$$\sqrt{x^2 + y^2} = 2\sqrt{x^2 + (y-1)^2}$$

Squaring and simplifying, we get

$$3x^2 + 3y^2 - 8y + 4 = 0$$

which is recognizable as the equation of a circle. After completing the square, we get

$$x^2 + \left(y - \frac{4}{3}\right)^2 = \left(\frac{2}{3}\right)^2$$

so the circle has center at (0, 4/3) and radius 2/3.

20. Let α and β be complex constants, and let $z = x + iy$. Show that

$$\text{Im}(\alpha z + \beta) = 0$$

is the equation of a straight line. Can every straight line be expressed by such an equation?

Hyperbolic Functions

Discussion: The functions $\cosh(t) = (e^t + e^{-t})/2$ *and* $\sinh(t) = (e^t - e^{-t})/2$ *are important in hyperbolic geometry. These exercises outline a few of their properties.*

21. Prove the following hyperbolic identities:
   (a) $\cosh(t)^2 - \sinh(t)^2 = 1$
   (b) $\cosh(t)' = \sinh(t)$ and $\sinh(t)' = \cosh(t)$

22. Show that if $x = \cosh(t)$ and $y = \sinh(t)$, then the point (x, y) lies on the branch of the hyperbola $x^2 - y^2 = 1$ where $x > 0$. Show that every point on this branch corresponds to exactly one value of t.

23. The hyperbolic functions have properties analogous to those of the trigonometric functions. What properties of sin and cos correspond to the properties of sinh and cosh in Exercises 21 and 22? Why are the trigonometric functions called the **circular** functions?

3 GEOMETRIC TRANSFORMATIONS

Complex Functions

A complex function is a function whose domain and range consist of complex numbers. Such a function, like all functions, accepts an input (in this case a complex number z) and produces an output (another complex number). In the usual notation for functions, the output of a complex function f, given input z, is written $f(z)$. Thus, for example, if $f(z) = 3z + i$, then $f(i) = 4i$ and $f(1 - i) = 3 - 2i$.

The functions we study have both an algebraic and a geometric aspect. The algebraic aspect is the formula used to compute the output of the function. The functions presented here are all rather simple algebraically: nothing much more complicated than linear and quadratic polynomials. What *is* difficult is to visualize these functions geometrically. It is necessary to adopt a transformation viewpoint.

Transformations

A **transformation** is a one-to-one function whose range and domain are the same set. Most of the functions we are interested in fit this description: The range and the domain are both the whole complex plane. Transformations can be visualized as moving or transforming each input point to its unique output point.

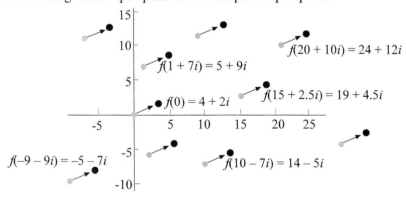

Figure 3.1 A flight of complex numbers

For example, consider $f(z) = z + (4 + 2i)$, which adds the constant $4 + 2i$ to the input z. Figure 3.1 shows the application of this function to a sample of points in the complex plane. As the figure shows, $f(z)$ moves all the points in the plane the same distance in the same direction. Figure 3.2 displays this another way: The complex plane is drawn twice, before being transformed (in gray) and after transformation (in black).

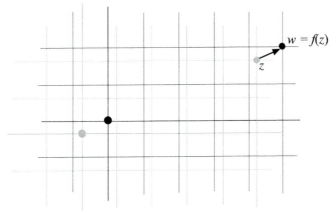

Figure 3.2 The whole plane moves at once!

This type of transformation is called a **translation**.

Examples of Transformations

Figures 3.3, 3.4 and 3.5 show the basic geometric transformations; translation, rotation, and homothetic transformations (that is, a stretching or shrinking of the plane), expressed as complex functions. Rotations and translations are considered Euclidean transformations because they transform every geometric figure into a congruent figure. Homothetic transformations, however, are not Euclidean, since they change the size of figures.

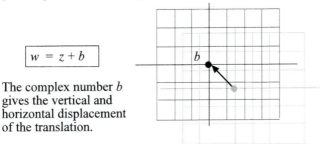

$$w = z + b$$

The complex number b gives the vertical and horizontal displacement of the translation.

Figure 3.3 Translation

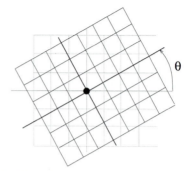

The real number θ is the
angle of rotation.

Figure 3.4 Rotation

The positive number k
gives the amount of:
   stretching (if $k > 1$)
or
   shrinking (if $k < 1$).

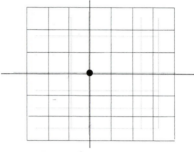

Figure 3.5 Homothetic transformation

A–Sketch the following transformations $w = f(z)$ using the
technique of Figure 3.1 or of Figure 3.2.
        (a) $w = z + 2$ (b) $w = iz$
        (c) $w = 5z$ (d) $w = z + i - 3$
    B–What is the type of each of these transformations?

Inversion

The transformation $w = Tz = 1/z$ (for $z \neq 0$) is called **inversion**.
It will be our most radical transformation. The geometric effect of
inversion is somewhat mysterious, so we investigate it in some detail.
Inversion is a non-Euclidean transformation.

In the first place, points *inside* the unit circle (that is, with $|z| < 1$)
are transformed to points *outside* the circle. This can be seen using
polar coordinates. If

$$z = re^{i\theta}$$

then

$$w = \frac{1}{z} = \frac{1}{r}e^{-i\theta}$$

Therefore, if $|z| < 1$, then $|w| = 1/r = 1/|z| > 1$. Our calculation with
polar form also shows that points in the upper half plane (that is,

with $0 \le \theta \le \pi$) exchange places with points in the lower half plane $(-\pi \le \theta \le 0)$.

What is remarkable is what inversion does to straight lines. For simplicity, let us consider lines parallel to the coordinates axes, as in Figure 3.6. Inversion turns these lines into circles passing through the origin! Specifically, each gray line, when inverted, becomes one of the black circles. Thus the gray network of coordinate lines inverts to the black network of circles. Furthermore, lines closer to the origin become bigger circles than lines farther from the origin.

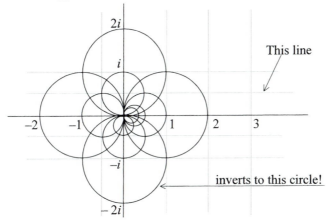

Figure 3.6 Inversion of the complex plane

C–Which line in Figure 3.6 inverts to the smallest circle there?

To confirm the inversion of straight lines analytically, let $z = x + ik$ be a point on the line $y = k$ (a line parallel to the x-axis). Applying inversion, we get

$$w = \frac{1}{z} = \frac{1}{x+ik} = \frac{x-ik}{x^2 + k^2}$$

Now, let $w = s + it$ (i.e., let s and t be the real and imaginary parts of w). Then

$$s = \frac{x}{x^2 + k^2} \quad \text{and} \quad t = \frac{-k}{x^2 + k^2}$$

These are parametric equations for s and t, with x as the parameter (k is constant). Together, they describe the curve into which the line $y = k$ is transformed by inversion. Notice that $ks + xt = 0$ or

$$x = \frac{-ks}{t}$$

Using this identity, x can be eliminated from the equation for s (or for t), getting

$$k^2s^2 + k^2t^2 + kt = 0$$

Completing the square gives

$$s^2 + \left(t + \frac{1}{2k}\right)^2 = \left(\frac{1}{2k}\right)^2$$

This *is* the equation of a circle: the circle with radius $1/2k$ and center at $(0, -1/2k)$. This proves that the inversion of a straight line parallel to the x-axis is a circle tangent to the x-axis at the origin. As we shall see later, the inversion of any straight line is a circle passing through the origin!

D–What value of k corresponds to the line indicated in Figure 3.6?

Conformality

A transformation is **conformal** when it preserves angles, that is, when the measure of an angle after transformation is the same as it was before transformation.   All the basic types of transformation introduced so far are conformal.

For rotations, translations, and homothetic transformations, conformality is obvious. This is indicated in Figure 3.7.   The gray triangles are either congruent or similar to the black, transformed triangles; hence, their angles are the same.

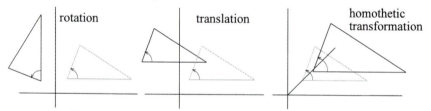

Figure 3.7 Conformality of familiar transformations

Inversion is conformal, too, but this is not obvious. In the first place, inversion turns segments of straight lines into arcs of circles, so it is necessary to specify what we mean by the angle between two circles. We adopt the standard definition that the angle between two curves is the angle between their tangents.   With this understanding, we can prove the following theorem:

Theorem *Inversion is conformal at every point in the complex plane except the origin.*

Proof: Let z be any point in the complex plane (except zero). Let two curves γ_1 and γ_2 pass through z (as in Figure 3.8). Let Γ_1 and Γ_2

be the inversions of the curves γ_1 and γ_2. Then Γ_1 and Γ_2 will pass through $z' = 1/z$. Let p and q be points on γ_1 and γ_2 near z on the same radial line from the origin. Then $p' = 1/p$ and $q' = 1/q$ will be on Γ_1 and Γ_2, respectively.

A straightforward computation shows that angles $\angle pzq$ and $\angle p'z'q'$ are equal. (See Exercise 9.) If we now let p and q approach z, then p' and q' approach z' (because inversion is continuous), and the secants approach tangents. In the limit, the angles have to be equal also. This proves that the angle between the tangents to two curves equals the angle between the tangents to the inversion of those curves, that is, inversion is conformal. ∎

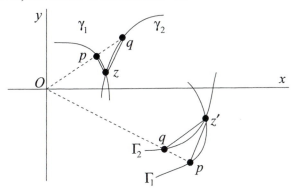

Figure 3.8 Proof that inversion is conformal

E–Draw a line parallel to the x-axis and a line parallel to the y-axis. Call the lines α and β. Draw the inversions of α and β. What should the angle between the inverted curves be, according to conformality? Is it?

Stereographic Projection

Our last function is not a transformation of the complex plane but rather maps a different set (the surface of a sphere) onto the complex plane.

Consider, in particular, the surface of the **unit sphere** in three-dimensional space. With (a, b, c) as coordinates, the unit sphere has the equation $a^2 + b^2 + c^2 = 1$. Stereographic projection maps this sphere onto its equatorial plane (the (a, b)-plane) by means of **central projection**. This means that the lines joining corresponding points all meet at a common point called the **center of projection**. For stereographic projection, the center of projection is the North Pole.

This process is illustrated in Figure 3.9. Each point $P = (a, b, c)$ on the sphere is connected by a straight line with the North Pole N. The stereographic projection of P is, then, the point $S(a, b, c)$ where this line meets the (a, b)-plane. The North Pole itself is the only point on the sphere that does not have a stereographic projection in the complex plane (because it does not determine a unique line with itself). We will do something about this soon.

At first, it may seem that stereographic projection wildly distorts geometric properties of the sphere and the plane. However, a number of significant geometric properties are actually preserved by stereographic projection, and the resulting mapping from sphere to plane is often used by cartographers to map the polar regions. For example, stereographic projection preserves angles, that is, it is conformal. The conformality of stereographic projection (proved shortly) means that the basic shape of geographic entities is preserved.

Stereographic projection is also important in non-Euclidean geometry. It stretches the sphere out onto the plane, permitting the study of spherical geometry (an important non-Euclidean geometry) to be carried out in the more convenient setting of the plane.

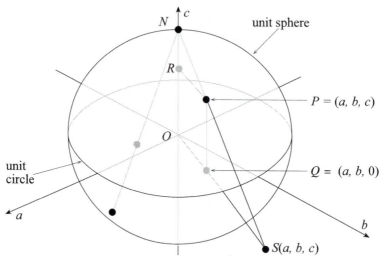

Figure 3.9 Stereographic projection

F–Draw the region on the unit sphere whose stereographic projection is the first quadrant of the (a, b)-plane.

A Formula for Stereographic Projection

To obtain a formula for stereographic projection, we use Figure 3.9. First note that $S(a, b, c)$ lies somewhere along the line from the origin O through the point $Q = (a, b, 0)$. The only question is how far? In other words,

$$S(a, b, c) = k\,(a, b, 0)$$

where k is an unknown scalar. Using the similar triangles ΔNSO and ΔNPR, we get

$$\frac{\overline{ON}}{\overline{RN}} = \frac{\overline{OS}}{\overline{RP}} = \frac{\overline{OS}}{\overline{OQ}}$$

But, since

$$\frac{\overline{OS}}{\overline{OQ}} = k \quad \text{and} \quad \frac{\overline{ON}}{\overline{RN}} = \frac{1}{1-c}$$

it follows that $k = 1/(1-c)$ so that

$$S(a,b,c) = \frac{1}{1-c}(a,b,0) = \frac{a+ib}{1-c}$$

Here we suddenly treat the (a, b)-plane as the complex plane. Thus, stereographic projection transforms the surface of the unit sphere to the complex plane so that the image of the point (a, b, c) is the complex number $z = x + iy$, where $x = a/(1-c)$ and $y = b/(1-c)$.

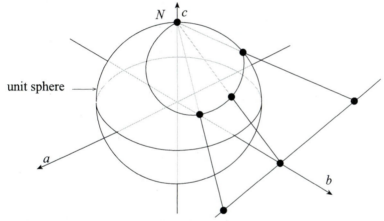

Figure 3.10 Stereographic projection turns circles through the North Pole into straight lines.

Like inversion, stereographic projection can turn circles into straight lines. Figure 3.10 shows how a circle on the sphere passing through the North Pole projects to a line in the complex plane. We know this must be so because all the lines of projection from N through points of the circle lie together in the plane of the circle, Therefore, the stereographic projection of the circle is the line of intersection of the plane of the circle and the (a, b)-plane.

Theorem *Stereographic projection is conformal.*

Proof. For simplicity, we only prove conformality when the image curves are straight lines. Suppose that two lines pq and pr in the complex plane are the projections of two circular arcs on the sphere meeting at points P and N. (See Figure 3.11.)

By symmetry, the angle between the circular arcs at P is the same as the angle between the circular arcs at N. The angle at N, by definition, equals the angle between the tangents to the arcs (in the plane tangent to the sphere at N). Finally, the angle between the tangents equals the angle between the parallel straight lines pq and pr in the complex plane below. (This is a theorem of solid Euclidean geometry: proposition 10 from the 11th book of Euclid.) ■

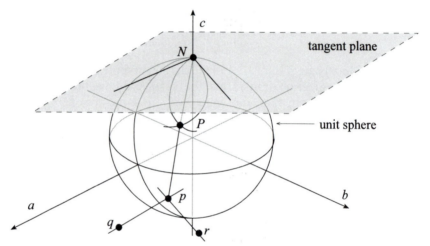

Figure 3.11 Stereographic projection is conformal.

The Point-at-Infinity

Stereographic projection stretches the unit sphere, *minus one point*, into the complex plane. The exceptional point is the North Pole. In order to remove this exception, it is customary to imagine a new point, called ∞, added to the complex plane. The new point ∞ is not combined *algebraically* with ordinary complex numbers, but, as the point-at-infinity, it occupies a perfectly definite *geometric* place as the limit of any sequence of finite points whose absolute values approach (real) infinity. It is actually possible to evaluate (at least some) functions at ∞ by taking limits. For example, if $f(z) = 1/z$, then $f(\infty) = 0$ and $f(0) = \infty$.

With the point-at-infinity added to the complex plane, stereographic projection establishes a one-to-one correspondence between the whole unit sphere (including the North Pole) and the complex plane (including ∞).

G–Is ∞ on the x-axis? Is it on the y-axis?

H–What other straight lines contain ∞?

Lifts

Since stereographic projection is a one-to-one correspondence between the unit sphere and the complex plane, any transformation or motion of one of these sets can be transferred (the technical term is **lifted**) to the other. For example, let $f(z)$ be a rotation of the complex plane about the origin. To lift f from the plane to the sphere, let P be any point on the sphere. We start by applying stereographic projection S to P (obtaining $z = S(P)$) and then apply the rotation f to z. This is illustrated in Figure 3.12. Start in the upper left of the figure and proceed counterclockwise around to the lower righthand corner to see the point P first projected into the plane and then rotated.

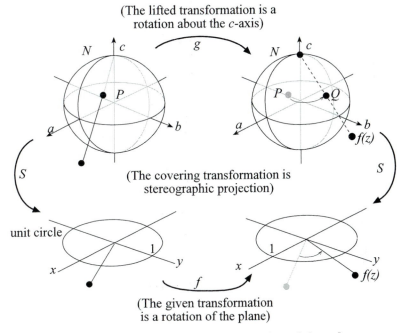

(The lifted transformation is a
rotation about the c-axis)

(The covering transformation is
stereographic projection)

(The given transformation
is a rotation of the plane)

Figure 3.12 Planar rotation lifts to rotation of the sphere.

Because stereographic projection is onto, the point $f(z)$ (in the plane) is the stereographic projection of some point Q on the sphere. (See Figure 3.12; upper right.) The combined transformation $P \to z \to f(z) \to Q$ is the rotation f transferred (or lifted) to the sphere. In this example the lifted motion $g: P \to Q$ happens also to be a rotation, namely, rotation about the c-axis.

Here is the formal definition of this situation:

Defintion *Let S be a continuous function from a set D onto a set R. We say that S is a **covering transformation** from D to R or that $D covers R via S.*

*Let f be a transformation from R to R. A transformation g from D to D is a **lift** of f if, for every z in D, S(g(z)) = f(S(z)).*

The functions in this definition can be displayed in a diagram:

$$D \xrightarrow{\ g\ } D$$

$$S \Big\downarrow \qquad\qquad \Big\downarrow S$$

$$R \xrightarrow[\ f\ ]{} R$$

The definition of lift simply says that applying these functions around the square from the upper-left corner to the lower-right corner in both directions, clockwise and counterclockwise, gives the same result. Figure 3.12 (with stereographic projection as the covering transformation) is laid out like this diagram. Note that the lifted transformation $P \to Q$ (in Figure 3.12) is the composition $Q = S^{-1}(f(S(P)))$, where S^{-1} is inverse stereographic projection.

Inversion is a more interesting transformation to lift. (See Figure 3.13.) Indeed, inversion is such a strange transformation of the plane that one might expect that its lift to the sphere would also be strange. To see what actually happens, however, let us apply stereographic projection to a point (a, b, c) on the sphere, obtaining

$$z = \frac{a+bi}{1-c}$$

Next, we apply inversion, to get

$$w = \frac{1}{z} = \frac{1-c}{a+bi} = \frac{(1-c)(a-bi)}{a^2+b^2}$$

But, because the point (a, b, c) lies on the unit sphere, $a^2 + b^2 = 1 - c^2$. Therefore,

$$w = \frac{(1-c)(a-bi)}{a^2+b^2} = \frac{(1-c)(a-bi)}{(1-c)(1+c)} = \frac{a-bi}{1+c}$$

This reveals that w is the stereographic projection of the point $(a, -b, -c)$. Hence, inversion in the complex plane lifts to the transformation of the sphere given by

$$g: (a, b, c) \to (a, -b, -c)$$

which is merely a 180° rotation of the unit sphere: about the a-axis!

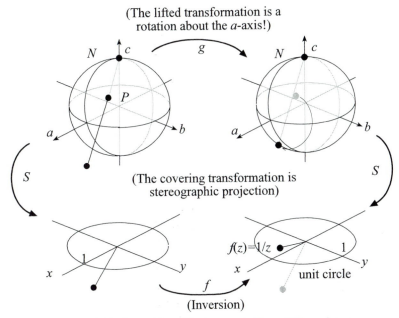

Figure 3.13 Inversion lifts to a rotation of the sphere.

Summary

Complex functions concisely express fundamental geometric transformations, including rotations, translations, homothetic transformations, and inversion. Stereographic projection completes the list of transformations needed to begin the study of non-Euclidean geometry. Stereographic projection introduces the point-at-infinity into the complex plane. All these transformations are conformal; that is, they preserve angles.

EXERCISES

Properties of Transformations

1. Let $f(z) = iz + 2$ and $g(z) = -iz + 5$. Sketch the composition transformation $w = f(g(z))$. What type of transformation is this?

2. Show that the composition of two translations is another translation. Show that the composition of two homothetic transformations (or two rotations) is likewise of the same type.

Example: Let $f(z) = z + b$ and $g(z) = z + d$ be two translations. Then the composition $f(g(z)) = f(z + d) = (z + d) + b = z + (d + b)$ is also a translation: by $(d + b)$.

3. Show that the general complex linear function $w = az + b$, where a and b are any complex numbers, is the composition of a

rotation, followed by a homothetic transformation followed by a translation.

Hint: Use the polar form of a.

4. Prove that the composition of conformal transformations is conformal.

Formulas

5. Find a formula for these rotations:
   (a) by 45° about the point i
   (b) by 90° about $2 + i$
   (c) by –60° about 3

Example: So far, we have only seen rotations about the origin. In order to rotate about another point, we must first translate that point to the origin, then rotate, then translate back. For (a): We translate i to the origin by using $f(z) = z - i$. At the origin, a rotation by $45°$ (= $\pi/4$) is given by the transformation

$$g(z) = e^{i\pi/4}z = \frac{(1+i)z}{\sqrt{2}}$$

The translation back to i is $f^{-1}(z) = z + i$. The desired rotation is the composition $f^{-1}(g(f(z)))$:

$$\frac{(1+i)(z-i)}{\sqrt{2}} + i = \frac{(1+i)z + i\sqrt{2} - i + 1}{\sqrt{2}}$$

6. Find a formula for these homothetic transformations:
   (a) a stretching of the plane from the point i by a factor of 3
   (b) a shrinking of the plane towards -2 by a factor of $1/2$

7. Find a formula for a transformation of the complex plane that carries the unit disk $\{z: |z| < 1\}$ to the disk $\{z: |z - 5| < 3\}$. Find a second transformation that does the same thing.

Inversion

8. The points $1, 1 + i, 1 - i, 1 + 2i$ all lie on a line. Plot them and their inversions. The inverted points lie a circle. What is the equation of this circle?

9. Complete the proof that inversion is conformal by showing that angle $\angle pzr$ equals angle $\angle p'z'q'$.

Hint: Use the formula in Exercise 17 of Chapter 2 plus the assumption that p and q lie on a line through the origin.

Stereographic Projection

10. Let P and Q be diametrically opposite points on the unit sphere, $a^2 + b^2 + c^2 = 1$. Let z and w be the stereographic projections of these points in the complex plane. Draw a picture of this situation. Explain why one of the two points z and w is always above the x-axis and the other below the x-axis, except when they are both on the x-axis. Explain why one is always inside the unit circle and the other outside, except when both are on the unit circle.

11. Let P and Q be diametrically opposite points on the unit sphere, $a^2 + b^2 + c^2 = 1$. Let z and w be the stereographic projections of these points in the complex plane. Show that $z\overline{w} = -1$.

Note: This result is important in elliptic geometry.

12. Inverse stereographic projection. Solve the equation

$$x + iy = \frac{a + ib}{1 - c}$$

for a, b, and c in terms of x and y.

Hint: A complex equation, such as this, is really two real equations; one for the real parts and one for the imaginary parts. Remember that (a, b, c) lies on the unit sphere.

13. What transformation of the complex plane lifts to the rotation of the unit sphere by $180°$ about the b-axis?

14. Let $f(z) = z^2$. Let α and β be lines parallel to the coordinate axes. Find the curves into which α and β are transformed by f. Do the transformed curves meet at right angles? Why do these curves meet twice?

Comment: This is a messy calculation. A rough idea of what is going on can be obtained simply by plotting some points.

The Point-at-Infinity

15. For these complex functions, find $f(\infty)$:

(a) $f(z) = \dfrac{1}{z + 3}$

(b) $f(z) = \dfrac{2z - 1}{z + i}$

(c) $f(z) = \dfrac{2z - i}{iz + i}$

(d) $f(z) = \dfrac{2z - 5}{i - 3}$

4 THE *ERLANGER PROGRAMM*

The *Erlanger Programm* is a tool for describing geometries. It makes possible the uniform development and comparison of different geometries. With only a few exceptions, all geometries fit into the *Erlanger Programm* mold.

Comparing two geometries means determining to what extent they overlap or, indeed, whether the concepts of one geometry are entirely contained inside another. A key question, therefore, is how to decide whether a particular geometric idea belongs to one geometry instead of another. This was a question Klein and other mathematicians faced in the geometrically fertile period of the late nineteenth century. In this chapter, we see how the *Erlanger Programm* solves these organizational problems, provides a framework for classifying geometric ideas, geometry by geometry, and, most significantly, provides a technique of proof uniformly applicable to all theorems of all geometries.

Congruence

Klein made congruence the central unifying concept of his approach to geometry. Of course, congruence already plays a large part in Euclidean geometry. In fact, it often appears that proving the congruence of triangles is all that goes on in Euclidean geometry!

What is the meaning of congruence? It is that congruent figures have identical geometric properties. For example, in the usual development of Euclidean geometry, two figures are called congruent whenever corresponding Euclidean measurements in the two figures always agree: Corresponding sides have the same length, corresponding angles have the same measure, and so on. Congruent figures have the same geometric properties, so that any statement true about one figure is automatically true of any congruent figure. In Euclidean geometry, congruence, thus, depends on measurement: Measurement comes first, and congruence is simply the word used to describe the situation when corresponding measurements taken in two figures are always the same.

Under the *Erlanger Programm*, this relationship is reversed; congruence *precedes* measurement. Geometry begins with a naked plane (or some other set) without axioms, postulates or, in particular, any ideas of measurement. All that this set comes with is congruence;

that is, a way is given to tell when two subsets of the geometry are congruent. (How this is done is described shortly.) These two things—a set, plus a notion of congruence—make a geometry. Geometries can be classified by the different ways they define congruence. And a particular measurement (length or area, for example) is introduced into a geometry only if it is proved in advance that the measurement yields equal results when applied to congruent figures.

Summary

In both the traditional and the modern Kleinian view of geometry, it is correct to say that any statement true about one figure is always true about any congruent figure. In the traditional view, measurement comes first and two figures are called congruent *because they have the same measurements*. In the *Erlanger Programm*, congruence comes first and a measurement is studied only *because it gives the same result for congruent figures*.

Congruence and Transformation

To use Klein's ideas, it is necessary to be able to describe when two figures are congruent without reference to preexisting geometrical ideas. Klein did this using the concept of **superposition**: Two figures are congruent when one can be moved so as to coincide with the other. The notion of superposition goes back to Euclid but was used by him with reluctance, as he probably realized that his postulates provided no basis for moving figures around. By the late nineteenth century, however, the idea of motion had been placed on a rigorous foundation. In fact, it is nothing more than the idea of a function or transformation. Thus, describing a particular notion of congruence is the same thing as specifying a collection of functions to act as congruence transformations. Then, given two figures A and B, A and B are congruent (written $A \cong B$) if, and only if, $A = f(B) = \{f(p): p$ is in $B\}$ where f is one of these congruence transformations.

Not just any collection of functions will do as congruence transformations, however. By universal agreement, the relation of congruence must possess these three properties:

(a) (reflexivity) $A \cong A$ for any figure A.
(b) (symmetry) If $A \cong B$, then $B \cong A$.
(c) (transitivity) If $A \cong B$ and $B \cong C$, then $A \cong C$.

A–Name an important Euclidean geometric relationship besides congruence that is reflexive, symmetric, and transitive.

Definition of a Geometry

These common-sense properties of congruence follow if the collection of congruence motions has these three properties:

(a) The identity function $f(z) \equiv z$ is a congruence transformation.
(b) If $f(z)$ is a congruence transformation, then f is invertible and f^{-1} is also a congruence transformation.
(c) If $f(z)$ and $g(z)$ are congruence transformations, then so is the composition $f(g(z))$.

This leads to the following definitions:

Definition *Let S be a nonempty set. A **transformation group** is a collection G of transformations $T: S \rightarrow S$ such that*
   (a) G contains the identity,
   (b) the transformations in G are invertible and their inverses are in G, and
   (c) G is closed under composition.

Definition *A **geometry** is a pair (S, G) consisting of a nonempty set S and a transformation group G. The set S is the **underlying space** of the geometry. The set G is the **transformation group** of the geometry.*

Definition *A **figure** is any subset A of the underlying set S of a geometry (S, G). Two figures A and B are **congruent** if there is a transformation T in G such that $T(A) = B$, where $T(A)$ is defined by the formula $T(A) = \{Tz : z \text{ is a point from } A\}$*

B–Explain how congruence (as defined using a transformation group) is reflexive, symmetric, and transitive.

Examples of Geometries

The most important part of a geometry is the transformation group G. It is G which determines the character of the geometry, in ways that will become clear as we study more and more examples.

Euclidean Geometry

Let the underlying set be the complex plane $\mathbf{C}$ and the transformation group be the set $\mathbf{E}$ of transformations of the form

$$Tz = e^{i\theta}z + b$$

where θ and b are constants (θ real, b complex). The transformation T is called a **rigid motion** because it preserves Euclidean distances between points. (See Exercise 1.) A rigid motion is composed of a rotation followed by a translation. A rigid motion may also be obtained by a rotation and translation applied in the opposite order; that is,

$$Tz = e^{i\theta}(z + a)$$

where

$$a = be^{-i\theta}$$

The pair (**C**, **E**) models **Euclidean geometry**. (More precisely, this is Euclidean geometry without reflections. For more on reflections, see Chapter 21.)

This example requires careful treatment. We cannot use it to develop Euclidean geometry here. That would be circular reasoning since we have *assumed* Euclidean geometry as a basis for work in the complex plane. However, such an approach to the study of Euclidean geometry is *possible*. All it requires is the development of the theory of complex numbers without the use of geometry, a somewhat lengthy process. It is fair, however, to exhibit the pair (**C**, **E**) as an example of how the *Erlanger Programm* is used to set up and describe a geometry.

C–Show that (**C**, **E**) is a geometry; that is, prove that **E** is a transformation group of **C**.

Translational Geometry

We *can* use the *Erlanger Programm* to introduce non-Euclidean geometries! For example, let **T** be the set of all translations T of the complex plane $Tz = z + b$, where b is a complex constant. The pair (**C**, **T**) is a geometry which we shall call **translational geometry**.

To be certain that (**C**, **T**) is a geometry, one must verify the requirements of the definition; that is, one must verify that **T** is a transformation group. Since the elements of **T** are transformations of **C**, all that is really necessary is to verify the defining conditions:

(a) The identity transformation is a translation.
(True: Choose $b=0$.)
(b) The inverse of a translation is a translation.
(True: $T^{-1}z = z - b$ is a translation.)
(c) Translations are closed under composition.
(True: See Exercise 2 of Chapter 3.)

Accordingly, translational geometry *is* a geometry. Of course, translational geometry is only a *toy* geometry; that is, it has only very simple properties. For example, in translational geometry triangles that are normally considered congruent in Euclidean geometry can fail to be congruent. Figure 4.1 gives an example.

The triangle $\Delta p'q'r'$ is a rotated version of Δpqr. But the transformation group of translational geometry does not contain rotations, so these triangles are not congruent in translational geometry. Only two triangles that are translations of each other are congruent in translational geometry. Congruence is much simpler in translational geometry than in Euclidean geometry.

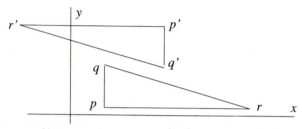

Figure 4.1 These are *not* congruent in translational geometry.

D–Note that **T** is a subgroup of **E**. (That is, translations are rigid motions.) Find another geometry whose underlying space is the complex plane and whose transformations are rigid motions.

The Trivial Geometry

Let S be any set and let G be the smallest transformation group of S. What is this smallest transformation group? The three properties (a), (b), and (c) in the definition of a transformation group require only that G contain the identity transformation. Thus, the identity alone forms a very small transformation group, and the resulting geometry is called the **trivial** geometry.

In the trivial geometry, two different figures are *never* congruent because the only congruence transformation is the identity, and the identity doesn't move any figure to any place different. Even two distinct points are incongruent! From the point of view of congruence, the trivial geometry is the simplest possible geometry, useful only as an extreme example.

Invariants

The patient reader may reasonably be wondering just how the transformation group approach to geometry really describes geometric ideas. This is what we now explain.

According to the *Erlanger Programm*, geometry is the study of those statements that, if true of one figure, are automatically true of all congruent figures. To formulate such statements, we need concepts and measurements that are unaltered when congruence transformations are applied. Such concepts and measurements are called invariants.

Definition *Let (S, G) be a geometry.   Let D be a set of figures from (S, G), i.e., the members of D are subsets of S.   Let T be a*

transformation from G. The set D is **invariant** *if, for every member B of D, T(B) is also in D.*
   A function f defined on D is called **invariant** *if*

$$f(T(B)) = f(B).$$

In other words, a *set of figures* is invariant if, together with any one figure *B*, it contains all figures congruent to *B*. A *function* is invariant if it gives the same value for congruent figures.

Example 1: Triangles

Let *D* be the set of all triangles in the Cartesian plane. (By a **triangle**, we mean the set consisting of three noncollinear points, plus the line segments connecting them.) In Euclidean geometry, *D* is an invariant set: A rigid motion always maps a triangle to another triangle. However, in a geometry that happened to include inversion among its transformations, *D* would *not* be an invariant set, because inversion converts the straight sides of a triangle into arcs of circles.

Example 2: Area, Perimeter, and Distance Sum

For any triangle Δ, let $a(\Delta)$ be the area of Δ. This defines an invariant function in Euclidean geometry, since congruent triangles have the same area. Similarly, the perimeter $p(\Delta)$ is a Euclidean invariant. On the other hand, if we let $d(\Delta)$ be the sum of the distances from the vertexes of Δ to the origin, perhaps calling this the **distance sum** of Δ, then *d* is *not* an invariant function, since this sum is often different for congruent triangles.

Example 3: Lines Parallel to the *x*-Axis

Let *E* be the set of all lines in the Cartesian plane parallel to the *x*-axis. This is *not* an invariant set in Euclidean geometry, since, in Euclidean geometry, there are transformations (rotations, for instance) that take a line that is parallel to the *x*-axis and turn it into one that is not parallel to the *x*-axis. On the other hand, *E* *is* an invariant set in translational geometry.

Example 4: Invariants for the Trivial Geometry

Every set of figures and every possible function is invariant under the identity transformation, since the identity doesn't change anything. Thus, to study the trivial geometry thoroughly, one must study *every* set of figures and *every* function. From the point of view of invariants, the trivial geometry is maximally complicated, since the identity transformation does not at all reduce the set of figures and functions that must be studied.

According to the *Erlanger Programm, the proper subject matter of a geometry is its invariant sets and the invariant functions on those sets.* Hence, the concept of a triangle and, more generally, the whole subject of triangles belongs to Euclidean geometry because the set D of all triangles is invariant in this geometry. Likewise, area and perimeter are properly associated with triangles in Euclidean geometry because $a(\Delta)$ and $p(\Delta)$ are invariant functions. The distance sum, $d(\Delta)$, of a triangle, however, is not a Euclidean concept.

In translational geometry, triangle, area, and perimeter are all legitimate concepts. In addition, translational geometry allows the concept of the direction of a line. For example, the set E of lines parallel to the x-axis (that is, with zero slope) is invariant in this geometry. In fact, the idea of slope, so important in analytic geometry and the calculus, is characteristic of (or proper to) translational geometry, *not* Euclidean geometry. The concept of slope plays no role in Euclidean geometry because (for example) the set E of lines parallel to the x-axis is not invariant there and, more generally, because the rotations of Euclidean geometry change slopes.

The *Erlanger Programm* does not forbid the use of noninvariant figures and/or functions in geometry. Such use may be interesting, practical, and even essential. The *Erlanger Programm* only insists that statements identified as belonging to a particular geometry be expressed exclusively in terms invariant under the transformation group of that geometry.

Summary

Given a geometry, it is invariance, determined by congruence, as specified by a transformation group, that establishes what ideas, quantities, figures, and functions are appropriate, characteristic, important, proper, and, in a word, deserving of study in that geometry. Additionally, invariance is important because, as long as a statement only mentions figures that form invariant sets and only talks about invariants of these figures, then it is guaranteed that, if the statement is true of just one figure, then it will be true of all congruent figures. The theorems of a geometry are those statements that, if true of one figure, are true of all congruent figures.

E–Consider the pair $(\mathbf{C}, \mathbf{R})$, where $\mathbf{R}$ is the set of all rotations of the plane $\mathbf{C}$ about the origin. Show that this *is* a geometry (another toy geometry which we shall call **rotational** geometry).

F–Find a set of figures in the plane that is an invariant set for rotational geometry but is not invariant for Euclidean geometry.

G–Find a function that is an invariant for rotational geometry but is not invariant for Euclidean geometry.

Geometric Proof

Systematic adoption of the *Erlanger Programm* means that all ideas introduced into a geometry must be carefully examined for invariance. Whenever a particular type of figure is defined (a triangle, for example), then the set of all such figures should be invariant. If a dimension of a figure (width or height, for example) or some other computed quantity (area, say) is defined, then the resulting function should be invariant.

One benefit of this point of view is that it clearly classifies and categorizes the subject matter of different geometries. Much more important, however, is the fact that the *Erlanger Programm* provides a powerful proof technique applicable to all theorems of all geometries. (*All* theorems of *all* geometries? YES!)

This works as follows: Let (S, G) be a geometry, and suppose that W is a statement that we wish to prove. Let F be a figure in S to which the statement W applies. Suppose further that the statement W is correctly posed according to the *Erlanger Programm*. This means that all measurements and other quantities mentioned in W are invariants. The effect of this is that if we prove that W is true for some figure $T(F)$ (where T is a transformation from the group G), then W must be true for F, since the measurements associated with $T(F)$ are the *same* as for F. Therefore, to prove W for F, we need only find a transformation T in G so that the proof is particularly simple for $T(F)$. If we can complete the proof in this case, then we're done. Typically, T is chosen so that $T(F)$ is in some special relation to a coordinate system, and W is proven for $T(F)$ by direct computation using coordinates.

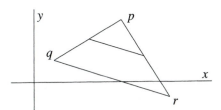

Figure 4.2 A typical triangle

Here is an example from Euclidean geometry:

Theorem *The line joining the midpoints of two sides of a triangle is parallel to and half the length of the third side.*

Proof: Let Δpqr be any triangle. (See Figure 4.2.) Our goal is to apply Euclidean transformations so that the triangle is placed in such a position that the proof is easy. We first apply a rotation so that the side qr is parallel to the x-axis. Then a translation will put the line segment qr on the x-axis in such a way that p lies on the y-axis.

Figure 4.3.) It is important to point out that rotation and translation are Euclidean transformations, so that we are using transformations appropriate to the geometry in which we are working.

In this position, the midpoints mentioned in the theorem are easily calculated. They are $s = (p + q)/2$ and $t = (p + r)/2$. (Remember that p, q, and r are in the complex plane, and so are complex numbers!) Hence,

$$s - t = \frac{p+q}{2} = \frac{p+q}{2} = \frac{1}{2}(q - r)$$

from which it is clear that st, the line joining the midpoints, is parallel to and half the length of the base qr. Since parallelism and length are invariant concepts in Euclidean geometry, confirmation of the theorem in this special case proves it in general. ■

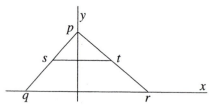

Figure 4.3 The typical triangle in a standard position

The foregoing is the pattern of most proofs in this book, regardless of the geometry being considered. As long as the conclusion of the theorem is expressed in invariant terms (using invariant functions and figures), the proof can be confined to computation of a well-chosen special case. Such proofs are often wonderfully easy. (Of course, if you like hard work, you can use an ill-chosen special case instead!)

Abstract Geometries and their Models

Let D_1 and D_2 be two disks: $D_1 = \{z: |z| < 1\}$ and $D_2 = \{z: |z - 5| < 3\}$. Here D_1 is the so-called (open) **unit disk**, the disk of radius 1 centered at the origin, and D_2 is a disk of radius 3 centered at 5. Let R_1 and R_2 be the groups of rotations of these two disks about their centers. Then (D_1, R_1) and (D_2, R_2) are two (toy) geometries that are really the same: The underlying set is a disk and the group of transformations consists of rotations. Every concept, figure, or invariant of one geometry has a corresponding concept, figure, or invariant in the other. This is captured in the following definition:

Definition *Two geometries* (S_1, G_1) *and* (S_2, G_2) *are* **models** *of the same* **abstract geometry** *if there is an invertible covering* *transformation μ from S_1 to S_2 so that every transformation in G_1 is a lift of a transformation of G_2 and, conversely, every transformation in G_2 is a lift of a transformation of G_1. Alternatively, the geometries (S_1, G_1) and (S_2, G_2) are called **isomorphic** (meaning of the same form) and the covering transformation μ is called an **isomorphism**.*

The situation described in the definition is depicted in Figure 4.4. For (S_1, G_1) and (S_2, G_2) to be models of the same abstract geometry, it must be that for every transformation T_2 in G_2, the composition $T_1 = \mu^{-1}T_2\mu$ is in G_1 and conversely (with G_1 and G_2 switched).

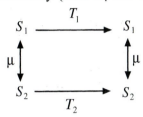

Figure 4.4 A lifting transformation connecting two models

What *is* an abstract geometry really? The odd thing is that we are not in a position to say. One thing, at least, is clear, however: Whatever an abstract geometry is, it has *models*. Furthermore, the only way to study an abstract geometry is to study its models. But, since all concepts, figures, and invariants appearing in one model of an abstract geometry can be lifted to all other models of the abstract geometry, these concepts, figures and invariants can be considered as belonging to the abstract geometry itself.

Sometimes one model of a geometry is preferred over other models for historical reasons, for convenience, or both. This is the case with Euclidean geometry, which is invariably studied through its model in the plane **C**. Even Euclidean geometry, however, has interesting alternative models. For example, consider the function

$$\mu(z) = \frac{z}{\sqrt{1+|z|^2}}$$

which maps the whole complex plane to the inside of the unit circle. (See Exercise 14.) Figure 4.5 shows how a Cartesian coordinate grid and the graph of the function $y = .5\cos(10x)$ appears when squashed into the unit circle by μ. Using μ (and its inverse), we can lift the rigid motions, invariants, and all the rest of Euclidean geometry from the plane to the unit disk, and construct a model of Euclidean geometry there.

The usual preference for the plane model of Euclidean geometry over other models is a matter of convenience. It stems from the intended application of this geometry to the study of real-world situations that are flat, or approximately flat, like the surface of the earth over a small region. The realization by mathematicians and scientists that *all* applications of mathematics are matters of convenience, that mathematics only *models* but never *is* reality, was a consequence of the discovery of non-Euclidean geometries described in Part II of this book. Klein's *Erlanger Programm*, invented in response to this realization, provides the means to organize our knowledge of all geometries.

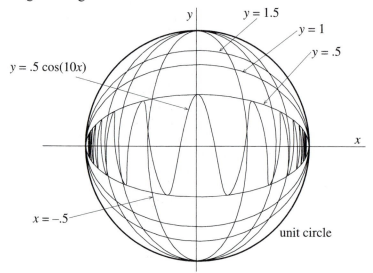

Figure 4.5 A model for Euclidean geometry inside a disk

EXERCISES

Invariant Functions

1. Prove that the distance function

$$d(z_1, z_2) = |z_1 - z_2|$$

is invariant under any Euclidean transformation of the two variables z_1 and z_2. This demonstrates that distance is a Euclidean concept according to the *Erlanger Programm* and justifies the term **rigid motion** for Euclidean transformations.

Hint: Let T be a rigid motion, say, $Tz = e^{i\theta}z + b$. The problem is to show that $d(T(z_1), T(z_2)) = d(z_1, z_2)$. Use properties of the modulus.

2. Prove that the vector function

$$\mathbf{v}(z_1, z_2) = z_1 - z_2$$

(so called because it represents a vector from the point z_2 to the point z_1) is invariant under translations, but *not* rotations. Therefore, **v** is an invariant for translational geometry, but not Euclidean geometry.

3. Prove that the **direction number** of a vector, defined by

$$a(z_1, z_2) = \frac{\mathbf{v}(z_1, z_2)}{|\mathbf{v}(z_1, z_2)|} = \frac{z_1 - z_2}{|z_1 - z_2|}$$

is invariant in translational geometry.

4. Prove that the function of four variables

$$f(z_1, z_2, z_3, z_4) = \frac{a(z_1, z_2)}{a(z_3, z_4)}$$

is invariant under Euclidean transformations (translations *and* rotations). Show that $f(z_1, z_2, z_3, z_4) = 1$ if, and only if, the line segment from z_1 to z_2 is parallel to the line segment from z_3 to z_4. This proves that parallelism is an invariant in Euclidean geometry (although slope is not).

Note: I know you already know that parallelism is a Euclidean concept, but this exercise gives the flavor of an Erlanger Programm *approach to this fact.*

Proofs

5. Prove (in Euclidean geometry): The diagonals of a rectangle bisect each other.

Hint: Use a Euclidean transformation to place the rectangle with one vertex at the origin. Complete the proof in the spirit of the Erlanger Programm; *like the example in the text.*

6. Prove (in Euclidean geometry): The medians of a triangle intersect in a point that is two-thirds the length of each median from each vertex.

7. Prove (in Euclidean geometry): The area of a triangle whose sides have lengths a, b, and c is given by Heron's formula,

$$A = \sqrt{s(s-a)(s-b)(s-c)}$$

where $s = (a + b + c)/2$.

Hint: Place the triangle as in Figure 4.3. Then prove the square *of Heron's formula rather than the formula itself (to avoid square roots of square roots). This is a messy computation, but it works out! A computer algebra system might be of assistance, if one is available.*

Invariant Sets

8. An invariant set is **minimal** if no proper subset of it is invariant. Prove that any two figures in a minimally invariant set are congruent.

Hint: Use proof by contradiction. Let E be an invariant set. If E contains two figures A and B that are not congruent, then consider the set formed by deleting B and all figures congruent to B from E .

9. Let ø be the straight line in the plane with the equation $2x + y = 1$. What is the minimal invariant set containing ø in Euclidean geometry? In translational geometry? In rotational geometry?

10. Let C be the set of all circles in the plane. What are the minimal invariant subsets of C in Euclidean geometry? In translational geometry? In rotational geometry?

Example: Let L be the set of all straight lines in the plane. Because all straight lines are congruent in Euclidean geometry, L is minimally invariant in Euclidean geometry. However, not all lines are congruent in translational geometry; only lines with the same slope are. Therefore, in translational geometry, minimally invariant sets of lines are sets of lines with the same slope. In rotational geometry, minimally invariant sets of lines are sets of lines that are the same distance from the origin.

11. Let S be the set of all squares in the plane. What are the minimal invariant subsets of S in Euclidean geometry? In translational geometry? In rotational geometry?

Subgeometries and Abstract Geometries

12. Let (S', G') and (S, G) be two geometries. If G' is a subset of G and S' is a subset of S, then (S', G') is called a **subgeometry** of (S, G). For example, translational geometry and rotational geometry are subgeometries of Euclidean geometry. Explain how an invariant for (S, G) (set or function) can be used to define an invariant for (S', G').

Note: Thus, the concepts of a geometry automatically appear in all its subgeometries.

13. Show that every abstract geometry has an infinite number of different models.

Hint: The definition of a geometry requires that the underlying set S be non-empty.

14. Find the inverse of the transformation

$$\mu(z) = \frac{z}{\sqrt{1 + |z|^2}}$$

Show that μ maps the whole complex plane onto the inside of the unit disk.

15. Show that there is a model for Euclidean geometry on the unit sphere minus the north pole. In this model, what figure on the sphere corresponds to a straight line in the plane?

16. Find another subgeometry of Euclidean geometry besides translational geometry and rotational geometry. Describe the transformations of your geometry. Give examples of sets of figures and numerical functions that are invariants for your geometry, but not for Euclidean geometry.

17. Invent a geometry of your own. Choose a set and a group of transformations of that set. Prove that your chosen transformations form a group. Define at least one figure and one invariant.

PART II

PLANE GEOMETRY

GEOMETRY AND ART: PLATE II

Stephan Lochner: *Madonna in the Rose Garden*. Tempera on wood (c. 1430-1435) 20"x16". Wallraf-Richartz Museum, Köln.

Like most pre-Renaissance art, this charming painting by the German artist Stephan Lochner makes no attempt to portray three-dimensional space realistically. The world of the figures represented here is simply a plane, delicately ornamented according to geometric principles. The rose arbor itself, for example, is just a rectangular arrangement of lines dressed with leaves and flowers. The painting as a whole has a carefully planned geometric balance (reflection symmetry) and makes heavy use of repeating decorative patterns.

> I am becoming more and more convinced that the necessity of our geometry cannot be proved, at least not by human reason nor for human reason. Perhaps in another life we will be able to obtain insight into the nature of space, which is now unattainable. Until then we must place geometry not in the same class with arithmetic, which is purely *a priori*, but with mechanics.
>
> –Carl Frederich Gauss (1817), in a letter to a friend (quoted in Kline [E1])

This part begins the study of non-Euclidean geometries.

Two preliminary chapters, Chapters 5 and 6, introduce Möbius geometry, a complex plane geometry that includes Euclidean and non-Euclidean geometries as subgeometries. The background provided by this study makes possible a concise analytic treatment of non-Euclidean geometry.

Chapters 7–10 present hyperbolic geometry, the geometry almost discovered by Saccheri and the non-Euclidean geometry most strikingly different from Euclidean geometry.

Chapter 11 presents spherical geometry, the other non-Euclidean geometry.

Chapter 12 introduces absolute geometry, which consists exactly of the results held in common by the three principal plane geometries: hyperbolic geometry, spherical geometry, and Euclidean geometry.

To the reader interested in cosmology and the geometry of the universe, this may seem like an awful lot of time devoted to two-dimensional geometries. It is time well spent, however, since many questions about (and features of) non-Euclidean geometries appear initially in these plane geometries.

5 MÖBIUS GEOMETRY

This chapter presents a very general geometry. The group of transformations used in Möbius geometry is so comprehensive that it includes the transformation groups of all the principal non-Euclidean geometries (and Euclidean geometry). The non-Euclidean geometries we study later are *subgeometries* of Möbius geometry; the concepts, results, and invariants of Möbius geometry appear in them also.

Definition *Let* **C**⁺ *be the complex plane including* ∞, *the point-at-infinity, and let* **M** *be the set of transformations of the form*

$$w = Tz = \frac{az + b}{cz + d}$$

*where a, b, c, and d are complex constants, and the quantity ad – bc (the **determinant** of T) is not zero. Such a transformation is called a* **Möbius transformation**.
The pair **(C⁺, M)** *models* **Möbius geometry**.

Möbius transformations include all the transformations that we have hitherto encountered: rotations, translations, rigid motions, homothetic transformations, and inversion. Conversely, a Möbius transformation can be written in terms of these transformations:

$$w = Tz = \frac{a}{c} - \frac{ad - bc}{c^2}\left(\frac{1}{z + \dfrac{d}{c}}\right) \quad \text{if } c \neq 0$$

Thus when c is not zero, the Möbius transformation T consists of a translation (by d/c), followed by inversion, followed by a homothetic transformation and a rotation (multiplication by the coefficient $-(ad - bc)/c^2$), followed by another translation (by a/c). (When $c = 0$, T is the combination of a translation, a rotation, and a homothetic transformation. See Exercise 3 of Chapter 3).

A–Use long division to verify the preceding formula for T.

B–Choose a, b, c, and d, to show that a Möbius transformation can be a rotation, translation, homothetic transformation, or inversion.

Is Möbius Geometry a Geometry?

To show that Möbius geometry is a geometry, we must verify that the Möbius transformations **M** have the properties required of a transformation group. To do this, we compute the composition of two transformations T and S. If

$$Tz = \frac{az+b}{cz+d}, \quad Sz = \frac{ez+f}{gz+h}$$

then

$$T(S(z)) = \frac{a\left(\dfrac{ez+f}{gz+h}\right)+b}{c\left(\dfrac{ez+f}{gz+h}\right)+d} = \frac{(ae+bg)z+(af+bh)}{(ce+dg)z+(cf+dh)}$$

This shows that the composition of two Möbius transformations is another Möbius transformation. Note that the rule for the composition (or multiplication) of Möbius transformations corresponds to the usual rule for the multiplication of matrices:

$$\begin{pmatrix} a & b \\ c & d \end{pmatrix}\begin{pmatrix} e & f \\ g & h \end{pmatrix} = \begin{pmatrix} ae+bg & af+bh \\ ce+dg & cf+dh \end{pmatrix}$$

The inverse transformation, $z = T^{-1}w$, is

$$z = T^{-1}w = \frac{dw-b}{-cw+a}$$

which is also a Möbius transformation.

Fixed Points

A fixed point of a transformation T is a point z such that $Tz = z$; that is, the point z is not moved by the transformation. In the case of a Möbius transformation, this means that z satisfies the equation

$$z = Tz = \frac{az+b}{cz+d}$$

or

$$c z^2 + (d-a)z - b = 0 \qquad (*)$$

If c is *not* zero, (*) is a quadratic equation. Quadratic equations have one or two roots (see Exercise 15 in Chapter 2), so T has one or two fixed points.

If c *is* zero, then T has ∞, at least, as a fixed point. (See Exercise 7.) If also $a \neq d$, then (*) is linear and has the finite root $b/(d-a)$.

Then T has two fixed points. In either case, T has only one or two fixed points.

A final possibility is that c is zero and $a = d$. In this case, $b = 0$ also. Then T is the identity transformation, and *every* point is a fixed point.

In summary:

Lemma *If T is not the identity transformation, then T has only one or two fixed points. A Möbius transformation with three or more fixed points must be the identity.*

The next theorem is crucial. It describes how much freedom of movement is possible using Möbius transformations. Similar theorems are important in every geometry we study.

Theorem *(The Fundamental Theorem of Möbius Geometry)* *There is a unique Möbius transformation taking any three distinct complex numbers z_1, z_2, z_3 to any other three distinct complex numbers w_1, w_2, w_3.*

Proof: I claim that all we have to do is find a Möbius transformation that sends the three given numbers z_1, z_2, z_3 to the three particular numbers 1, 0, and ∞. In other words, all we need do is find a Möbius transformation T such that

$$Tz_1 = 1, \quad Tz_2 = 0, \quad Tz_3 = \infty$$

If this is possible for *any* given numbers z_1, z_2, z_3, then, using w_1, w_2, w_3 in place of z_1, z_2, z_3, there will also exist a transformation S such that

$$Sw_1 = 1, \quad Sw_2 = 0, \quad Sw_3 = \infty$$

Then the composition $U = S^{-1}T$ is the transformation we seek since

$$
\begin{aligned}
U(z_1) &= S^{-1}T(z_1) = S^{-1}(1) = w_1 \\
U(z_2) &= S^{-1}T(z_2) = S^{-1}(0) = w_2 \\
U(z_3) &= S^{-1}T(z_3) = S^{-1}(\infty) = w_3
\end{aligned}
$$

We next prove that the transformation U is unique. Thus, suppose V is a second transformation sending z_1 to w_1, z_2 to w_2, and z_3 to w_3. Then $V^{-1}U$ has three fixed points. According to the preceding lemma, $V^{-1}U$ must be the identity transformation. Therefore, $V = U$.

Finally, we must find the transformation T used to construct U, the transformation mapping z_1, z_2, z_3 to 1, 0, and ∞. This is

$$Tz = \frac{z - z_2}{z - z_3} \frac{z_1 - z_3}{z_1 - z_2} \qquad (**)$$

C–Check that $Tz_1 = 1$, $Tz_2 = 0$, and $Tz_3 = \infty$.

This completes the proof of the theorem. ∎

Corollary *All figures consisting of three distinct points are congruent in Möbius geometry.*

The quantity opposite Tz in formula (**) is so important that it gets a separate name and definition:

Definition *The **cross ratio** is the following function of four complex variables:*

$$(z_0, z_1, z_2, z_3) = \frac{z_0 - z_2}{z_0 - z_3} \frac{z_1 - z_3}{z_1 - z_2}$$

If z_1, z_2, and z_3 are held constant, then, as a function of z_0, the cross ratio is the unique Möbius transformation sending z_1 to 1, z_2 to 0, and z_3 to ∞.

Using cross ratios, we can summarize the proof of the fundamental theorem as follows: Given six complex numbers z_1, z_2, z_3 and w_1, w_2, w_3, the transformation sending z_1 to w_1, z_2 to w_2, and z_3 to w_3 is found by solving the equation

$$Tz = (z, z_1, z_2, z_3) = (w, w_1, w_2, w_3) = Sw$$

for w in terms of z.

D–Prove the corollary.

E–According to the fundamental theorem, Möbius geometry has enough transformations to map any three points to any other three points. What analogous result is true in translational geometry?

Invariants of Möbius Geometry

We now study Möbius geometry itself. According to the *Erlanger Programm*, properties of a geometry appear as invariants of the group of motions of the geometry. Thus, we proceed by listing the important invariants of Möbius geometry. (Remember that every Möbius invariant is also invariant in all the non-Euclidean geometries we study later.)

Angle Measure

The first thing to point out is that Möbius transformations are conformal. Hence, angle measure is an invariant of Möbius geometry.

Angle measure is one feature, therefore, that Euclidean geometry and the non-Euclidean geometries have in common.

Cross Ratio

The cross ratio (as a function of four variables) is an invariant of Möbius geometry. Here is a complete statement and proof:

Theorem *Consider the figure* $\{z_0, z_1, z_2, z_3\}$ *consisting of four distinct points from the complex plane, and let T be any Möbius transformation. Then*

$$(Tz_0, Tz_1, Tz_2, Tz_3) \;=\; (z_0, z_1, z_2, z_3)$$

Proof: (This proof is from Ahlfors [A1].) Let $Sz = (z, z_1, z_2, z_3)$. In other words, S is the Möbius transformation sending $z_1 \to 1$, $z_2 \to 0$, and $z_3 \to \infty$. Therefore, the transformation ST^{-1} takes $Tz_1 \to 1$, $Tz_2 \to 0$, and $Tz_3 \to \infty$. But, the transformation taking $Tz_1 \to 1$, $Tz_2 \to 0$, and $Tz_3 \to \infty$ is also given by the cross ratio: (z, Tz_1, Tz_2, Tz_3). Furthermore, by the Fundamental Theorem, this transformation is unique, so that,

$$ST^{-1}z \;=\; (z, Tz_1, Tz_2, Tz_3)$$

Therefore,

$$(z_0, z_1, z_2, z_3) \;=\; Sz_0 \;=\; ST^{-1}(Tz_0) \;=\; (Tz_0, Tz_1, Tz_2, Tz_3) \quad \blacksquare$$

Since it is an invariant, the cross ratio measures something about the configuration of four distinct points that has meaning in Möbius geometry. The next theorem tells part of the story.

Theorem *The cross ratio* (z_0, z_1, z_2, z_3) *is real if, and only if, the four points* z_0, z_1, z_2, z_3 *lie on a Euclidean circle or straight line.*

Proof: More precisely, we will prove that z is on the Euclidean circle or straight line determined by the three points z_1, z_2, and z_3 if, and only if, the cross ratio (z, z_1, z_2, z_3) is real. Therefore, let $Tz = (z, z_1, z_2, z_3)$. Since T is a Möbius transformation, T can be written as a fraction

$$Tz = \frac{az+b}{cz+d}$$

for some complex constants a, b, c, and d. When is Tz real? To find out, let's solve the equation

$$\frac{az+b}{cz+d} = Tz = \overline{Tz} = \frac{\bar{a}\bar{z}+\bar{b}}{\bar{c}\bar{z}+\bar{d}}$$

Cross multiplying gives

$$(a\bar{c} - c\bar{a})z\bar{z} + (a\bar{d} - c\bar{b})z + (b\bar{c} - d\bar{a})\bar{z} + (b\bar{d} - d\bar{b}) = 0 \qquad (***)$$

The question is: What z's satisfy (***)? Dividing by the leading coefficient yields

$$z\bar{z} + \left(\frac{a\bar{d} - c\bar{b}}{a\bar{c} - c\bar{a}}\right)z + \left(\frac{b\bar{c} - d\bar{a}}{a\bar{c} - c\bar{a}}\right)\bar{z} + \left(\frac{b\bar{d} - d\bar{b}}{a\bar{c} - c\bar{a}}\right) = 0$$

and completing the square (in a crazy, complex sort of way: see Exercise 16 in Chapter 2) gives

$$\left[z - \left(\frac{d\bar{a} - b\bar{c}}{a\bar{c} - c\bar{a}}\right)\right]\left[\bar{z} - \overline{\left(\frac{d\bar{a} - b\bar{c}}{a\bar{c} - c\bar{a}}\right)}\right] = \frac{d\bar{b} - b\bar{d}}{a\bar{c} - c\bar{a}} + \left|\frac{d\bar{a} - b\bar{c}}{a\bar{c} - c\bar{a}}\right|^2$$

so that

$$\left|z - \left(\frac{d\bar{a} - b\bar{c}}{a\bar{c} - c\bar{a}}\right)\right|^2 = \left|\frac{ad - bc}{a\bar{c} - c\bar{a}}\right|^2$$

This is the equation of a circle.

However, this solution of (***) is only valid if the denominator, $a\bar{c} - c\bar{a}$, is not zero. If $a\bar{c} - c\bar{a}$ is zero, then, letting $\alpha = a\bar{d} - c\bar{b}$ and $\beta = bd$, we find that equation (***) takes the form

$$\text{Im}(\alpha z + \beta) = 0 \qquad (****)$$

which is the equation of a straight line. (See Exercise 20 in Chapter 2.) ∎

F–Check equation (****).

Clines

The preceding theorem indicates that the cross ratio distinguishes circles and straight lines from other figures. Therefore, let us adopt this definition:

Definition *A subset C of the complex plane is a **cline** if C is a Euclidean circle or straight line.*

The word "cline" is a contraction of "'circle" and "line." Since the cross ratio is an invariant function, we expect that the set of clines is invariant in Möbius geometry:

Theorem *If C is a cline, and T is a Möbius transformation, then T(C) is a cline.*

Proof: Let z_1, z_2, and z_3 be three distinct points on C. According to the previous theorem, a fourth point, z, is on C if, and only if, the cross ratio (z, z_1, z_2, z_3) is real. Now, the cross ratio is invariant, so $(z,$

$z_1, z_2, z_3)$ is real if, and only if, (Tz, Tz_1, Tz_2, Tz_3) is real. But (Tz, Tz_1, Tz_2, Tz_3) is real if, and only if, Tz is on the circle or straight line passing through the points $Tz_1, Tz_2,$ and Tz_3. This proves that z is on C if, and only if, Tz is on the circle (or straight line) passing through $Tz_1, Tz_2,$ and Tz_3. Thus, $T(C) = \{Tz : z$ is on $C\}$ is a cline; viz., the cline passing through the points $Tz_1, Tz_2,$ and Tz_3. ∎

Note that if C is a circle, $T(C)$ need not be a circle, and if C is a straight line, $T(C)$ need not be a straight line. In Möbius geometry, circles and straight lines are two manifestations of a *single* geometric figure. All circles and straight lines are congruent to each other; the latter just happen to pass through ∞! Thus, the two pictures in Figure 5.1 are congruent in Möbius geometry. The only difference between them (from a Möbius point of view) is that the point where the three clines meet is an ordinary point in (a), but is ∞ in (b).

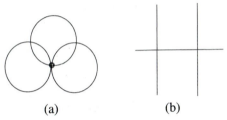

(a)   (b)

Figure 5.1 These are congruent in Möbius geometry.

G–Are all clines congruent in Möbius geometry?

Symmetry

Invariance of the cross ratio leads to another important concept:

Definition *Let C be a cline passing through the three distinct points $z_1, z_2,$ and z_3. Two points z and z^* are called **symmetric** with respect to C if*

$$\left(z^*, z_1, z_2, z_3\right) = \overline{\left(z, z_1, z_2, z_3\right)}$$

The symmetric point z^* is uniquely determined by z and this equation. (See Exercise 21.) Furthermore, if C happens to be a straight line, z^* is the usual point that is mirror symmetric to z across C. (See Exercise 22.)

Symmetry is an invariant of Möbius geometry. Thus, let z and z^* be symmetric with respect to C and let T be a Möbius transformation. We will show that Tz^* and Tz are symmetric with respect to the cline $T(C)$. This follows from the computation

$$(Tz^*, Tz_1, Tz_2, Tz_3) = (z^*, z_1, z_2, z_3)$$

$$= \overline{(z, z_1, z_2, z_3)} = \overline{(Tz, Tz_1, Tz_2, Tz_3)}$$

because $T(C)$ is the cline passing through the points Tz_1, Tz_2, and Tz_3.

To gain some feeling for the meaning of symmetry in the case of a circle, let C be the circle with center a and radius R:

$$C = \left\{ z : |z-a|^2 = R^2 \right\} = \left\{ z : \overline{z} - \overline{a} = \frac{R^2}{z-a} \right\} \qquad (+)$$

Let z_1, z_2, and z_3 be points on C and z be a point not on C. Let z^* be the point symmetric to z. Then

$$\left(z^*, z_1, z_2, z_3 \right) = \overline{\left(z, z_1, z_2, z_3 \right)}$$

$$= \overline{\left(z-a, z_1-a, z_2-a, z_3-a \right)}$$

$$= \left(\overline{z-a}, \overline{z_1-a}, \overline{z_2-a}, \overline{z_3-a} \right)$$

Here we are using the fact that the cross ratio is invariant under the Möbius transformation $z \to z - a$ (a translation) and the obvious fact that

$$\overline{(a,b,c,d)} = (\overline{a}, \overline{b}, \overline{c}, \overline{d})$$

Now, z_1, z_2, and z_3 are on C. Therefore, using (+) we get

$$\left(z^*, z_1, z_2, z_3 \right) = \left(\overline{z} - \overline{a}, \frac{R^2}{z_1-a}, \frac{R^2}{z_2-a}, \frac{R^2}{z_3-a} \right)$$

Next, applying the inversion $z \to R^2/z$, and another translation (namely, $z \to z + a$), we get

$$\left(z^*, z_1, z_2, z_3 \right) = \left(\frac{R^2}{\overline{z}-\overline{a}}, z_1-a, z_2-a, z_3-a \right) = \left(\frac{R^2}{\overline{z}-\overline{a}} + a, z_1, z_2, z_3 \right)$$

As a result, we conclude that

$$z^* = \frac{R^2}{\overline{z}-\overline{a}} + a = a + \left(\frac{R^2}{|z-a|^2} \right)(z-a)$$

and

$$|z^*-a||z-a| = R^2$$

It follows, from these formulae, that z^* lies along the same radius from a as z does, and that the product of the distances of z^* and z

from a is R^2. The geometric relationship between z and z^* is depicted in Figure 5.2.

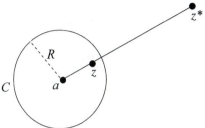

Figure 5.2 Two points symmetric with respect to a circle

Summary

A Möbius transformation is the composition of all the fundamental types of transformations we have studied: rotations, translations, homothetic transformations, and inversions. These transformations can map any three points to any other three points. In the resulting geometry, Möbius geometry, there is one important figure, the cline, and two important invariants, angle measure and the cross ratio. Symmetry, a final important concept, is based on the cross ratio.

EXERCISES

Cross Ratios

1. Verify that the cross ratio $Tz = (z, z_1, z_2, z_3)$ as a function of its first variable (supposing the other three are distinct constants) is a Möbius transformation that sends z_1 to 1, z_2 to 0, and z_3 to ∞.

2. When one of the variables in the cross ratio is ∞, the value of the cross ratio can be determined by taking a limit. Thus,

$$(\infty, z_1, z_2, z_3) = \frac{z_1 - z_3}{z_1 - z_2}$$

since $(z - z_2)/(z - z_3)$ tends to 1 as z tends to ∞.
Find (z_0, ∞, z_2, z_3), (z_0, z_1, ∞, z_3), and (z_0, z_1, z_2, ∞).

Möbius Transformations: Formulas

3. Find a Möbius transformation:
   (a) sending 1 to 4, 0 to i, and ∞ to -1
   (b) sending 0 to 0, i to 1, and $-i$ to 2
   (c) sending 1 to 2, 2 to 3, and 3 to -1

Hint: Use cross ratios, as suggested on page 60.

4. Find a Möbius transformation that takes the circle $|z| = 1$ to the straight line $x + y = 1$.

Hint: Just map three convenient points from the circle to three points on the straight line.

5. What Möbius transformation represents a rotation of $90°$ about the point 1?

6. Find the fixed points of these transformations

$$\frac{2z}{3z-1}, \qquad \frac{3z-2}{2z-1}, \qquad \frac{-2}{z+1}, \qquad \frac{-iz}{(1-i)z-1}$$

7. Let $Tz = (az + b)/(cz + d)$. Show that $T(\infty) = a/c$, if c is not zero, while $T(-d/c) = \infty$. What happens if c is zero?

8. Find all Möbius transformations
   (a) with the fixed points 1 and -1
   (b) with just one fixed point at -1

Möbius Transformations: Theory

9. Let T and S be Möbius transformations. Prove that $|TS| = |T||S|$, where $|T|$ is the determinant of the transformation T.

10. Show that if the determinant of a Möbius transformation were allowed to equal to zero, the transformation would be constant. Why must constant functions be excluded from transformation groups?

11. Prove that a Möbius transformation with a single fixed point at ∞ is a translation.

12. Show that every Möbius transformation can be written in a form with determinant 1. Is this form unique?

Hint: What happens to a Möbius transformation and its determinant when the numerator and denominator are multiplied by the same quantity?

13. Prove that every Möbius transformation is conformal.

Fundamental Theorems

14. The Fundamental Theorem of Möbius geometry states that Möbius transformations are capable of mapping any three points to any other three points. State and prove an analogous Fundamental Theorem for Euclidean geometry

15. State and prove a Fundamental Theorem for translational geometry. What about rotational geometry?

Congruence in Möbius Geometry

16. Prove that all clines are congruent in Möbius geometry.

17. Some parts of Figure 5.3 are congruent in Möbius geometry. Which?

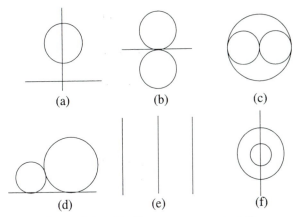

(a)   (b)   (c)

(d)   (e)   (f)

Figure 5.3 A Möbius congruence puzzle

Symmetry

18. Let C be the unit circle. Find the points z^* symmetric with respect to C for $z = 1$, $1/2$, i, $i/2$, $1 + i$, and $(1 + i)/2$. Draw C and all the points z and z^*.

19. Let C be the circle with center i and radius 2. Find the points z^* symmetric with respect to C when $z = 0$, 1, 2, $\sqrt{3}$, and 3. Draw C and all the points z and z^*.

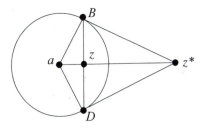

Figure 5.4 Construction of z^*

20. Here is a Euclidean construction for the symmetric point z^* when C is a circle: Let C have center a and radius R. First connect a and z. If z is inside C, draw the perpendicular to the line connecting a and z at z. The perpendicular will intersect C in two points B and D. Construct the tangents to C at these points. These tangents intersect at z^*. If z is outside C, reverse this process, constructing the tangents first and then connecting them. This is illustrated in Figure 5.4.

Prove that z^* is the symmetric point to z with respect to C.

Hint: Use similar triangles.

21. Let C be a circle and z any complex number. Prove that the point z^* symmetric to z with respect to C is unique.

22. Prove that when C is a straight line, the symmetric point z^* is the mirror symmetric point to z across C.

Hint: Consider first the special case where C is the real axis. Use invariance to extend the result from this particular line to all straight lines.

23. Discuss the relationship between symmetry with respect to the unit circle and inversion. Are they the same transformation?

Discussion: The answer is no. On the other hand, symmetry with respect to the x-axis is the same as a familiar operation on complex numbers. What is that operation?

Symmetry and Orthogonality

24. Let C be a straight line. Let z and z^* be distinct symmetric points with respect to C. Prove that any circle C' **orthogonal** (i.e., perpendicular) to C and passing through z must also pass through z^*. Conversely, prove that any circle that passes through z and z^* is orthogonal to C.

Hint: The problem is to prove that C′ passes through z*. (See Figure 5.5.) Use the Erlanger Programm: Show that the figure can be transformed to the case where b is at ∞. What do C and C′ become?

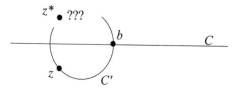

Figure 5.5 Problem 24

25. Let C be a cline. Let z and z^* be distinct symmetric points with respect to C. Prove that any cline C' that is orthogonal to C and passes through z must also pass through z^*. Conversely, prove that any cline that passes through z and z^* is orthogonal to C.

Hint: Use the Erlanger Programm. Consider the case where C is a straight line. Use invariance to extend this special case to all clines. This proof is a good example of the power of the Erlanger Programm. Try to prove this result using Euclidean geometry alone without using Möbius transformations!

6 STEINER CIRCLES

This chapter continues the study of Möbius geometry. Steiner circles are families of circles (families of clines, really) that serve as coordinate systems in Möbius geometry and other geometries. The chapter shows how they are used to classify Möbius transformations into different types: elliptic, parabolic, hyperbolic, and loxodromic. Steiner circles provide a strange kind of graph paper perfect for visualizing Möbius transformations.

Families of clines

Let p and q be two points in the complex plane. There are *two* important families of clines associated with these points. In the first place, there is the family of all clines passing *through p and q*. Most of these clines are circles, but one is a straight line. The whole family constitutes the **Steiner circles of the first kind** with respect to the points p and q (see Figure 6.1).

In order to study these clines further, imagine them subjected to the Möbius transformation

$$w = Sz = \frac{z - p}{z - q}$$

which sends p to zero and q to ∞. The result is shown in Figure 6.1.

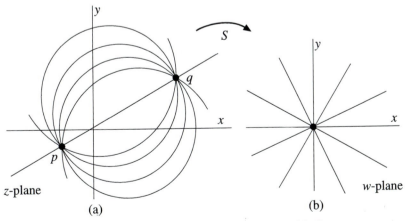

Figure 6.1 Steiner circles of the first kind

In the w-plane, the Steiner circles (of the first kind) become the family of all clines through 0 and ∞, that is, the family of all straight lines through the origin.

In the w-plane, let us now add the orthogonal (i.e., perpendicular) clines: the circles with center at the origin and equations $|w| = k$ (k a constant). The picture in the w-plane now consists of both the straight lines through the origin and the concentric circles with center at the origin. This looks like polar coordinate graph paper. (See Figure 6.2.) Every circle meets every straight line and *vice versa*, and they always meet at right angles.

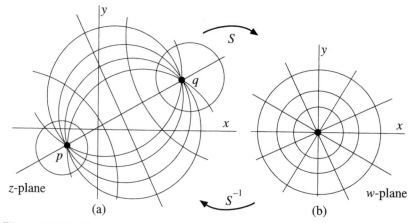

Figure 6.2 Steiner circles of the first and second kinds with respect to two general points and with respect to 0 and ∞

Finally, we pull the circles back to the z-plane via the inverse transformation S^{-1}. The resulting curves in the z-plane are the **Steiner circles of the second kind** (also called **circles of Apollonius**). Since Möbius transformations preserve clines, the Steiner circles of the second kind really *are* circles (instead of some other curve). Furthermore, since the transformations S and S^{-1} are conformal, the two families of circles in the z-plane meet at right angles, the same as those in the w-plane. Since the equation for the concentric circles in the w-plane is $|w| = k$, the Steiner circles of the second kind have the equation

$$\frac{|z - p|}{|z - q|} = k$$

The whole picture in the z-plane (including both families of circles) is a generalization of polar coordinates to Möbius geometry.

A–"Through every point in the plane there passes one and only one Steiner circle of the first kind." This statement has two exceptions. What are they? What about Steiner circles of the second kind?

Normal Form of a Möbius Transformation

Let T be a Möbius transformation with two fixed points p and q. We will analyze T using the Steiner circles with respect to p and q. Steiner circles are relevant because (a) T fixes the points p and q and (b) T maps clines to clines. Therefore, T maps each cline passing through p and q to another such cline. (See Exercise 11.) In other words, the Steiner circles of the first kind with respect to p and q are invariant under T. The Steiner circles of the second kind are also invariant. (See Exercise 13.) Our goal is to watch how T moves these circles around. This will give a graphic picture of the action of T on the whole plane.

To accomplish this task, we take a detour through the auxiliary w-plane using the transformation $w = Sz = (z - p)/(z - q)$ introduced in the previous section. Consider S and S^{-1} as a pair of covering transformations between the w-plane and the z-plane. (The definition of "covering transformation" is in Chapter 3.) Using S^{-1} we can lift T from the z-plane to a transformation R of the w-plane by setting

$$Rw = STS^{-1}w$$

Because T has p and q as fixed points, R has fixed points 0 and ∞. (See Exercise 10.) If we write R as a fraction

$$Rw = \frac{aw + b}{cw + d}$$

(as we can any Möbius transformation), the fact that 0 is a fixed point means that $b = 0$, while the fact that ∞ is fixed means that $c = 0$. Thus,

$$Rw = \frac{a}{d}w = \lambda w$$

where $\lambda = a/d$. In other words, Rw is a very simple transformation, namely, multiplication of w by a constant. This very specific form for R gives information about the transformation T in several ways.

In the first place,
$$ST = RS$$
or

$$\frac{Tz - p}{Tz - q} = \lambda \frac{z - p}{z - q}$$

This equation is called the **normal form** of T. It is a formula for T in terms of just three constants: p, q (T's fixed points) and λ. The normal form of T is more informative than the form $(az + b)/(cz + d)$, since the constants p, q, and λ are directly connected to the action of T whereas a, b, c, and d are not. That exactly three constants can specify T is a reflection of the Fundamental Theorem of Möbius Geometry.

Next, we write

$$T = S^{-1}RS$$

expressing the fact that, as R is the lift of T to the w-plane, conversely T is the lift of R back to the z-plane. This suggests that T be understood as the result of three operations: first S (sending the fixed points to 0 and ∞), then R (multiplication by a constant), and last S^{-1} (putting the fixed points back at p and q). This view makes the nature of R and the constant λ central to understanding T.

In particular, there are three essentially different types of transformation depending on the value of λ.

Case 1. Elliptic Transformations

Suppose that λ has modulus 1. Then $\lambda = e^{i\theta}$ for some real constant θ. In this case, R is a rotation about the origin, and T is called an **elliptic** transformation. The action of T (corresponding to the rotating action of R) is to move points *along* the Steiner circles of the second kind (the circles of Apollonius) in a swirling motion around the fixed points. (See Figure 6.3.)

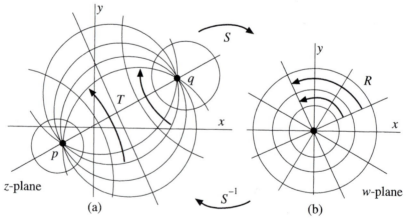

Figure 6.3 Graphing an elliptic transformation

Case 2. Hyperbolic Transformations

Suppose that λ is real and positive, that is, $\lambda > 0$. Then R is a homothetic transformation and T is called a **hyperbolic** transformation. The action of T is to move points along the Steiner circles of the first kind in a kind of flow from one of the fixed points toward the other. (See Figure 6.4.)

Case 3. Loxodromic Transformations

If λ is a general complex quantity, then $\lambda = ke^{i\theta}$, where neither $k = 1$ nor $\theta = 0$. In this case, T is called **loxodromic**. The action of T is a combination of the motions of an elliptic and a hyperbolic transformation.

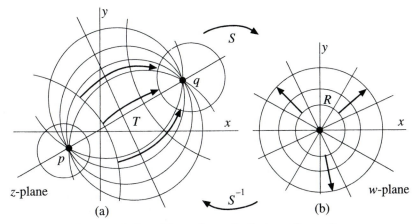

Figure 6.4 Graphing a hyperbolic transformation

Conclusions

(1) Together, the two kinds of Steiner circles constitute generalized polar coordinate systems for Möbius geometry.

(2) Any Möbius transformation with the two fixed points p and q must transform each Steiner circle with respect to p and q to another Steiner circle of the same type with respect to p and q.

(3) The simplest types of transformation, elliptic and hyperbolic, move points along one or another of the two families of Steiner circles.

(a) Elliptic transformations move points along Steiner circles of the second kind (see Figure 6.5a). (The **orbits** of T lie on the circles of Apollonius.) In the special case where one of the fixed points is ∞, an elliptic transformation is a rotation.

(b) Hyperbolic transformations move points along Steiner circles of the first kind. (See Figure 6.5b.) In case one of the fixed points is ∞, a hyperbolic transformation is a homothetic transformation.

(4) A more complicated transformation, the loxodromic, combines the preceding two motions.

(5) The normal form of a transformation is an expression of the relationship between the transformation and the Steiner circle coordinate system determined by its fixed points.

B–Draw a picture like the ones in Figure 6.5 illustrating the motion of a loxodromic transformation about its fixed points p and q.

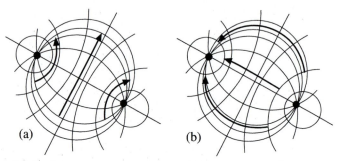

(a)   (b)

Figure 6.5 Graph of (a) an elliptic transformation and (b) a hyperbolic transformation

Parabolic Transformations

In addition to elliptic, hyperbolic, and loxodromic transformations, there are Möbius transformations with only *one* fixed point. These are called **parabolic**. Their behavior can be understood with the aid of a third type of Steiner circle.

Let $w = Tz$ be a transformation whose one fixed point is p. Let S now be the transformation

$$w = Sz = \frac{1}{z - p}$$

simply sending p to ∞. Like T, the composition $R = STS^{-1}$ has just one fixed point, now at ∞. (See Exercise 10.) Therefore, R is a translation (See Exercise 11 of Chapter 5); that is,

$$Rw = w + \beta$$

for some constant β. The transformation T is completely determined by the constants p and β. Thus, since $ST = RS$, we get the formula

$$\frac{1}{Tz - p} = \frac{1}{z - p} + \beta$$

which is the **normal form** of a parabolic transformation.

To visualize R and T, draw the family of parallel lines in the w-plane oriented in the direction given by β (see Figure 6.6). R translates every point in the w-plane along one of these lines.

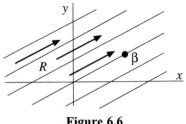

Figure 6.6

If we now apply the transformation S^{-1}, we obtain a family of circles in the z-plane. Since the given lines (in the w-plane) meet only at infinity, the circles (in the z-plane) meet only at the point p, where they must be mutually tangent. The resulting circles are called **degenerate Steiner circles**. (See Figure 6.7.)

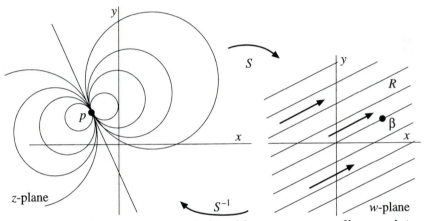

Figure 6.7 Degenerate Steiner circles with respect to an ordinary point [and with respect to ∞]

In the w-plane, we now add a perpendicular family of lines. The two families of lines in the w-plane can be thought of as coordinate lines for a Cartesian coordinate system that has been rotated in order to study the transformation R. (See Figure 6.8.) Note that R moves points along one set of lines, while each line in the other set is transformed to another line from that set.

Applying S^{-1} to this picture, we get two perpendicular families of degenerate Steiner circles in the z-plane. (See Figure 6.8.) These families of circles can be thought of as curves of a **generalized Cartesian coordinate system**. In this example, the system is arranged in order to study the parabolic transformation T that transforms each Steiner circle into another Steiner circle in the same family and actually moves points along the circles of one of the families.

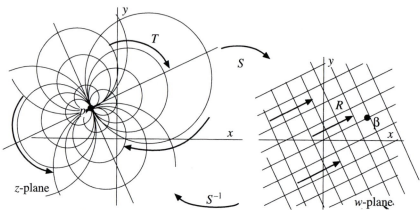

Figure 6.8 Graphing a parabolic transformation

Summary

Steiner circles provide families of curves that generalize Cartesian and polar coordinate systems to Möbius geometry. Generalization of the Cartesian system uses two families of degenerate Steiner circles (mutually tangent circles). This generalization is used to study transformations with just one fixed point (parabolic transformations). The generalization of polar coordinates uses two families of circles: the Steiner circles of the first and second kinds with respect to two points. This generalization is used to study transformations with two fixed points (elliptic, hyperbolic, and loxodromic transformations).

EXERCISES

Normal Forms

1. Let

$$\frac{Tz-p}{Tz-q} = \lambda\,\frac{z-p}{z-q}$$

be the normal form of a Möbius transformation with two fixed points. Prove that $\lambda = (Tz, z, p, q)$, where z is any complex number.

Note: This provides a convenient means for computing λ.

Hint: Solve the normal form for λ.

2. Let

$$\frac{1}{Tz-p} = \frac{1}{z-p} + \beta$$

be the normal form of a parabolic transformation whose fixed point p is not ∞. Let z_0 be the point that T sends to ∞. Prove that

$$\beta = -\frac{1}{z_0 - p} = \frac{1}{T(\infty) - p}$$

Note: This provides a means for computing β.

3. Analyze each of the following Möbius transformations by finding the fixed points, finding the normal form, and sketching the appropriate coordinate system of Steiner circles indicating the motion of the transformation.

(a) $Tz = \dfrac{z}{2z - 1}$ (b) $Tz = \dfrac{3z - 4}{z - 1}$

(c) $Tz = \dfrac{z}{-z + 2}$ (d) $Tz = \dfrac{-z}{(1 + i)z - i}$

Example: Analysis of the Möbius transformation $Tz = -z/(2z - 3)$. To find the fixed points, we solve the equation

$$z = \frac{-z}{2z - 3}$$

or

$$z^2 - z = 0$$

The roots are 0 and 1. Next, λ is found using the result of Exercise 1 using $z = 2$ (any z except 0 or 1 would do): $\lambda = (Tz, z, 0, 1) = (-2, 2, 0, 1) = 1/3$. Therefore, the transformation is hyperbolic with normal form

$$\frac{Tz}{Tz - 1} = \frac{1}{3}\frac{z}{z - 1}$$

A graph of the action of T looks roughly like Figure 6.9.

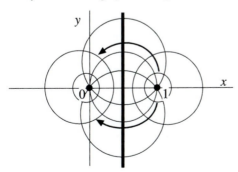

Figure 6.9 Example for Exercise 3

4. Find the normal form of the Euclidean transformation $Tz = e^{i\theta}z + b$.

Möbius Transformations: Classification

5. When is the Euclidean transformation $Tz = e^{i\theta}z + b$ elliptic? hyperbolic? parabolic? loxodromic?

6. What kind of transformation (elliptic, parabolic, hyperbolic, or loxodromic) is inversion? What is its normal form?

7. Find formulas in the form $(az + b)/(cz + d)$ for Möbius transformations satisfying the following descriptions:

   (a) an elliptic transformation with fixed points at i and $-i$ and $T(\infty) = 0$

   (b) a parabolic transformation with fixed point $1 + i$ and $T(\infty) = 100$

   (c) a hyperbolic transformation with fixed points 5 and 10 and $T(\infty) = 20$.

Hint: Use normal forms.

Möbius Transformations: Theoretical Results

8. For a hyperbolic or a loxodromic transformation, the limit

$$\lim_{n \to \infty} T^n z$$

has the same value regardless of the value of z, with one exception. What is this limit and what is the exceptional z?

9. A transformation such that $T^2 = I$ (the identity transformation) is called an **involution**. Prove that an involutory Möbius transformation must be elliptic. What must λ be for such a transformation?

Hint: Given the normal form of T, what is the normal form of T^2?

10. Let R, S, and T be transformations such that $T = S^{-1}RS$. Prove that z is a fixed point of T if ,and only if, Sz is a fixed point of R.

11. Let the Möbius transformation T take z_1 to w_1 and z_2 to w_2. Prove that T transforms any Steiner circle of the first kind with respect to z_1 and z_2 into a Steiner circle of the first kind with respect to w_1 and w_2.

12. Prove that the only circles perpendicular to all Steiner circles of the first kind with respect to given points p and q are the Steiner circles of the second kind with respect to p and q.

Hint: This is a theorem of Möbius geometry. Give a proof in the spirit of the Erlanger Programm by placing the points p and q in the best possible location.

13. Let the Möbius transformation T take z_1 to w_1 and z_2 to w_2. Prove that T transforms any Steiner circle of the second kind with respect to z_1 and z_2 into a Steiner circle of the second kind with respect to w_1 and w_2.

Theorems of Möbius Geometry

14. Let C_1 and C_2 be two nonintersecting clines. Prove that there is a unique pair of points that are simultaneously symmetric to both C_1 and C_2.

Hint: Consider first the special case where one of the clines is a straight line.

15. Let C_1 and C_2 be two clines. Find and describe all clines perpendicular to C_1 and C_2 under the following circumstances:

    (a) C_1 and C_2 intersect in two points

    (b) C_1 and C_2 are tangent

    (c) C_1 and C_2 do not intersect

Hint: Consider special cases where the answer, based on Euclidean considerations, is more or less obvious. Then express your answer in invariant terms that make sense when the clines are in general position.

Example: (a) In Möbius geometry, two intersecting clines are congruent to two intersecting straight lines, since one of the points of intersection can be moved to ∞. From Euclidean geometry, we know that a circle orthogonal to a straight line has that straight line as diameter. Therefore, a circle orthogonal to two intersecting lines has both lines as diameters and the intersection of the lines as its center. (Draw a picture!)

In invariant terms, the circle is a circle of Apollonius with respect to the intersection point and ∞. (Problem 13 proved that circles of Apollonius are invariant.) Therefore, the clines orthogonal to two intersecting clines are the circles of Apollonius with respect to the two points of intersection of the given clines.

Note: This exercise is important to hyperbolic geometry.

16. Given two nonintersecting clines and a third that intersects the first two, prove that there is a unique fourth cline perpendicular to the three given ones.

7 HYPERBOLIC GEOMETRY

At last we are ready to study hyperbolic geometry, the first non-Euclidean geometry to be discovered, the revolutionary geometry of Gauss, Bolyai, and Lobatchevsky.

After so much time floating around in the rarefied atmosphere of Möbius geometry, hyperbolic geometry will seem very concrete, almost familiar, with straight lines, distances, circles, and triangles very much like Euclidean geometry. However, the theory of parallels is completely altered from Euclidean geometry, and it is remarkable how this difference creates strange consequences throughout hyperbolic geometry.

We eventually present two models of hyperbolic geometry. The first uses the **unit disk D** as its underlying set. **D** is defined as $\{z: |z| < 1\}$. The second uses the **upper half plane U** $= \{z: z = x + iy, y > 0\}$. This chapter introduces the disk model. The half plane model appears in Chapter 9.

For the disk model, the group of transformations consists of all Möbius transformations that map **D** onto itself. Thus, hyperbolic geometry is a natural geometry from a mathematical point of view: The underlying set is a disk (the most natural bounded shape imaginable), and the group of transformations is as large as possible, consistent with this choice of set (the group of *all* Möbius transformations of the disk).

Therefore, the first question is: Which Möbius transformations map the unit disk onto itself? To answer this question, let T be a Möbius transformation, and suppose that $|Tz| < 1$ whenever $|z| < 1$. In particular, T maps some point z_0 to zero, where z_0 is inside the unit disk; that is, $|z_0| < 1$. The symmetric point $1/\bar{z}_0$ (symmetric with respect to the unit circle) is sent to ∞, and some third point is sent to 1. Hence,

$$Tz = \alpha \left(\frac{z - z_0}{z - \dfrac{1}{\bar{z}_0}} \right) = \lambda \left(\frac{z - z_0}{1 - \bar{z}_0 z} \right)$$

where the factor α is determined by the point that T sends to 1 and $\lambda = -\bar{z}_0 \alpha$. Moving to the boundary of **D**, if $|z| = 1$, then, because T is continuous, $|Tz| = 1$ also. This means that

$$1 = |Tz| = |\lambda| \left| \frac{z - z_0}{1 - \bar{z}_0 z} \right| = |\lambda| \left| \frac{z - z_0}{\bar{z}z - \bar{z}_0 z} \right| = |\lambda| \frac{|z - z_0|}{|\bar{z} - \bar{z}_0||z|} = |\lambda|$$

where we have used the fact that $1 = \bar{z}z$. Since $|\lambda| = 1$, $\lambda = e^{i\theta}$ for some θ, and so

$$Tz = e^{i\theta} \frac{z - z_0}{1 - \bar{z}_0 z}$$

where $|z_0| < 1$. This, therefore, is the form that T must have in order to map the unit disk onto itself. Conversely, if a transformation T is given by the preceding equation, it is easy to show that T does map the unit disk onto itself. (See Exercise 1.) Note that T is specified by just two constants: z_0 and λ.

Definition *Let* **D** *be the unit disk in the complex plane; that is,* **D** = $\{z: |z| < 1\}$. *Let* **H** *be the set of transformations of* **D** *of the form*

$$Tz = e^{i\theta} \frac{z - z_0}{1 - \bar{z}_0 z} \qquad (+)$$

where $|z_0| < 1$. *The pair* (**D**, **H**) *models* **hyperbolic geometry**.

The set **D** will be called the **hyperbolic plane**. The group **H** is the **hyperbolic group**. Since **H** is a subgroup of **M** and **D** is a subset of **C**$^+$, hyperbolic geometry is a subgeometry of Möbius geometry. Therefore, every statement true in Möbius geometry is also true in hyperbolic geometry. However, every figure in hyperbolic geometry has fewer congruent figures than in Möbius geometry (because hyperbolic geometry has fewer transformations than Möbius geometry does).

A–Compare the formula for a transformation from the hyperbolic group with the formula for a Euclidean rigid motion (given in Chapter 4). How many constants appear in the two formulas and how are they placed?

Straight Lines

In Möbius geometry, we studied just one figure: the cline (Euclidean circles + Euclidean straight lines). In hyperbolic geometry it is no longer the case that all clines are congruent. For example, the unit circle is unique: *No* other circle or Euclidean straight line is congruent to the unit circle. (Of course, the unit circle is not really in the hyperbolic plane. We shall see later that it acts as a line at infinity.)

A second example of a special class of circles is provided by the circles that happen to be perpendicular to the unit circle. (See Figure

7.1.) These play the role of straight lines in hyperbolic geometry, as the following definition and theorem explain.

Definition *A **hyperbolic straight line** is a circle or Euclidean straight line in the complex plane that intersects the unit circle at a right angle.*

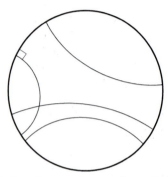

7.1 The hyperbolic plane and some hyperbolic straight lines

Theorem *In hyperbolic geometry, all hyperbolic straight lines are congruent. Two points in the hyperbolic plane determine a unique hyperbolic straight line.*

Proof: Symmetry is the key to this proof. Let T be a transformation of hyperbolic geometry, and let z be any point in the hyperbolic plane. Let z^* be the point symmetric to z with respect to the unit circle. Since symmetry is preserved by all Möbius transformations, the points Tz and Tz^* will be symmetric with respect to the circle obtained by applying T to the unit circle. But T maps the unit circle to itself. This proves the following:

Lemma *Each transformation of hyperbolic geometry maps each pair of points symmetric with respect to the unit circle to another pair of points symmetric with respect to the unit circle.*

Now, let C be any hyperbolic straight line, and let z_0 be any point of the hyperbolic plane on C. Since C is a cline that is perpendicular to the unit circle, C must also pass through the point z_0^* that is outside the unit circle and symmetric to z_0 with respect to the unit circle. (See Exercise 25 in Chapter 5.)

Let T be a transformation from the hyperbolic group sending z_0 to 0 as in equation (+). The symmetric point z_0^* will be sent to ∞. Therefore, $T(C)$ is a *Euclidean straight line* perpendicular to the unit circle and passing through its center. In other words, $T(C)$ is a *diameter* of the unit circle. Finally, choose θ so that T will rotate C, making $T(C)$ become the x-axis. The action of T is illustrated in Figure 7.2.

This proves that every hyperbolic straight line C is congruent to a specific standard straight line: the x-axis. Therefore, all hyperbolic straight lines are congruent. Note that the constants z_0 and θ allow us exactly this much freedom of movement: to choose one point to send to the origin and then rotate.

Next, let z_1 and z_2 be any two points in the hyperbolic plane. To prove that there is a unique hyperbolic straight line through z_1 and z_2, we may place z_1 and z_2 in any congruent position convenient to us. Using the same freedom of motion as in the first part of the proof, we can place z_1 at the origin and then rotate z_2 to a point somewhere along the positive x-axis.

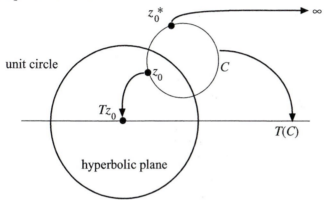

Figure 7.2 Putting a straight line in a standard position in hyperbolic geometry

Now, every hyperbolic straight line passing through 0 also passes through the symmetric point ∞ with respect to the unit circle. In other words, every hyperbolic straight line passing through 0 ($= z_1$) is an ordinary Euclidean straight line.

From Euclidean geometry, there is a unique Euclidean straight line passing through z_1 ($= 0$) and z_2 (on the positive x-axis). (This line is, in fact, the x-axis.) Since, in the case $z_0 = 0$, the Euclidean straight lines through z_1 are the same as the hyperbolic straight lines, this proves the theorem. ∎

B—Compare the freedom of movement permitted by the hyperbolic group with that of the Euclidean group.

Euclid's Postulates

Let us now survey Euclid's postulates from the point of view of hyperbolic geometry:

POSTULATE 1. *Two points determine a straight line.* We have just proved this.

POSTULATE 2. *A line can be produced indefinitely in either direction.* When we introduce the hyperbolic distance function (in Chapter 9), this axiom will be true because the unit circle will be infinitely far away from the points inside the unit disk, and so hyperbolic straight lines will be infinitely long.

POSTULATE 3. *A circle can be described with any center and radius.* This will also prove true by virtue of the infinite remoteness of the unit circle.

POSTULATE 4. *All right angles are congruent.*   The transformations of the hyperbolic group are conformal, so that this axiom is also true in hyperbolic geometry. (See Exercise 8.)

POSTULATE 5. (Playfair's Axiom) *Through a point not on a line there is a unique line parallel to the given line.*   This is complicated enough to receive separate treatment.

Parallelism

In Euclidean geometry, parallel lines are defined to be lines that never meet no matter how far they are extended. Adopting the same definition in hyperbolic geometry, we see that parallelism occurs very often. Two different types of parallelism may be distinguished:

Definition *The points on the unit circle are called **ideal points**.*
  *Two hyperbolic lines are called **parallel** if they do not intersect inside **D** but do share one ideal point. Two hyperbolic lines are called **hyperparallel** if they do not intersect inside **D** and do not have an ideal point in common.*

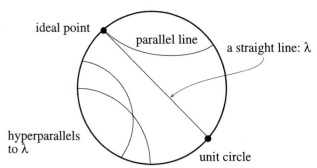

Figure 7.3 Types of parallelism

Figure 7.3 shows a hyperbolic straight line λ, a line parallel to λ, and several lines hyperparallel to it.   Notice that two lines hyperparallel to a given line can intersect, a situation that never happens in Euclidean geometry.

direction) is less than 180°, the given lines may be hyperparallel, may be parallel, or may intersect (in that direction).

7. Show that every acute angle is an angle of parallelism for some point and some hyperbolic straight line.

Results of Hyperbolic Geometry

8. Prove that all right angles are congruent in hyperbolic geometry.

Hint: Give an Erlanger Programm *proof. Use a transformation from* **H** *to put all right angles in a standard position.*

9. Given a point and a hyperbolic line not passing through it, prove that there is a perpendicular hyperbolic line from the point to the given hyperbolic line in hyperbolic geometry.   Is this perpendicular unique?

Hint: Give an Erlanger Programm *proof. Use a transformation from* **H** *to put the given point in the most advantageous place in* **D**.

10. Given an acute angle, show that *not* every point inside the angle lies on a segment of a hyperbolic straight line joining points on the sides of the angle.

11. Prove that two hyperparallel lines have a unique common perpendicular.

Hint: See Exercise 16 in Chapter 6.

12. Prove that the sum of the angles of a triangle in hyperbolic geometry is less than 180°.

8 CYCLES

In the last chapter, we studied the role of two special clines in hyperbolic geometry: (a) the *unit circle* (which we still claim acts as a "line at infinity") and (b) *clines perpendicular to the unit circle* (which act as straight lines). In this chapter, we consider the role in hyperbolic geometry played by the remaining clines. The three possibilities are given in the following definition and displayed in Figure 8.1.

Definition *Let C be the portion of a Euclidean circle (or straight line) inside the unit disk. Suppose that C is not perpendicular to the unit circle. Then C is called a* **cycle**. *If C is entirely contained in* **D**, *then C is a* **hyperbolic circle**. *If C is tangent to the unit circle, then C is a* **horocycle**. *Finally if C intersects the unit circle (at an angle other than a right angle), C is a* **hypercycle**.

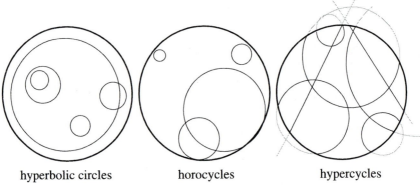

| hyperbolic circles | horocycles | hypercycles |

Figure 8.1 Examples of cycles

It is amazing, but true, that each type of cycle is closely connected with one of the fundamental types of Möbius transformation: elliptic, parabolic, or hyperbolic. We will explore this connection now. First, here is a plausible result:

Lemma *Let T be a Möbius transformation, C a circle, and z a point. If T(C) = C, and z is a fixed point for T, then the point z* symmetric with respect to C is also fixed by T.*

Proof: See Exercise 4. ■

We apply this lemma to a transformation that leaves the unit circle invariant, in other words, a transformation T from the hyperbolic group **H**. If T has a fixed point inside the unit circle, then, according to the lemma, the symmetric point outside is also a fixed point. Thus, there are just three possible locations for the fixed points of transformations from the hyperbolic group: (1) one fixed point is *inside* the unit circle and a second one is *outside*, (2) there are two fixed points *on* the unit circle, or (3) there is only one fixed point (which must be *on* the unit circle).

Let us examine these three cases in more detail.

Elliptic Transformations and Hyperbolic Circles

An elliptic transformation T has two fixed points: one inside the unit circle and the other the symmetric point outside the circle. It is not possible for the fixed points to be *on* the unit circle; that would make the unit circle a Steiner circle of the first kind with respect to the fixed points, but the Steiner circles of the first kind are *not* invariant for an elliptic transformation.

Let p and p^* be the fixed points of T, $|p| < 1$ and $|p^*| > 1$. Any circle passing through p and p^* is perpendicular to the unit circle (see Exercise 25 in Chapter 5) and, therefore, is a hyperbolic straight line. Thus, the Steiner circles of the first kind with respect to p and p^* are hyperbolic straight lines; in fact, they constitute the family of *all* hyperbolic straight lines passing through the point p.

On the other hand, the unit circle is a Steiner circle of the second kind with respect to p and p^*. The other circles of Apollonius with respect to p and p^* (at least those inside the unit circle) are hyperbolic circles. The transformation T rotates points around these circles, creating a circulatory motion about the fixed point p. For this reason, T is called a **hyperbolic rotation**. (See Figure 8.2.)

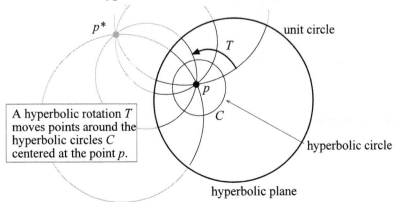

A hyperbolic rotation T moves points around the hyperbolic circles C centered at the point p.

Figure 8.2 An elliptic transformation is like a Euclidean rotation.

Conversely, let C be a hyperbolic circle; in other words C is a Euclidean circle that just happens to be inside the unit circle. Then the family of all circles perpendicular to C *and* perpendicular to the unit circle is a family of Steiner circles of the first kind with respect to two points p and p^* symmetric with respect both to the unit circle and C, where $|p| < 1$ and $|p^*| > 1$. (See Exercise 15c in Chapter 6.) The point p is called the **center** of C and the Steiner circles of the first kind are called **diameters** of C.

Parabolic Transformations and Horocycles

If T is a parabolic transformation, then it has just one fixed point, and that fixed point must be *on* the unit circle. Let p be this fixed point. Since p is an ideal point in hyperbolic geometry, the family of all hyperbolic straight lines through p is a family of parallel lines. From a Euclidean point of view, these hyperbolic straight lines are the family of Euclidean circles, mutually tangent at p and perpendicular to the unit circle. This is a family of degenerate Steiner circles.

On the other hand, the perpendicular family of degenerate Steiner circles includes the unit circle and a number of horocycles inside the unit circle. All these horocycles share the same ideal point p. The transformation T moves points along the horocycles. The hyperbolic straight lines through p are the **diameters** of the horocycles; the transformation T is called a **parallel displacement**. (See Figure 8.3.) Parallel displacement is a weird kind of translation in which points move toward a particular ideal point-at-infinity and also away from the same point!

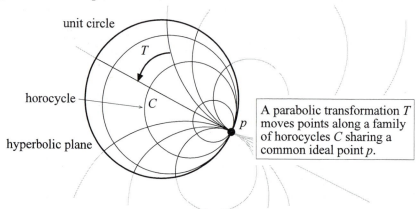

unit circle

horocycle

hyperbolic plane

A parabolic transformation T moves points along a family of horocycles C sharing a common ideal point p.

Figure 8.3 A parabolic transformation is like a Euclidean translation.

The horocycle is a unique and fascinating curve. Euclidean geometry has nothing like it. Horocycles are not straight lines; they are curved. But, like straight lines, horocycles are self-congruent to themselves at every point. (See Exercise 7.)

Hyperbolic Transformations and Hypercycles

If T is hyperbolic, then it has two fixed points, p and q, say, and these fixed points are *on* the unit circle. Thus the unit circle is a Steiner circle of the first kind with respect to p and q. This makes sense, since a hyperbolic transformation does leave invariant the Steiner circles of the first kind with respect to the transformation's fixed points. In addition to the unit circle, the family of Steiner circles of the first kind with respect to p and q includes one hyperbolic straight line: the hyperbolic straight line determined by p and q. The other Steiner circles of the first kind are hypercycles.

On the other hand, the Steiner circles of the second kind with respect to p and q are perpendicular to all the Steiner circles of the first kind, including the unit circle. This make them a family of mutually hyperparallel hyperbolic straight lines.

Conversely, let C be a hypercycle. Then C intersects the unit circle in two points, say, p and q. C is a Steiner circle of the first kind with respect to these two points, the unit circle is another, and the hyperbolic straight line determined by the two ideal points is a third. It will turn out, after we introduce the hyperbolic distance function, that the perpendicular distance from C to the straight line connecting p and q is constant. For this reason, hypercycles are also called **equidistant curves**. A hyperbolic transformation T with the fixed points p and q moves points along the equidistant curves from one fixed point toward the other. Such a transformation is called a **hyperbolic translation**. (See Figure 8.4.) The unique straight line, among the Steiner circles of the first kind connecting p and q, is called the **axis of translation**. The Steiner circles of the second kind are the family of all hyperbolic straight lines perpendicular to the axis of translation.

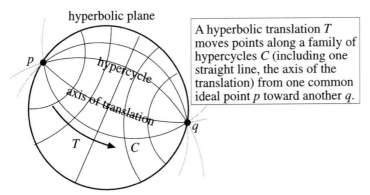

A hyperbolic translation T moves points along a family of hypercycles C (including one straight line, the axis of the translation) from one common ideal point p toward another q.

Figure 8.4 A hyperbolic transformation is a second kind of translation.

Summary

If we loosely define a **cycle** as a curve in (any) geometry traced out when a point is repeatedly transformed by a single congruence transformation, then, since there are three types of transformation in hyperbolic geometry, there are three types of cycles.

A **hyperbolic circle** is the curve, C, traced out by a point in the hyperbolic plane that is repeatedly subjected to an elliptic transformation. Such a transformation has one fixed point p inside the unit disk, and another symmetrically placed outside. The transformation is called a hyperbolic rotation and does indeed rotate points in the hyperbolic plane around the fixed point p. The curve C (see Figure 8.2) surrounds the fixed point p, which is called its center. In the terminology of Möbius geometry, a hyperbolic circle C is a circle of Apollonius with respect to p and p^*; in Euclidean terms, it is a circle inside the unit disk; in hyperbolic terms, it is simply a curve that is like what is called a circle in Euclidean geometry.

A **horocycle** is the curve traced out by a point in the hyperbolic plane that is subjected to a parabolic transformation. Such a transformation has one fixed point p on the unit circle. The transformation is called a parallel displacement, and, under its influence, points move into the unit disk away from the ideal point p and then back out to p. (See Figure 8.3.) In the terminology of Möbius geometry, a horocycle is a degenerate Steiner circle with respect to the point p; in Euclidean terms, it is a circle tangent to the unit circle; in hyperbolic terms, it is a unique curve unlike anything in Euclidean geometry.

A **hypercycle** is the curve traced out by a point in the hyperbolic plane that is subjected to a hyperbolic transformation. Such a transformation has two fixed points p and q, both on the unit disk. The transformation is called a hyperbolic translation, and, under its influence, points move into the unit disk from one ideal point, p or q, toward the other. The cycle traced out by this motion is a curve (see Figure 8.4), whereas in Euclidean geometry the analogous motion would create a straight line. In the terminology of Möbius geometry, a hypercycle is a Steiner circle of the first kind with respect to p and q; in Euclidean terms, it is an arc of a circle not perpendicular or tangent to the unit circle; in hyperbolic terms, it is a curve that stays a constant distance from a straight line.

EXERCISES

Cycles in General

1. Show that there exists a tangent hyperbolic straight line at every point on a hyperbolic circle, horocycle, or hypercycle. Show that this tangent is perpendicular to the diameter at that point.

Hint: Use the Erlanger Programm. *Place the desired point of tangency at a convenient point of the hyperbolic plane.*

2. Show that every hyperbolic triangle has a circumscribed cycle. Show by examples that this cycle can be a hyperbolic circle, horocycle, or hypercycle.

Hint: This exercise just calls for drawing some pictures.

3. A **fundamental arc** on a cycle is an arc such that the tangent at one end is parallel to the diameter at the other end. Draw examples of fundamental arcs on (a) a circle in Euclidean geometry, (b) a hyperbolic circle, and (c) a horocycle.

Cycles and Transformations

4. Prove the lemma at the beginning of the chapter.

5. Prove that a transformation from the hyperbolic group cannot be loxodromic.

6. Find formulas for transformations from the hyperbolic group satisfying the following conditions:
    (a) a hyperbolic rotation about the point $1/2$
    (b) a parallel displacement with ideal point -1
    (c) a hyperbolic translation from $-i$ to i

Hint: Use normal forms. In each case, there are actually many solutions; for example, there are hyperbolic rotations of all angles about $1/2$. Try to find a formula for the most general transformation of each type.

Horocycles

7. Prove that any two horocycles are congruent and that, furthermore, a congruence transformation can be chosen that maps any point on one horocycle to any point on another.

Note: Applied to a single horocycle, this exercise says that a horocycle is self-congruent to itself at every point. Do the other cycles have this property?

8. How many horocycles are determined by two points?

9. Is it possible to drop a perpendicular horocycle from a point to a given horocycle? If so, how many?

10. Do hyperbolic circles have tangent horocycles as well as tangent straight lines at every point?

11. Develop a theory of parallelism for horocycles. For example, can you distinguish different types of parallelism for horocycles as we have for straight lines? Is Playfair's form of the parallel postulate (see Chapter 1) true for horocycles?

12. In what ways are horocycles like straight lines? Write a brief essay on this question, considering both similarities and differences. Use the results of Exercises 7–11.

9 Hyperbolic Length

This chapter introduces hyperbolic distance. With this step, hyperbolic geometry becomes at once both more familiar and more strange: More familiar, because distance is a well-known concept from Euclidean geometry that obeys many of the same rules in non-Euclidean geometry as in Euclidean geometry; more strange, not only because the customary mensuration formulae from Euclidean geometry are altered almost beyond recognition in hyperbolic geometry, but also because distance measurement is fundamentally linked to angle measurement in hyperbolic geometry in a way totally foreign to Euclidean geometry.

Let γ be a curve in the hyperbolic plane. To determine the length of γ, let γ be given parametrically; that is, $\gamma = (x(t), y(t))$, where x and y are described by separate functions $x(t)$ and $y(t)$ of the parameter t which ranges over an interval $[a, b]$ of the reals. (See Figure 9.1.)

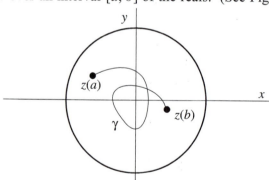

Figure 9.1 A curve in the hyperbolic plane

The functions $x(t)$ and $y(t)$ can be regarded as describing a *complex* quantity $z(t) = x(t) + iy(t)$. In other words, we consider a parametric curve γ as given by a complex function $\gamma = z(t)$ of a real variable. For example, $z(t) = a + re^{it}$ (where $0 \le t \le 2\pi$) is a parametrization of a circle. We assume that the functions $x(t)$ and $y(t)$ are differentiable. so that the curve γ is **smooth**.

A–If $\gamma = z(t) = a + re^{it}$ $(0 \le t \le 2\pi)$ is a circle, what is its center and what is its radius?

B–What are the functions $x(t)$ and $y(t)$ of this parametrization?

C-Is this a smooth parametrization?

Definition *In the hyperbolic plane, the **length** of a smooth curve γ with parametrization $z(t) = x(t) + iy(t)$ (where $a \leq t \leq b$) is given by*

$$\ell(\gamma) = 2 \int_a^b \frac{|z'(t)|\, dt}{1 - |z(t)|^2}, \quad \text{where } z'(t) = x'(t)) + iy'(t)$$

Let z_1 and z_2 be two points in the hyperbolic plane. The distance from z_1 to z_2, written $d(z_1, z_2)$, is defined to be the length of the hyperbolic straight-line segment between the two points.

Only *positive* quantities appear under this integral sign, so the length of a curve is a positive, real number. The numerator,

$$|z'(t)|dt = \sqrt{x'(t)^2 + y'(t)^2}\, dt$$

is the usual Euclidean integrand for arc length (usually called ds). However, the denominator has a profound effect on the integral. As $z(t)$ approaches the unit circle, $1 - |z(t)|^2$ approaches zero, so that the integrand approaches infinity. This has the effect, as we shall soon see, of making the unit circle infinitely distant from the points inside it. However, we must prove that length is invariant before making any calculations.

Theorem *Let T be a transformation of hyperbolic geometry, and let γ be a smooth curve. Then $\ell(T(\gamma)) = \ell(\gamma)$.*

Proof: Let the transformation T be

$$w = Tz = e^{i\theta}\frac{z - z_0}{1 - \bar{z}_0 z}$$

and let γ be described by a parametrization $z(t)$. Then $T(\gamma)$ has the parametrization

$$w(t) = T(z(t)) = e^{i\theta}\frac{z(t) - z_0}{1 - \bar{z}_0 z(t)}$$

with

$$w'(t) = T'(z(t))z'(t) = e^{i\theta}\frac{\left(1 - |z_0|^2\right)}{\left(1 - \bar{z}_0 z(t)\right)^2} z'(t)$$

This formula is a consequence of the chain rule. Now we can compute

$$\ell(T(\gamma)) = 2\int_a^b \frac{|w'(t)|}{1-|w(t)|^2} \, dt = 2\int_a^b \frac{\left| e^{i\theta} \dfrac{(1-|z_0|^2)z'(t)}{(1-\bar{z}_0 z(t))^2} \right|}{1 - \left| e^{i\theta} \dfrac{z(t)-z_0}{1-\bar{z}_0 z(t)} \right|^2} \, dt$$

Using the fact that $|e^{i\theta}| = 1$, and converting to a simple fraction gives

$$\ell(T(\gamma)) = 2\int_a^b \frac{(1-|z_0|^2)|z'(t)|dt}{\left|1 - \bar{z}_0 z(t)\right|^2 - |z(t)-z_0|^2}$$

$$= 2\int_a^b \frac{|z'(t)|dt}{1-|z(t)|^2} = \ell(\gamma)$$

The last step uses the fact (and what a fact!) that,

$$|1 - \bar{z}_0 z|^2 - |z - z_0|^2 = (1-\bar{z}_0 z)(1-z_0\bar{z}) - (z-z_0)(\bar{z}-\bar{z}_0)$$

$$= (1 - \bar{z}_0 z - z_0\bar{z} + |z_0|^2 |z|^2) - (|z|^2 - z_0\bar{z} - \bar{z}_0 z + |z_0|^2)$$

$$= (1-|z_0|^2)(1-|z|^2)$$

This proves the theorem. ∎

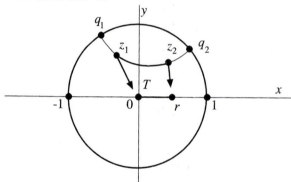

Figure 9.2 Putting a line segment in a standard position

Distance Formulas

To derive a formula for the distance between two points, $d(z_1, z_2)$, it is sufficient, since hyperbolic distance is an invariant, to place the points z_1 and z_2 in a convenient location. For this purpose, we apply the transformation

$$Tz = e^{i\theta} \frac{z-z_1}{1-\bar{z}_1 z}$$

so that z_1 goes to zero and further choose the rotation $e^{i\theta}$ so that $r = Tz_2$ ends up on the positive x-axis (as shown in Figure 9.2). The length of the straight line from 0 to r is easily computed to be

$$d(0,r) = \int_0^r \frac{2dt}{1-t^2} = \int_0^r \left(\frac{1}{1+t} + \frac{1}{1-t} \right) dt = \ln\left(\frac{1+r}{1-r} \right)$$

from which it follows that

$$d(z_1,z_2) = \ln\left(\frac{1+\left|\dfrac{z_2-z_1}{1-\bar{z}_1 z_2}\right|}{1-\left|\dfrac{z_2-z_1}{1-\bar{z}_1 z_2}\right|} \right) \qquad (*)$$

since

$$r = \left| \frac{z_2-z_1}{1-\bar{z}_1 z_2} \right|$$

This is rather a forbidding formula. A nicer distance formula is

$$d(z_1, z_2) = \ln((z_1, z_2, q_2, q_1)) \qquad (**)$$

where q_1 and q_2 are the ideal points on the hyperbolic straight line connecting z_1 and z_2. To be specific, q_1 is the ideal point reached by starting at z_2, going to z_1, and continuing on to the unit circle (q_1 is the ideal point at z_1's end of the straight line determined by z_1 and z_2). The proof of (**) is in Exercise 1.

D□What happens to $d(0, r)$ as r tends toward the unit circle?

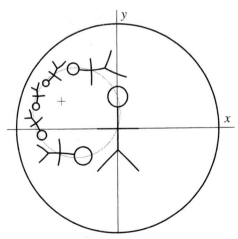

Figure 9.3 A stick figure rotated about a point

Some intuitive feel for hyperbolic distance can perhaps be gained from Figure 9.3. where several stick figures are depicted, all of which are transforms of each other by hyperbolic rotation. All the stick figures in the figure are congruent in hyperbolic geometry. Their decreasing Euclidean size indicates the way that distances are scaled approaching the boundary in the unit circle model of hyperbolic geometry. The orbits of the hyperbolic rotation used are hyperbolic circles, one of which is drawn in the figure.

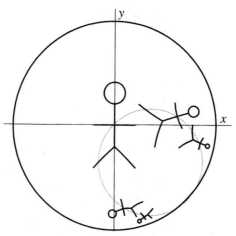

Figure 9.4 A stick figure displaced around a horocycle

For comparison, Figures 9.4 and 9.5 illustrate the two other types of hyperbolic motion. In Figure 9.4, the stick figures have been moved by a parallel displacement whose fixed point is the ideal point $e^{-i\pi/6}$. The orbits of this motion are horocycles one of which is drawn in the figure.

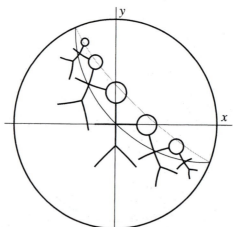

Figure 9.5 A stick figure translated along a hypercycle

In Figure 9.5, the stick figures differ by a hyperbolic translation whose axis (indicated by the gray hyperbolic line) connects the ideal points 1 and $e^{2\pi i/3}$. The orbits of this motion are hypercycles, one of which is drawn in the figure.

Figures 9.3, 9.4 and 9.5 were drawn with the aid of the computer algebra system Mathematica. All the stick figures are actually exactly the same size in hyperbolic geometry. In fact, they are congruent by the transformations indicated. They look different only because our model crams the whole infinite hyperbolic plane into a finite Euclidean circle. The Euclidean plane would experience similar distortion, if we did the same thing to it. (See Figure 4.5.)

Fundamental Properties of Distance

The following theorem lists a number of basic properties of hyperbolic distance. The Euclidean distance function satisfies exactly the same properties.

Theorem *Let z_1, z_2, and z_3 be points in the hyperbolic plane. Then*
  (1) $d(z_1, z_2) \geq 0$
  (2) $d(z_1, z_2) = d(z_2, z_1)$
  (3) *If z_1, z_2, and z_3 are collinear (in that order), then*
$$d(z_1, z_3) = d(z_1, z_2) + d(z_2, z_3)$$

Proof: (1) is true because the integral that defines distances (and lengths) has a nonnegative integrand. (2) can be proven several ways. It follows, for example, from the formula

$$d(z_1, z_2) = \ln(z_1, z_2, q_2, q_1)$$

since

$$(z_1, z_2, q_2, q_1) = (z_2, z_1, q_1, q_2)$$

Finally, (3) can also be proven by manipulating cross ratios. (See Exercise 3). ∎

The next theorem is extremely important, but its proof is difficult.

Theorem *Let z_1, and z_2 be points in the hyperbolic plane. Then the shortest curve connecting z_1 and z_2 is a hyperbolic straight line.*

Discussion: To give a complete proof would take us too far afield (into the calculus of variations). The following argument is intended to make the conclusion plausible.

First we apply a transformation so that z_1 is sent to 0 and z_2 becomes a point r on the positive x-axis. In these circumstances, let γ be any curve connecting 0 and r, perhaps such as depicted in Figure

9.6. Our goal is to discover what curve γ minimizes the integral in the quantity

$$\ell(\gamma) = 2\int_a^b \frac{|z'(t)|dt}{1-|z(t)|^2}$$

where γ is parametrized by $z(t)$.

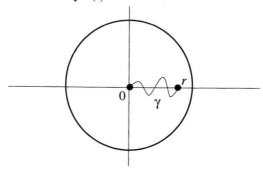

Figure 9.6 Another curve in the hyperbolic plane

At this point our argument becomes less precise. To minimize the integral, we must make the denominator as large as possible and the numerator as small as possible. Not much can be done about the denominator, except to say that γ should stay away from the unit circle. Apart from this conclusion, it appears that the integral will be minimized when the numerator

$$|z'(t)| = \sqrt{x'(t)^2 + y'(t)^2}$$

is minimized. Now, $x'(t)$ can only be minimized so much, because the curve γ simply *must* move in the x-direction; it must go from 0 to r. But movement in the y-direction can be eliminated. Thus, to minimize the integral in $\ell(\gamma)$, we *can* make $y'(t) = 0$. This strongly suggests that the minimizing curve is the interval from 0 to r on the x-axis, in other words, the hyperbolic straight line connecting 0 and r. ∎

Corollary (Triangle Inequality) For any three points z_1, z_2, and z_3 in the hyperbolic plane,

$$d(z_1, z_3) \leq d(z_1, z_2) + d(z_2, z_3)$$

E–Deduce the corollary from the preceding theorem.

The Formula of Lobatchevsky

The previous section showed that in many ways hyperbolic distance has properties analogous to Euclidean distance in Euclidean geometry. The formula of Lobatchevsky, however, embodies something new and unprecedented. In Euclidean geometry, there is no connection between angle measure and distance measure; however, in hyperbolic geometry, angle measure and distance measure are linked.

Theorem *Let the point p be given at the hyperbolic distance d from a hyperbolic straight line. Let θ be the angle of parallelism of p with respect to this line. Then*

$$e^{-d} = \tan\left(\frac{\theta}{2}\right)$$

Proof: Applying a hyperbolic transformation, we can put the point p at the origin and arrange the given hyperbolic straight line symmetrically about the x-axis, that is, like the line qrs in Figure 9.7.

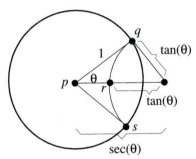

Figure 9.7 Proof of Lobatchevsky's formula

Note that r is the Euclidean distance from p to the line qrs. Now, we already know that

$$d = \ln\left(\frac{1+r}{1-r}\right)$$

On the other hand, using Euclidean geometry, we see that

$$r = \sec(\theta) - \tan(\theta) = \frac{1 - \sin(\theta)}{\cos(\theta)}$$

Therefore,

$$e^{-d} = \frac{1-r}{1+r} = \frac{\cos(\theta) + \sin(\theta) - 1}{\cos(\theta) - \sin(\theta) + 1}$$

$$= \frac{\cos^2(\theta) + 2\cos(\theta)\sin(\theta) + \sin^2(\theta) - 1}{\cos^2(\theta) + 2\cos(\theta) + 1 - \sin^2(\theta)}$$

where the last result comes from multiplying numerator and denominator by $(\cos(\theta) + \sin(\theta) + 1)$. Using double-angle formulae, we get

$$e^{-d} = \frac{2\cos(\theta)\sin(\theta)}{2\cos^2(\theta) + 2\cos(\theta)} = \frac{\sin(\theta)}{\cos(\theta)+1} = \frac{2\sin\left(\dfrac{\theta}{2}\right)\cos\left(\dfrac{\theta}{2}\right)}{\left(2\cos^2\left(\dfrac{\theta}{2}\right)-1\right)+1} = \tan\left(\dfrac{\theta}{2}\right)$$

This proves the theorem. ■

The Half Plane Model

There is really nothing special about the unit disk. Hyperbolic geometry can be modeled inside any cline. An alternative model that actually makes some computations easier is obtained by choosing the inside of a straight line. What is the inside of a straight line? From the point of view of Möbius geometry, it is a half plane.

*Definition The **upper half plane** is the subset $\mathbf{U} = \{z\colon \mathrm{Im}(z) > 0\}$ of the complex plane. Let $\overline{\mathbf{H}}$ be the group of transformations of $\mathbf{U}$ of the form*

$$w = Tz = \frac{az+b}{cz+d}$$

*where a, b, c, and d are all real and $(ad - bc) > 0$. The pair $(\mathbf{U}, \overline{\mathbf{H}})$ models **hyperbolic geometry**.*

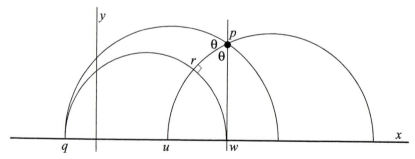

Figure 9.8 Life in the hyperbolic half plane

The geometry $(\mathbf{U}, \overline{\mathbf{H}})$ is a model of the same abstract geometry as $(\mathbf{D}, \mathbf{H})$. (See Exercise 13.) *Both* are models of hyperbolic geometry. Figure 9.8 gives an idea of what hyperbolic geometry looks like in the half plane.

F–What curves should play the role of straight lines in the half plane model of hyperbolic geometry?

G☐Consider the point p and line qrw (in Figure 9.8) to be given, while pru is the perpendicular from p to qrw. What is the angle θ called?

Distance in the Half Plane

In the upper half plane, the length of a curve γ is given by

$$\ell(\gamma) = \int_a^b \frac{|z'(t)|\,dt}{y(t)}$$

where γ is given parametrically by $z(t) = x(t) + iy(t)$ for $a < t < b$. This can be established in two steps.

In the first place, $\ell(\gamma)$ is an invariant. To prove this, let

$$w = Tz = \frac{az+b}{cz+d}$$

be a Möbius transformation of the upper half plane. By dividing numerator and denominator by the same constant, if necessary, we can assume that $ad \,\square\, bc = 1$. Now let $w(t) = T(z(t)) = X(t) + iY(t)$ be the transformation of γ by T. Then

$$w'(t) = \frac{z'(t)}{(cz(t)+d)^2}$$

and

$$2iY(t) = w(t) - \overline{w(t)} = \frac{az(t)+b}{cz(t)+d} - \frac{a\overline{z(t)}+b}{c\overline{z(t)}+d}$$

After a little computation, it appears that

$$2iY(t) = \frac{z(t) - \overline{z(t)}}{|cz(t)+d|^2} = \frac{2iy(t)}{|cz(t)+d|^2}$$

Therefore,

$$\frac{|w'(t)|\,dt}{Y(t)} = \frac{|z'(t)|\,dt}{y(t)}$$

from which $\ell(T(\gamma)) = \ell(\gamma)$ follows immediately.

In the second place, we check that the formula at the top of this page gives the same straight line distance between two points as in the circle model. To do this, let p_1 and p_2 be two points in the upper half plane. By applying a transformation from $\mathbf{H}$, p_1 and p_2 can be placed in a standard position. In the half plane model, we have the same flexibility as in the circle model of hyperbolic geometry: A point can be moved to any other single point, and, in addition, a (hyperbolic) rotation can be applied about that point. In the half point with specially useful Euclidean coordinates (like the origin in the circle model). A standard place to put points is along the y-axis. Having done this (with p_1, say, as shown in Figure 9.9), we can then rotate about the point y_1. The effect of a hyperbolic rotation about y_1 is that any real point can be set to ∞. By placing r at ∞, p_2 must also end up on the y-axis, since the hyperbolic straight line γ now is a Euclidean straight line.

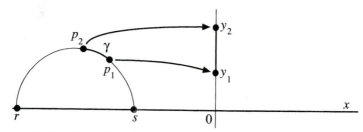

Figure 9.9 Putting a line in standard position in the half plane model

In this position, the segment between y_1 and y_2 can be parametrized by $z(t) = it$ ($y_1 < t < y_2$), so the length of the segment is

$$\ell(\gamma) = \int_{y_1}^{y_2} \frac{dt}{t} = \ln\left(\frac{y_2}{y_1}\right)$$

while the cross ratio is

$$(y_1, y_2, \infty, 0) = \frac{y_2}{y_1}$$

Thus,

$$\ell(\gamma) = \ln\big((y_1, y_2, \infty, 0)\big) = \ln\big((p_1, p_2, r, s)\big)$$

the same formula as in the circle model.

EXERCISES

Theoretical Properties of Distance

1. Derive formula (**) for $d(z_1, z_2)$ from formula (*). (The formulas are on page 97.)

Hint: Both distance and the cross ratio are invariants.

2. What happens to the distance $d(z_1, z_2)$ as z_2 approaches the unit circle?

3. Prove part (3) of the theorem on properties of distance.

4. Prove that a hyperbolic circle is the locus of points that are a fixed distance from its center.

Hint: In other words, prove that all radii of a hyperbolic circle have the same length.

5. Let C be a hypercycle, and let λ be the hyperbolic straight line that shares the same ideal points as C. Prove that the perpendicular distance from C to λ is the same at every point of C. This explains the term **equidistant curve** used for hypercycles.

6. **The chain rule.** Let $z(t)$ be a complex-valued function of a real variable. Let $w(t)$ be the composition of $z(t)$ with a Möbius transformation T; that is, $w(t) = T(z(t))$. Prove that $w'(t) = T'(z(t))\, z'(t)$.

Hint: Let Tz = (az + b)/(cz + d). Let z(t) = x(t) + iy(t) and let w(t) = X(t) + iY(t). Express X and Y in terms of a, b, c, d, x, and y. Differentiate.

Bisectors

7. Prove that every angle in hyperbolic geometry is bisected by a unique hyperbolic straight line that is the locus of points equidistant from the sides of the angle.

Hint: Show that there are points equidistant from the sides of an angle and that these points lie along a straight line.

8. Show that the bisectors of the angles of a hyperbolic triangle meet in a single point. This point is called the **incenter** of the triangle. Show that every triangle has an inscribed circle.

9. Prove that a triangle has a circumscribed hyperbolic circle if the perpendicular bisectors of the sides of the triangle meet at a point. Show, by examples, that the perpendicular bisectors need not meet at a point.

The Half Plane Model

10. Show that any Möbius transformation that maps the upper half plane onto itself can be expressed in the form

$$Tz = \frac{az+b}{cz+d}$$

with real coefficients, such that $ad \,\square\, bc > 0$. Use this idea to verify that **H** is a transformation group.

Hint: First explain that continuity implies that a Möbius transformation mapping the upper half plane to itself must map real numbers to real numbers. Then use the fact that 0, 1, and ∞ are real.

11. Draw examples of hyperbolic circles, horocycles, and hypercycles in the half plane model for hyperbolic geometry.

12. What kind of curve in hyperbolic geometry is a straight line parallel to the x-axis (in the half plane model)? What kind of curve is a straight line whose slope is neither zero nor ∞?

13. Show that $(\mathbf{D}, \mathbf{H})$ and $(\mathbf{U}, \overline{\mathbf{H}})$ are models of the same abstract geometry.

Hint: Use a Möbius transformation from $\mathbf{D}$ to $\mathbf{U}$ as a lifting transformation.

Mensuration Formulae

14. Given a hyperbolic circle, find a formula for its circumference in terms of the hyperbolic length of its radius.

Hint: Place the center of the circle at the origin. Let C be its circumference, R its hyperbolic radius, and r its Euclidean radius. Integrate to find C in terms of r. Substitute to replace r by R. Try to get the answer in the form $C = 2\pi\sinh(R)$.

15. Given a point and a line, Lobatchevsky's formula indicates that the angle of parallelism is determined by the distance from the point to the line. The distance d such that the angle of parallelism is $\pi/4$ is called **Schweikart's constant.** Find Schweikart's constant.

16. The Pythagorean Theorem. Let a, b, and c be the hyperbolic lengths of the three sides of a right triangle, where c is the side opposite the right angle. Prove that $\cosh(c) = \cosh(a)\cosh(b)$.

Question: Why is this the Pythagorean theorem?

Answer: Because it connects the length of the hypotenuse of a right triangle with the lengths of the legs.

Hint: Place the right angle at the origin. Unfortunately, this is another monster computation.

17. Find the length of a fundamental arc on a horocycle. Prove that all fundamental arcs on a horocycle are congruent.

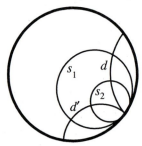

Figure 9.10 Common radii of two horocycles

18. Figure 9.10 shows two segments s_1 and s_2 of horocycles sharing a common ideal point. Show that $d = d'$ and $s_2 = s_1 e^{-d}$, where all these are hyperbolic distances.

Drawing Instruments for the Hyperbolic Plane

Discussion: In the Euclidean world, a straightedge is simply a piece of straight line. A straightedge suffices to draw all straight lines because two pieces of straight line (of the same length) are congruent. A more complicated tool is needed for circles because pieces of circles of the same length (but of different radii) are not congruent.

*A Euclidean **compass** is simply a device for holding two points (one of which has a pencil attached) a fixed distance apart. A compass suffices to draw all circles because a circle is the locus of all points a fixed distance from a given point.*

19. In Euclidean geometry, a straightedge is used to draw straight lines and a compass is used to draw circles. Explain what hyperbolic straightedges and compasses are and how they would be used in a hyperbolic world.

20. Design an instrument that would be used to draw horocycles in a hyperbolic world.

21. Design an instrument that would be used to draw hypercycles in a hyperbolic world.

22. You are standing on a hyperbolic world. You are in a corn field bounded on one side by a straight road. You are looking toward the point where the road passes closest to you. You want to know the distance d from you to the road. So, you carefully turn your head to look up the road toward the point where it passes out of sight toward infinity. Let θ be the angle through which you have just turned your head. Determine d from θ.

10 AREA

In this chapter we come to some of the most amazing results of hyperbolic geometry. Formulas for areas in hyperbolic geometry are startlingly different from those in Euclidean geometry.

To understand how to compute areas in hyperbolic geometry, let us review their computation in Euclidean geometry. There the area of a region is calculated by combining integration in two perpendicular directions into a double integral. In rectangular coordinates, this looks like Figure 10.1(a); in polar coordinates, like Figure 10.1(b).

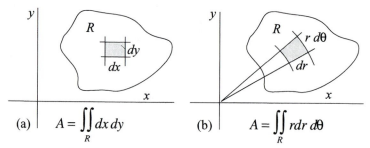

Figure 10.1 Euclidean areas are calculated by double integration.

The same idea leads to double integrals for area in hyperbolic geometry. (See Figure 10.2.) Here is the formal definition:

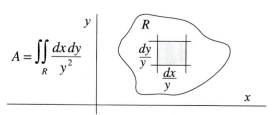

Figure 10.2 Hyperbolic areas are too.

Definition *The **area** of a region R in the hyperbolic half plane is given by the formula*

$$A = \iint_R \frac{dx\,dy}{y^2}$$

Areas of Triangles

As an application of this definition, we will find a formula for the area of a triangle.

First we consider **doubly asymptotic triangles**, that is, triangles with two ideal vertexes. Examples are given in Figure 10.3.

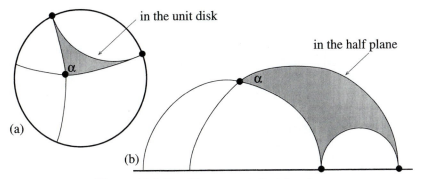

Figure 10.3 Doubly asymptotic triangles

At an ideal vertex, the angle is zero, so a doubly asymptotic triangle has just one nonzero angle (α in the figure). Although all three sides of the triangle are infinitely long, the area of such a triangle is finite! To compute this area, we place the triangle in the half plane with one ideal point at ∞, the other at -1, and the third vertex somewhere along the unit circle. (See Figure 10.4.)

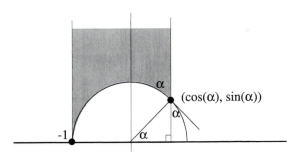

Figure 10.4 Calculation of the area of a doubly asymptotic triangle

Then, by integration, we obtain

$$A = \iint \frac{dx\,dy}{y^2} = \int_{-1}^{\cos(\alpha)} \int_{\sqrt{1-x^2}}^{\infty} \frac{dy\,dx}{y^2}$$

$$= \int_{-1}^{\cos(\alpha)} \frac{dx}{\sqrt{1-x^2}} = \frac{\pi}{2} - \arccos(x) \Big|_{-1}^{\cos(\alpha)} = \pi - \alpha$$

Using this formula, we find that the area of a trebly asymptotic triangle, the largest triangle in the hyperbolic plane, is only π. (See Figure 10.5.)

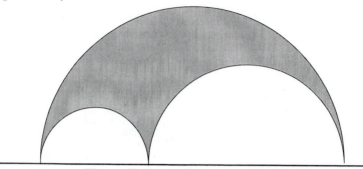

Figure 10.5 A trebly asymptotic triangle

A–Draw a quadrilateral in the hyperbolic plane with four ideal vertexes. What is the area of this figure?

More Triangular Areas

To determine the area of an arbitrary triangle, consider Figure 10.6. Here Δpqr is a given triangle with angles α, β, and γ. In addition, q^* is the ideal point along the ray from p to q, r^* is the ideal point along the ray from q to r, and p^* is the ideal point along the ray from r to p. The figure contains the original triangle, plus three doubly asymptotic triangles. All are contained inside a trebly asymptotic triangle.

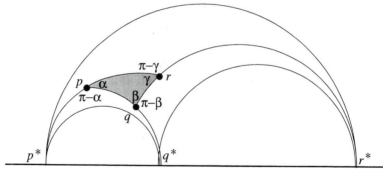

Figure 10.6 Area of a triangle in hyperbolic geometry

The areas of the triangles in Figure 10.6 are related by the equation

$$\Delta p^*q^*r^* \;=\; \Delta pqr + \Delta pp^*q^* + \Delta qq^*r^* + \Delta rr^*p^*$$

or, in terms of area,

$$\pi \;=\; A + [\pi - (\pi - \alpha)] + [\pi - (\pi - \beta)] + [\pi - (\pi - \gamma)]$$

where A, the area of Δpqr, is the only area we don't yet know. Solving for A gives

$$A = \pi - (\alpha + \beta + \gamma)$$

The quantity $(\alpha + \beta + \gamma)$ is the sum of the angles of Δpqr. The difference $\pi - (\alpha + \beta + \gamma)$ is called the **angular defect**. Thus, we have proved the following theorem:

Theorem *The area of a triangle equals its angular defect.*

This is an amazing result. In Euclidean geometry, area depends on length measurements. For example, the length of the base and height of a triangle determines its Euclidean area. But in hyperbolic geometry, the area of a triangle depends on angle measurements. Furthermore, because the area is positive, we have also proven:

Theorem *The sum of the angles of a triangle in hyperbolic geometry is less than π radians (180°).*

These are some of the most interesting results of hyperbolic geometry because they differ so much from our experience with Euclidean geometry. The term "defect" reflects this strangeness and expresses the attitude of the early explorers of hyperbolic geometry. The implication of this choice of language is that hyperbolic triangles must be defective because their angle sum is less than 180°.

Perhaps the most remarkable feature of hyperbolic geometry that these theorems display is the connection between angle measure and area or linear measure. In Euclidean geometry, we have free choice of linear measure. It doesn't matter what units are used: inches, centimeters, or whatever. On the other hand, there *is* a natural unit of angle measure: the radian. Radian measure is natural because it is related to natural measurements on the circle (laying a radius around the circumference), and the use of radian measure simplifies many formulas, notably those that use trigonometric functions.

Hyperbolic geometry is different. For both linear and area measures there are natural units, for example, the length of a fundamental arc and the area of a trebly asymptotic triangle. Furthermore, length and area measurements are linked to angle measurements by Lobatchevsky's formula, and the formula for the area of a triangle.

B–What is a formula for the area of a quadrilateral in hyperbolic geometry? What about the area of a general polygon?

C–Are rectangles possible in hyperbolic geometry?

Area in the Circle Model

In the circle model, polar coordinates are most convenient. Starting with the general differential of length

$$\frac{2|z'(t)|dt}{1-|z(t)|^2}$$

and letting $z = re^{i\theta}$, we obtain two differentials in perpendicular directions by taking the parameter t equal first to r, and then to θ (each time holding the other coordinate constant).

Here is what we get,

with respect to r: with respect to θ:

$$\frac{2|e^{i\theta}|dr}{1-r^2}=\frac{2\,dr}{1-r^2}\qquad\qquad\frac{2|ire^{i\theta}|d\theta}{1-r^2}=\frac{2r\,d\theta}{1-r^2}$$

where we are using

$$z'=\frac{dz}{dr}=e^{i\theta}\quad\text{and}\quad z'=\frac{dz}{d\theta}=ire^{i\theta}$$

Putting the two differentials together, we get,

$$dA=\frac{4r\,dr\,d\theta}{(1-r^2)^2}$$

This leads to the following definition:

Definition The **area** of a figure R in the hyperbolic plane (unit disk model) is defined by

$$A=\iint_{R}\frac{4r\,dr\,d\theta}{(1-r^2)^2}$$

D–In hyperbolic geometry the largest triangle has area π. Is there a largest circle, and if so, what is its area?

Similarity

In both hyperbolic and Euclidean geometry, two triangles are **similar** when their angles have the same measures. In Euclidean geometry, such triangles can be of different sizes, and the theory of similarity is an important part of Euclidean geometry. It is the basis, for example, of trigonometry and the application of trigonometry to surveying. In hyperbolic geometry, however, similar triangles cannot be of different sizes: They must have the same area since area is determined by angle measure. In fact, as the next theorem shows, there is no theory of similarity in hyperbolic geometry, because similar triangles are congruent!

Theorem *If corresponding angles are equal in two triangles Δpqr and $\Delta p'q'r'$, then the triangles are congruent.*

Proof: By applying a hyperbolic transformation to each triangle, we can place corresponding vertexes r and r' at the origin, as in Figure 10.7. Then, since the angle at r equals the angle at r', we can place the edges rp and $r'p'$ along the same straight line and similarly the edges rq and $r'q'$ along the same straight line. (See Figures 10.7 and 10.8.)

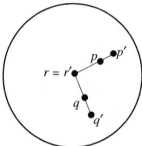

Figure 10.7 These aren't similar triangles!

Under these circumstances we wish to prove that $d(r, p) = d(r', p')$ and $d(r, q) = d(r', q')$, where d is hyperbolic distance. It will follow immediately that the third sides $d(p, q)$ and $d(p', q')$ are also equal. We prove these equalities by showing that any other result is impossible.

Suppose, for example, that $d(r, p) < d(r', p')$ and that $d(r, q) < d(r', q')$. This possibility is illustrated in Figure 10.7. Then the area of Δpqr would be less than the area of $\Delta p'q'r'$. But this is impossible.

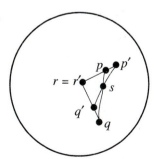

Figure 10.8 These aren't similar either.

Suppose next that $d(r, p) < d(r', p')$ but $d(r, q) > d(r', q')$. This arrangement is illustrated in Figure 10.8. Then the lines pq and $p'q'$

would intersect at a point s. (Technically this requires the Intermediate Value Theorem.) Now, $\angle sq'q + \angle sq'r = \pi$ and $\angle sq'r = \angle sqr$. So $\angle sq'q + \angle sqq' = \pi$. Thus, the triangle $\triangle sqq'$ would have an angle sum greater than π. This is impossible.

The preceding paragraphs dispose of two possibilities. All other cases are treated by similar arguments. ■

E–List all possible cases that need to be covered in the proof of the theorem on similarity. Verify that the two cases discussed in detail are characteristic of all these cases.

Life in the Hyperbolic Plane

In *Flatland* [B1], Edwin Abbott portrays the mathematical and spiritual adventures of the inhabitant of a two-dimensional Euclidean universe. The narrator, whose name is A. Square, describes in detail his world, all the inhabitants of which are regular geometric shapes ranging from segments of straight lines through equilateral triangles to regular polygons with various numbers of sides.

Let us imagine, instead, life in a hyperbolic plane inhabited perhaps by creatures like the stick figures in Figures 9.3, 9.4, and 9.5. This would be simultaneously richer and poorer than life in the Euclidean plane. For one thing, in addition to the straight lines and circles familiar from the Euclidean plane, hyperbolic Flatland would be enriched by the presence of equidistant curves and horocycles. Horocycles are particularly fascinating to contemplate. Like straight lines, all horocycles are congruent to each other, yet they are not straight! To what use, architectural or aesthetic, would these curves be put? On the other hand, life in hyperbolic Flatland is rendered poorer by the absence of squares and rectangles. These elementary building blocks are used in Euclidean worlds for almost every kind of construction. What would replace them?

Furthermore, life in a hyperbolic plane is made immensely more difficult by the absence of similarity. This implies, for example, the impossibility of making scale models of objects. Exact maps, which are possible in a flat Euclidean two-dimensional world, would be impossible. All of Euclidean trigonometry, with its applications to surveying and astronomy, disappears, since it is based on similar triangles.

Actually, the differences between life in the hyperbolic plane and life in the Euclidean plane would likely be evident only in activities and events on a large scale, since over short distances hyperbolic geometry is approximately Euclidean. (See Chapter 12.) Over longer distances some of the difficulties described can be dealt with by devising non-Euclidean solutions. (See the exercises.) In particular, there is a non-Euclidean theory of trigonometry that applies to the hyperbolic plane.

EXERCISES

Proofs

1. Justify the positioning of the doubly asymptotic triangle in the half-plane model used to calculate its area at the beginning of the chapter.

2. (For multivariable calculus students)   Prove that the double integral formula for area (in the half-plane model) is an invariant in hyperbolic geometry.

Hint: Use the theorem about changing variables in a double integral (proved in multivariable calculus). You will have to compute the Jacobian of a Möbius transformation T from the hyperbolic group!

Measuring Areas

3. What areas are possible for a Saccheri quadrilateral?

Hint: The Saccheri quadrilateral is defined on page 12.

4. A **sector** of a horocycle is the figure enclosed by the horocycle and two diameters (see Figure 10.9). Find a formula for the area of a sector of a horocycle in terms of the hyperbolic length d of the horocyclic arc between the two diameters.

Hint: Apply the half-plane area formula with the horocycle passing through ∞.

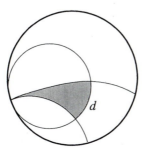

Figure 10.9 A sector of a horocycle

5. Prove that the area A of a hyperbolic circle whose radius has hyperbolic length R is given by the formula

$$A = 4\pi\left(\sinh\left(\frac{R}{2}\right)\right)^2$$

Compare this with the corresponding Euclidean formula.

Hint: Use the same strategy as for Exercise 14 in Chapter 9.

a sequence of circles with center on the x-axis and passing through the origin as the radius approaches infinity is the y-axis.

In hyperbolic geometry, what curve (or curves) can be obtained as the limit of a sequence of circles of larger and larger radius? Can a straight line be such a limit?

Hyperbolic Polygons

7. Are there parallelograms in hyperbolic geometry? Are there rectangles? If so, what are their areas?

8. The narrator of *Flatland* is a Euclidean square. Explain why this shape is impossible in the hyperbolic plane. Suppose, instead, that there is an inhabitant of hyperbolic Flatland who is an equiangular, four-sided figure with area 1. Explain how this figure exists and is uniquely determined (except for congruent figures).

9. What figure does an *n*-gon of area 1 approach in the limit as *n* tends to infinity?

10. Suppose we had a supply of blocks in the shape of the four-sided figure described in Exercise 8. Would these be suitable building blocks for the construction, for example, of hyperbolic houses? Design a set of hyperbolic building blocks.

Equidistant Curves

11. Two curves are **equidistant** if the perpendicular distance between them is constant. In Euclidean geometry, what curve is equidistant from a straight line? What curve is equidistant from a circle?

12. In Euclidean Flatland, the edges of roads (also railroad tracks) are parallel lines. Explain why, except for short distances, this doesn't work in hyperbolic Flatland.

13. In hyperbolic geometry, what curve is equidistant from a hyperbolic circle? What curve is equidistant from a horocycle? Would horocycles make suitable railroad curves?

11 ELLIPTIC GEOMETRY

Spherical geometry, the natural geometry of the surface of a sphere, is another non-Euclidean geometry. In this chapter, spherical geometry, also called **elliptic geometry**, will be modeled in the complex plane using stereographic projection to connect plane and sphere. This enables us to compare the two non-Euclidean geometries (elliptic and hyperbolic) with Euclidean geometry (in this context called **parabolic** geometry) and study all three in a unified setting.

To begin, let us find all Möbius transformations T that preserve diametrically opposite points; that is, if z_1 and z_2 are complex numbers that correspond (via stereographic projection) to the ends of a diameter of the unit sphere, then so do Tz_1 and Tz_2. These are the transformations of the complex plane that correspond (via stereographic projection) to rigid motions of the sphere.

Recall that two points z_1 and z_2 in the complex plane correspond to the endpoints of a diameter of the unit sphere if they satisfy the equation

$$z_1 \bar{z}_2 = -1$$

(See Exercise 11 of Chapter 3.) Therefore, let

$$w = Tz = \frac{az+b}{cz+d}$$

be a Möbius transformation, and suppose that $z_1 \bar{z}_2 = -1$. Then also,

$$Tz_1 \overline{Tz_2} = -1$$

or, in other words,

$$Tz_1 = \frac{-1}{\overline{Tz_2}} = \frac{-1}{\overline{T\left(\dfrac{-1}{\bar{z}_1}\right)}}$$

Substituting for T gives

$$\frac{az+b}{cz+d} = -\frac{\overline{c\left(\dfrac{-1}{\bar{z}}\right)+d}}{\overline{a\left(\dfrac{-1}{\bar{z}}\right)+b}} = -\frac{\bar{d}z - \bar{c}}{\bar{b}z - \bar{a}}$$

Now, let us assume that the coefficients a, b, c, d are such that the transformation T has determinant 1. This form is essentially unique. (See Exercise 12 of Chapter 5.) Therefore, the last equation implies that $b = -\bar{c}$ and $a = \bar{d}$, so that T takes the form

$$Tz = \frac{az + b}{-\bar{b}z + \bar{a}} \quad \text{where } |a|^2 + |b|^2 = 1$$

or, with an adjustment of constants,

$$Tz = e^{i\theta} \left(\frac{z - z_0}{\bar{z}_0 z + 1} \right) \quad \text{where } e^{i\theta} = \frac{a}{\bar{a}} \text{ and } z_0 = \frac{-b}{a}$$

Definition *The set **S** of all Möbius transformations just described is called the **elliptic group**. The pair $(\mathbf{C}^+, \mathbf{S})$ models **elliptic geometry**.*

A–Compare the two formulas for a transformation from the elliptic group with the formulas in Chapter 7 for transformations from the hyperbolic group. What similarities do you observe? What differences?

Double versus Single Elliptic Geometry

What curves should be considered the straight lines of elliptic geometry? On the surface of the sphere, curves that give the shortest distance between two points are arcs of **great circles**, the circles on the sphere whose center is the center of the sphere. When these circles are projected onto the complex plane, they become circles in the plane with the following property:

Definition *In the model $(\mathbf{C}^+, \mathbf{S})$ of elliptic geometry, a **great circle** is a circle C in the complex plane such that, whenever C contains a point z then C also contains the diametrically opposite point. An **elliptic straight line** is an arc of a great circle.*

Having defined elliptic straight lines, we turn to the consideration of which Euclidean postulates hold in elliptic geometry. The startling fact is that the very first postulate is false: two points do *not* always determine a unique straight line.

Diametrically opposite points are the problem. Such a pair of points (on the sphere or in the plane) determines many straight lines because, according to the proceeding definition, *every* straight line passing through one of the points must pass through the other. Figure 11.1 illustrates this; on the sphere, and also stereographically projected into the complex plane. The points z and z^d (likewise w and w^d) have many great circles in common.

of the pair $\{z, z^d\}$ contained in $\mathbf{D}^c$. (Note, however, that *both* w and w^d are in $\mathbf{D}^c$.) Since every point is represented there, the whole single elliptic plane $\mathbf{C}/d$ can be regarded as contained in $\mathbf{D}^c$. In other words, single elliptic geometry can be modeled in the set $\mathbf{D}^c$, with the advantage gained thereby that no identifications are needed, *except* for pairs of points around the boundary circle. The entire single elliptic plane (in this model) fits inside the closed disk. Figure 11.2 illustrates this another way. Note the curve connecting points p and q. Although, to Euclidean eyes, this curve appears to consist of two pieces, in elliptic geometry they are joined by the identification of the points r and r^d on the unit circle.

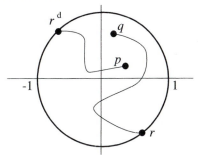

Figure 11.2 Disk model of the single elliptic plane

Thus, single elliptic geometry has a disk model much like the disk model of hyperbolic geometry. Conversely, hyperbolic geometry has a model based on the identification of points, like our first model of single elliptic geometry. In this model, the underlying set is the whole complex plane, but points *symmetric* with respect to the unit circle are identified. (See Figure 11.3.)

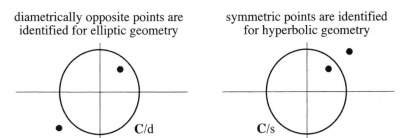

Figure 11.3 Models of single elliptic and hyperbolic geometry in the complex plane identify different pairs of points.

Distance and Area in Elliptic Geometry

As in hyperbolic geometry, basic measurements in elliptic geometry are given by integrals. For example, the arc length of a curve γ is given by

$$\ell(\gamma) = \int_a^b \frac{2|z'(t)|}{1+|z(t)|^2} \, dt$$

and the area of a plane region R is

$$A = \iint_R \frac{4r\,dr\,d\theta}{\left(1+r^2\right)^2}$$

Here are some applications of these formulas.

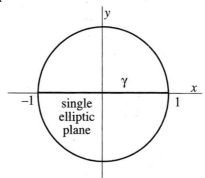

Figure 11.4 Standard position for an elliptic line

Length of a Straight Line

The most convenient placement of the line is along the x-axis. (See Figure 11.4.) Then the length of the whole line can be calculated

$$\ell(\gamma) = \int_{-1}^1 \frac{2dt}{1+t^2} = 2\arctan(t)\Big|_{-1}^1 = 4\arctan(1) = \pi$$

The surprise is that lines in elliptic geometry have finite length. On second thought, however, perhaps this is not so surprising, since an elliptic straight line is the stereographic projection of a great circle, which obviously has finite length.

Area of a Two-gon

A two-gon is a figure bounded by two straight lines and only one vertex. (See Figure 11.5.) There is nothing like it in Euclidean or hyperbolic geometry. Its area is computed as

$$A = 2\int_0^\alpha \int_0^1 \frac{4r\,dr\,d\theta}{\left(1+r^2\right)^2} = 2\int_0^\alpha \frac{-2}{1+r^2}\Big|_{r=0}^{r=1} d\theta = 2\alpha$$

Note that if $\alpha = \pi$, the two-gon is the whole plane. Thus, the single elliptic plane has a finite total area of 2π.

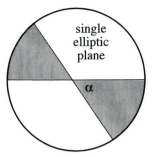

Figure 11.5 A two-gon

Area of a Triangle

To find the area of a triangle, say, Δpqr, extend all three sides to complete great circles, as in Figure 11.6.   Then the whole single elliptic plane is divided into four triangles: ΔP, ΔQ, ΔR, ΔS (all of which share the same three vertexes!). Thus,

$$\text{whole plane} = \Delta P + \Delta Q + \Delta R + \Delta S$$

Grouping these regions thus,

$$\text{whole plane} = (\Delta P + \Delta Q) + (\Delta P + \Delta R) + (\Delta P + \Delta S) - 2\Delta P$$

we see that the plane consists of three two-gons minus twice the triangle ΔP. If A represents the area of this triangle, then we have

$$2\pi = 2\alpha + 2\beta + 2\gamma - 2A$$

or

$$A = \alpha + \beta + \gamma - \pi$$

The quantity on the right-hand side of the last equation is called the **angular excess** of the triangle ΔP. This proves the following:

Theorem *The area of a triangle in single elliptic geometry is equal to its angular excess.*

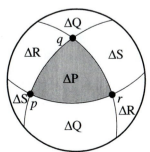

Figure 11.6 Finding the area of a triangle in elliptic geometry

Euclid's Axioms Again

It is relatively easy to prove many theorems of single elliptic geometry by adapting arguments used in the [open] disk model of hyperbolic geometry to the [closed] disk model of single elliptic geometry. In particular, the first four of Euclid's postulates are true in single elliptic geometry, and their proofs are similar to proofs of analogous results in hyperbolic geometry. (See Exercise 5.) Note that the phrase "a finite straight line can be produced to any length" (from Euclid's second postulate) must be interpreted carefully. In elliptic geometry, straight lines are **reentrant**; that is, if one follows a straight line long enough, one returns to the starting point. Hence, elliptic straight lines can be "produced to any length" but only by going around and around them!

D–What familiar curve that occurs in both hyperbolic and Euclidean geometry is reentrant?

E–Is a horocycle reentrant?

Parallelism

Let λ and μ be two lines in single *or* double elliptic geometry. By inverse stereographic projection, λ and μ correspond to two great circles on the surface of the unit sphere. Each great circle determines a plane passing through the origin in three-dimensional space. These planes intersect in a line: the common diameter of the two great circles. The endpoints of this diameter on the unit sphere are in both the lines λ and μ. Thus, in single or double elliptic geometry, two distinct lines always intersect in a point; therefore, there are *no* parallel lines.

Summary

When transferred to the complex plane (by stereographic projection), the geometry of a sphere is called double elliptic geometry. It is complicated by the fact that it does not satisfy the first of Euclid's postulates. However, by identifying diametrically opposite points, we get single elliptic geometry, in which the first four of Euclid's postulates are satisfied. Two models of single elliptic geometry were introduced in this chapter: a [closed] disk model and a model based on identification of points. There are many analogies with hyperbolic geometry, which also has two models based on these principles.

What Next?

There are too many points of similarity between elliptic and hyperbolic geometry to be explainable by coincidence: Both have disk models, both satisfy the first four Euclidean axioms, and both have similar integral formulae for length and area. While there are also many differences, these similarities strongly suggest that there is a common theory linking both non-Euclidean geometries. This theory, called **absolute geometry**, is the subject of the next chapter. Still more connections among the various non-Euclidean geometries are explored later: from a projective point of view in Chapter 16, from a solid geometrical perspective in Chapter 18, and from an axiomatic point of view in Chapter 26.

EXERCISES

Elliptic Transformations

1. Prove that **S** is a transformation group.

*Hint: At least two proofs are possible, one based directly on the formula for elliptic transformations and another based on the fact that **S** is the group of all Möbius transformations that preserve diametrically opposite points.*

2. Let T be a transformation from **S**. Prove that T is elliptic and lifts (by stereographic projection) to a rotation of the sphere.

Hint: Show that T has two fixed points that are diametrically opposite. Place one of these fixed points at the origin, and conclude that T is a rotation.

Straight Lines in Elliptic Geometry

3. Prove that the stereographic projection of a great circle on the unit sphere is an elliptic straight line in the complex plane. Conversely, show that inverse stereographic projection carries an elliptic straight line to a great circle on the sphere.

4. Prove that elliptic straight lines that happen to pass through the origin are also Euclidean straight lines. Show that these are the only elliptic straight lines that are also Euclidean straight lines.

5. Verify the first four Euclidean postulates in single elliptic geometry.

Hint: Imitate the corresponding proofs of these results in hyperbolic geometry. (See Chapter 7.)

6. Using an argument based directly on the disk model, prove that two lines are never parallel in single elliptic geometry.

7. What is the length of a straight line in double elliptic geometry? What is the area of a two-gon? What is the area of the whole double elliptic plane?

Cycles

8. Let C be the set of points in the single elliptic plane that are a fixed distance r from a fixed point z. The locus of all such points is called a **cycle** in elliptic geometry. The point z is called the **center** of the cycle, and r is its **radius**. Show that when

identifications are taken into account (see Figure 11.7), C consists of points from one or possibly two Euclidean circles.

9. Let C be an elliptic cycle with center z and radius r. For what values of r is C an elliptic straight line? For what value of r is C *not* a curve?

Hint: Think spherically.

10. Prove that every Euclidean circle in the complex plane is an elliptic cycle or an elliptic straight line.

Hint: Show that there is a transformation from **S** *that makes C symmetric about the origin.*

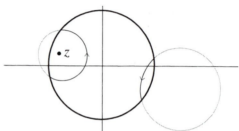

Figure 11.7 A cycle in elliptic geometry

Theorems of Elliptic Geometry

11. Do triangles in elliptic geometry have inscribed cycles?

12. Do triangles in elliptic geometry have circumscribed cycles?

13. What curve is equidistant from a straight line in elliptic geometry?

14. Prove that the sum of the angles of a triangle is greater than $180°$ in elliptic geometry.

15. State and prove the analog of the Pythagorean theorem in elliptic geometry.

Mensuration Formulas

16. Establish this formula for the distance between two points in elliptic geometry:

$$d(z_1, z_2) = 2\arctan\left(\left|\frac{z_1 - z_2}{1 + \overline{z_1}z_2}\right|\right)$$

17. Find a formula for the area of a cycle in elliptic geometry in terms of its radius.

Answer: $A = 4\pi\sin(R/2)^2$.

18. Find a formula for the circumference of a cycle in elliptic geometry in terms of its radius.

12 ABSOLUTE GEOMETRY

Elliptic, hyperbolic, and Euclidean (or **parabolic**) geometry are the most important plane geometries. In this chapter, we consider those theorems that are true in all three geometries. In other words, we ask: What results do the non-Euclidean geometries have in common—with each other and with Euclidean geometry? The collection of concepts common to all three principal plane geometries and the theorems true in all three goes by the name **absolute geometry**.

In the study of absolute geometry, we will encounter a series of concepts crucial to the possible application of geometry to physical space.

Models for Absolute Geometry

Absolute geometry is not actually a geometry! At least, it is not a geometry in the sense of the *Erlanger Programm*. The problem is that absolute geometry cannot be defined or understood on the basis of a single model. The very concept of absolute geometry demands the consideration of at least *three* models, one each for elliptic, parabolic, and hyperbolic geometry. Absolute geometry is an abstract geometry whose results appear in the models of several different geometries.

Since the consideration of a variety of models is essential to understanding absolute geometry, we begin the chapter by defining an *infinite* family of geometries A_k (one for each real number k). In the following definitions, all these geometries are laid out in one fell swoop—underlying sets (with identifications!), transformation groups, formulas for arc length and area—the whole schmeer:

Definition Let k be any real number. Let G_k be the set of Möbius transformations of the form

$$Tz = \frac{az + b}{-k\bar{b}z + \bar{a}} \quad \text{where } |a|^2 + k|b|^2 = 1$$

Let C_k be the set consisting of the complex plane (including the point at infinity) with points z_1 and z_2 identified if

$$kz_1\bar{z}_2 + 1 = 0$$

The geometry $\mathbf{A}_k = (\mathbf{C}_k, \mathbf{G}_k)$ *is* **absolute geometry** *with* **curvature** k.

Definition *In absolute geometry with curvature* k, *a* **straight line** *is a cline* C *such that if* C *passes through the point* z, *then* C *also passes through the point* $-1/k\bar{z}$.

Definition *The* **length** *of a parametric curve* γ *in absolute geometry with curvature* k *is given by the formula*

$$\ell(\gamma) = \int_a^b \frac{2|z'(t)|}{1+k|z(t)|^2}\, dt$$

The **area** *of a plane region* R *is given by*

$$A = \iint_R \frac{4r\, dr\, d\theta}{\left(1+kr^2\right)^2}$$

Casual inspection of these definitions reveals that all three classic geometries are here: $\mathbf{A}_1$ is elliptic geometry, $\mathbf{A}_{-1}$ is hyperbolic geometry, and $\mathbf{A}_0$ is Euclidean geometry (with a change in scale from the usual Euclidean geometry). What is not so obvious is that all the geometries $\mathbf{A}_k$ divide neatly into just three types:

(a) Euclidean geometry, $\mathbf{A}_0$, is a type all its own.

(b) for $k > 0$, the geometries $\mathbf{A}_k$ do not differ significantly from elliptic geometry. In all of them, there are no parallel lines, all transformations are elliptic, and the sum of the angles of a triangle is greater than π. Each of these geometries is equivalent to spherical geometry on a sphere of radius $R=1/\sqrt{k}$. (A sphere with radius R is said to have **curvature** $1/R^2$. This is the source of the term curvature for the parameter k.)

(c) for $k < 0$, the geometries $\mathbf{A}_k$ do not differ significantly from hyperbolic geometry. In all of them, there are parallel and hyperparallel lines, the group of transformations includes hyperbolic transformations, and the sum of the angles of a triangle is less than π.

These results are not difficult to prove, given all that we know already about elliptic and hyperbolic geometry. Some of the exercises at the end of the chapter ask you to verify them.

A–By comparing formulas, check that $\mathbf{A}_1$ is elliptic geometry, and $\mathbf{A}_{-1}$ is hyperbolic geometry.

B–Check that $\mathbf{A}_0$ is Euclidean geometry. What are the straight lines of $\mathbf{A}_0$?

Certain theorems are true in *all* the geometries $\mathbf{A}_k$. For example, all satisfy the first four Euclidean postulates. In addition, the longest side of any triangle is opposite the largest angle of the triangle. Results like these are theorems of absolute geometry. Absolute

The nature of absolute geometry can be put more poetically as follows: Suppose explorers from the Earth encounter (in outer space, perhaps) a delegation from another *universe*, a universe governed by a geometry with a different curvature than ours. Then the theorems of absolute geometry represent the geometrical common ground between the two delegations.

Put still another way, absolute geometry is the part of Euclidean geometry that is independent of the fifth postulate.

One remarkable fact needs to be noted concerning the geometries A_k. While on a large scale there are gross differences among them, on a small scale they are all approximately the same! For example, no matter what the curvature k, the sum of the angles of a small triangle (meaning small in area) is close to $180°$ in *all* the geometries A_k. This follows immediately from the formula

$$(\alpha + \beta + \gamma - \pi) \ = \ k\,A$$

where A is the area of the triangle. Other results of this nature are given in the exercises. We summarize the situation by saying that the geometries A_k (including elliptic and hyperbolic geometry) are locally Euclidean, meaning they are approximately Euclidean on a small scale.

Other Plane Geometries

Connections among the four two-dimensional (plane) geometries we have considered so far are represented in the following diagram where the geometry at the end of each arrow is a subgeometry of the geometry at the beginning of the arrow:

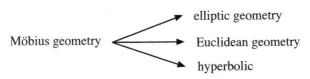

What other geometries are possible? The *Erlanger Programm* allows us to pose this question in precise technical terms. To find all possible geometries means to find and classify all possible transformation groups. Even in these terms, the problem is too broad. Therefore, a key word here is "classify." More interesting and useful than finding all geometries is finding all geometries of various basic types, for example, all *plane* geometries or all geometries that *satisfy the first four axioms of Euclid*. Even restricted in this or some similar way, the classification of all geometries is a tremendous task that is still the subject of active mathematical research. However, we can indicate how to begin to attack this problem.

We first point out some special properties of geometries.

Definition *A geometry (**S**, **G**) is*

(a) ***homogeneous*** *if, given any two points a and b in **S**, there is a transformation T in **G**, such that T(a) = b,*

(b) ***metric*** *if there is a real-valued invariant function of two variables, d(a, b), satisfying*

(1) *d(a, b) ≥ 0, and d(a, b) = 0 if, and only if, a = b,*

(2) *d(a, b) = d(b, a),*

(3) *d(a, b) ≤ d(a, c) + d(c, b),*

(c) ***isotropic*** *if the group of transformations **G** includes all rotations about every point in **S**.*

More informally, a geometry is

(a) *homogeneous* if all points are congruent; in other words, the geometry looks the same at every point;

(b) *metric* if the geometry has a distance function satisfying the fundamental properties given in Chapter 9; and

(c) *isotropic* if the geometry looks the same in every direction from each point.

Since Euclidean geometry, to which we are accustomed, has these properties, it seems natural to us that other geometries should have them. Particularly, if we are interested in geometries that can be used to describe or model the real world, these properties, for the moment, appear *necessary*, since to date all physical observations and all known physical laws suggest that the geometry of the universe *is* homogeneous, metric, and isotropic. In particular, as far as is presently known, the laws of physics are the same at every point in the universe and appear the same in every direction.

Thus, it is significant that the four plane geometries we have studied so far are all homogeneous. Möbius geometry, however, is *not* metric, although elliptic, hyperbolic and Euclidean geometry are both metric and isotropic. The really interesting fact (the proof of which is too involved for this book) is the following:

Theorem *The only two-dimensional geometries that are metric, homogeneous, and isotropic are elliptic geometry, parabolic (Euclidean) geometry, and hyperbolic geometry.*

Therefore, to the extent that we are interested only in geometries with these properties, we have already studied all possible *plane* geometries: they are the geometries $\mathbf{A}_k$. The other implication of this theorem is that, if we are interested in discovering new plane geometries, then we must relax one or another of the foregoing requirements. This is still of some practical interest, for although they may not be used to model the whole universe, nonisotropic and non-homogeneous geometries have applications on a smaller scale inside and outside mathematics.

One way to construct new geometries is to build them out of existing geometries. We have already encountered two fundamental geometry-building techniques: projection and identification. Stereographic projection enabled us to build double elliptic geometry in the plane. (More models of geometries built by *projection* appear in Chapters 18 and 19.) Identification enabled us to build single elliptic geometry in the plane and can also be used to build hyperbolic geometry. (Projective geometry, another geometry built using *identification*, appears in Chapter 13.)

Here is a third construction for new geometries:

Definition *Let* $\mathbf{Q}_1 = (\mathbf{S}_1, \mathbf{G}_1)$ *and* $\mathbf{Q}_2 = (\mathbf{S}_2, \mathbf{G}_2)$ *be two geometries. The **product** geometry is written* $\mathbf{Q}_1 \times \mathbf{Q}_2 = (\mathbf{S}_1 \times \mathbf{S}_2, \mathbf{G}_1 \times \mathbf{G}_2)$ *and defined as follows:*

(a) *The underlying set* $\mathbf{S}_1 \times \mathbf{S}_2$ *consists of all ordered pairs* (a, b) *of points where a is from* $\mathbf{S}_1$ *and b is from* $\mathbf{S}_2$. $\mathbf{S}_1 \times \mathbf{S}_2$ *is called the* **Cartesian product** *of the sets* $\mathbf{S}_1$ *and* $\mathbf{S}_2$.

(b) *The group of transformations* $\mathbf{G}_1 \times \mathbf{G}_2$ *is the Cartesian product of the sets* $\mathbf{G}_1$ *and* $\mathbf{G}_2$. *A typical member* $T = (T_1, T_2)$ *of* $\mathbf{G}_1 \times \mathbf{G}_2$ *acts on a point of* $\mathbf{S}_1 \times \mathbf{S}_2$ *according to the formula*

$$T(a, b) = (T_1 a, T_2 b)$$

C–Check that $\mathbf{G}_1 \times \mathbf{G}_2$ is a transformation group.

D–If $\mathbf{Q}_1$ is a k-dimensional geometry and $\mathbf{Q}_2$ is an h-dimensional geometry, what dimension is $\mathbf{Q}_1 \times \mathbf{Q}_2$?

To build *two*-dimensional geometries using products, we need *one*-dimensional geometries as building blocks. Here are some one-dimensional geometries for this purpose:

One-Dimensional Euclidean Geometry $\mathbf{E}^1$

The underlying set is the real line $\mathbf{R}$. The transformations are the translations T_k defined by

$$T_k x = x + k$$

where k is a real constant.

One-Dimensional Elliptic Geometry $\mathbf{S}^1$

The underlying set is the unit circle $\mathbf{S} = \{z : |z| = 1\}$ in the plane. The transformations are the rotations of the circle R_θ defined by

$$R_\theta z = e^{i\theta} z$$

where θ is a real constant.

As the exercises show, products of these geometries are not isotropic.

More generally, Riemann considered plane geometries that are not even homogeneous. In Riemannian geometry, the curvature of the geometry is given by a function $k(\mathbf{x})$ that varies from point to point. At points where $k(\mathbf{x})$ is negative, the geometry is qualitatively like hyperbolic geometry; at points where $k(\mathbf{x})$ is positive, the geometry is like elliptic geometry.

The two-dimensional geometry of the torus (the doughnut-shaped surface illustrated in Figure 12.1) is an example. At points on the inside of the torus, the curvature is negative; at points on the outside, curvature is positive. Along the circles at the top and bottom of the torus, curvature is zero. No transformation of the torus can map a point with negative curvature to one of positive curvature, therefore, the geometry of this surface (as of *most* surfaces) is not homogeneous.

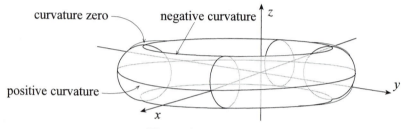

Figure 12.1 A torus

EXERCISES

The Geometries A_k

1. Verify that $\mathbf{A}_k$ is a transformation group. Check that the transformations in $\mathbf{A}_k$ map identified points to identified points; that is, if z_1 and z_2 are identified and T is in $\mathbf{A}_k$, then Tz_1 and Tz_2 are also identified.

2. Let $k \neq 0$. Let $\mathbf{D}[R] = \{z : |z| \leq R\}$ be the closed disk of radius R where $R{=}\sqrt{|k|}$. Show that if z_1 and z_2 are identified points in $\mathbf{A}_k$ then either z_1 and z_2 are both on the boundary of $\mathbf{D}[R]$ (i.e., on the circle of radius R), or just one of them is inside the disk and the other is outside. Explain how this means that a model for $\mathbf{A}_k$ can be based on $\mathbf{D}[R]$.

3. Let $k > 0$. In the geometry $\mathbf{A}_k$, show that there are no parallel lines and that the sum of the angles of a triangle is greater than π.

Hint: Show that diametrically opposite points on the circle of radius $R = 1/\sqrt{|k|}$ centered at the origin are identified. Show that every straight line connects such a pair of points.

4. Let $k < 0$. In the geometry $\mathbf{A}_k$, show that straight lines are arcs of circles perpendicular to the circle of radius $R = 1/\sqrt{|k|}$ centered at the origin. Show that the sum of the angles of a triangle is less than π.

Hint: Show that identified points in this geometry are symmetric with respect to this circle.

5. If A is the area of a triangle in $\mathbf{A}_k$, show that

$$kA = (\alpha + \beta + \gamma - \pi)$$

where α, β, and γ are the angles of the triangle. What result does this give when $k = 0$?

Hint: Consider separately three cases: k positive, negative, or zero. In each case, the calculation of the area of a triangle is more or less the same as in elliptic, hyperbolic, or Euclidean geometry (respectively). Use ideas from Chapter 11 (for elliptic geometry) and Chapter 10 (for hyperbolic geometry).

6. In the geometry $\mathbf{A}_k$, find formulas for

(a) the length of a straight line segment from z_1 to z_2
(b) the area of a circle given its radius
(c) the circumference of a circle given its radius

Hint: Consider three cases: k positive, negative, or zero. If k>0, try to use trigonometric functions in your formulas; if k<0, try to use hyperbolic functions. Use ideas from in Chapter 11 (for elliptic geometry), and Chapters 9 and 10 (for hyperbolic geometry).

7. The graph of the equation $x^2 + ky^2 = 1$ is a conic section: ellipse, degenerate parabola, or hyperbola. For what values of k does the graph take each shape?

Theorems of Absolute Geometry

8. Side-Angle-Side In absolute geometry, two triangles are congruent if two pairs of corresponding sides and the angle included between them are equal (SAS). Prove this.

Hint: Consider separately elliptic, hyperbolic and Euclidean geometry.

9. In absolute geometry, the longest side of any triangle lies opposite the largest angle. Prove this.

10. In absolute geometry, the length of any side of a triangle is less than the sum of the lengths of the other two sides. Prove this.

Absolute Geometries Are Locally Euclidean

11. Let C be a circle in the geometry $\mathbf{A}_k$. Let r be its radius (using the distance function of the geometry) and let A be its area. Prove that

$$\lim_{r \to 0} \frac{A}{\pi r^2} = 1$$

In other words, the area of small circles is approximately given by the Euclidean formula πr^2.

Hint: Use the formula for area you derived in Exercise 6. The limit is easy to evaluate if the area is expressed using hyperbolic and/or trigonometric functions (see Chapter 10, Exercise 5 and Chapter 11, Exercise 16).

12. Let C be a circle in the geometry $\mathbf{A}_k$. Let r be its radius and let D be its circumference. Prove that

$$\lim_{r \to 0} \frac{D}{2\pi r} = 1$$

In other words, the circumference of small circles is approximately given by the Euclidean formula $2\pi r$.

Hint: Use the formula for circumference you derived in Exercise 6. The limit is easy to evaluate if the area is expressed using hyperbolic and/or trigonometric functions.

Product Geometries

13. Prove that if $\mathbf{Q}_1 = (\mathbf{S}_1, \mathbf{G}_1)$ and $\mathbf{Q}_2 = (\mathbf{S}_2, \mathbf{G}_2)$ are homogeneous, then the product geometry $\mathbf{Q}_1 \times \mathbf{Q}_2$ is also homogeneous.

14. Prove that if $(\mathbf{S}_1, \mathbf{G}_1)$ and $(\mathbf{S}_2, \mathbf{G}_2)$ are metric geometries, then the product geometry also metric.

Hint: Consider the sum of the two metrics.

15. Verify that one-dimensional Euclidean geometry and one-dimensional elliptic geometry are geometries.

16. Consider the product geometry $\mathbf{E}^1 \times \mathbf{E}^1$. Is it homogeneous? metric?, isotropic? This geometry was already introduced in this book. Under what name?

Hint: This is not Euclidean plane geometry.

17. Consider the product geometry $\mathbf{E}^1 \times \mathbf{S}^1$. Is this geometry homogeneous?, metric?, isotropic? Show how it can be regarded as the geometry of an infinite cylinder.

18. What do you think should play the role of straight lines in $\mathbf{E}^1 \times \mathbf{S}^1$? Having made a choice, are any of Euclid's axioms valid in this geometry? What is the sum of the angles in a triangle?

19. Consider the product geometry $\mathbf{S}^1 \times \mathbf{S}^1$. Is this geometry homogeneous?, metric?, isotropic? This product is related to but *not* the same as the geometry of the torus. In what ways are they alike? In what ways are they different?

PART III

PROJECTIVE GEOMETRY

GEOMETRY AND ART: PLATE III

Albrecht Dürer: *Demonstration of Perspective Drawing of a Lute*. Woodcut (1525), 5 1/4x7 1/8 ". Kupferstichkabinett, West Berlin

Bernardo Bellotto. *Piazza San Marco, Venice*. Oil on canvas (ca. 1740) 136.2 x 323.5 cm.. The Cleveland Museum of Art 1995, Leonard C. Hanna Fund.

Albrecht Dürer, an outstanding artist of the Renaissance in northern Europe, was an early promoter of geometric perspective and projective geometry, the principles of which he learned in the course of several trips to Italy. He wrote a treatise on the application of geometry to art, from which the above woodcut is taken. The painting by Bernardo Bellotto shows the careful use of Renaissance ideas of geometric perspective in a later period for the purpose of realistic depiction of three-dimensional space.

... For what is the purpose of this conversation between us? Its purpose, as I understand it, is for me to explain to you, as briefly as possible, what I am–that is, what sort of a man I am, what I believe in, and what I hope for. And so I will just state here plainly and briefly that I accept God. But I must point out one thing: if God does exist and if He really created the world, then, as we well know, He created it according to the principles of Euclidean geometry and made the human brain capable of grasping only three dimensions of space. Yet there have been and still are mathematicians and philosoph-ers–among them some of the most outstanding–who doubt that the whole universe or, to put it more generally, all existence was created to fit Euclidean geometry; they even dare to conceive that two parallel lines that, according to Euclid, never meet on earth do, in fact, meet somewhere in infinity. And so, my dear boy, I've decided that since I'm incapable of understanding even that much, I cannot possibly understand about God. I humbly admit that I have no special talent for coping with such problems, that my brain is an earthly, Euclidean brain, and that therefore, I'm not properly equipped to deal with matters that are not of this world. And I would advise you too, Alyosha, never to worry about these matters, least of all about God–whether He exists or not. All such problems are quite unsuitable for a mind created to conceive only three dimensions. ...

–from *The Brothers Karamazov* (1880) by Fyodor Dostoyevsky, translated by Andrew H. MacAndrew. (Bantam Books, New York, 1981).

Projective geometry grew out of the work of visual artists who, particularly in Renaissance Italy, sought to understand how the eye perceives objects in three dimensions. They translated this understanding into rules that allowed them to produce increasingly accurate renderings of the three-dimensional world on the planar surface of their paintings.

Later, in the nineteenth century, projective geometry developed into a powerful mathematical tool with many applications, both geometric and nongeometric in nature. As such, projective geometry today is considered one of the most central of geometric theories.

In Chapters 13 and 14, we describe the fundamental concepts of the projective plane using ideas developed by Renaissance artists to help us visualize this rather peculiar geometry. Chapter 15 extends these planar ideas to other dimensions, and in Chapter 16 we set forth the relationship between projective geometry and other geometries, describing its fundamental importance in modern mathematics.

13 THE REAL PROJECTIVE PLANE

From the viewpoint of this book, projective geometry has some unique features. Like absolute geometry, it includes all the principal plane geometries (elliptic, parabolic, hyperbolic) as subgeometries. Unlike absolute geometry, projective geometry is a genuine geometry in itself (in the sense of the *Erlanger Programm*); that is, projective geometry is described completely by a single model. Unlike elliptic, parabolic, and hyperbolic geometry, however, projective geometry is not a metric geometry. Projective geometry does have coordinate systems, but they *cannot* be used to define a distance function!

Modeling Monocular Vision

Imagine a single eye located at the origin overlooking a three-dimensional scene (as in Figure 13.1).

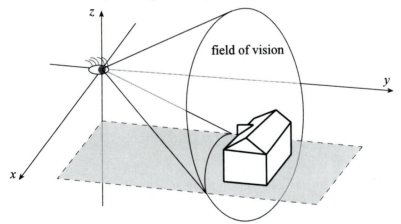

Figure 13.1 View of an eye at the origin

The vision of this eye is limited in two ways. In the first place, it only sees objects within its **field of vision**, a circular cone with vertex at the eye, opening out in the direction of the scene being viewed. For a human eye, the vertex angle of this cone is somewhere around 120°. This angle is determined by the size of the retina and its distance from the pupil. Similar mechanical considerations determine the size of the field of vision of artificial eyes (telescopes and cameras, for example). In the second place, the eye sees only one point in each direction; that is, it cannot see behind things. Therefore, for the purpose of modeling vision, all points along each ray emanating from the eye are identical because the eye sees only one of these points at a time. In mathematical terms, these points should be *identified*, just as we identified diametrically opposite points on the unit sphere for elliptic geometry.

The following definition embodies these limitations:

Definition *Three-dimensional Euclidean space* **V** *is the set of points* **p** = (x, y, z) *represented by three real coordinates. The **real projective plane** **P**$_2$ *is the set of all points in* **V** *except the origin, that is,*

$$\mathbf{P}_2 = \{\mathbf{p} = (x, y, z) : x, y, z \text{ all real, not all zero}\}$$

subject to the condition that each point is identified with all nonzero scalar multiples of itself, that is, **p** *is identified with* k**p** *for every nonzero real number k.*

Note that points in the real projective plane have three coordinates. These are so-called **homogeneous coordinates**, and **P**$_2$ is called the **homogeneous model** of the projective plane.

In Chapter 14, we will describe the transformations needed to complete the definition of the real projective plane as a geometry. Meanwhile, observe how neatly **P**$_2$ models an idealized field of vision. At first glance, it may appear that this field of vision (i.e., **P**$_2$) includes every point in space (except the origin, which, of course, is occupied by the eye itself). But, by identifying each point **p** = (x, y, z) with all *positive* scalar multiples k**p** (k > 0), we express mathematically the idea that the eye sees only *one* point in each direction. Additionally, by identifying the *negative* scalar multiples −k**p** with **p**, we express the idea that the eye sees either in front of, or behind itself, but *not* both. Thus, we regard **P**$_2$ as embodying an ideal field of vision with a vertex angle of 180°.

Although it seems paradoxical, a *point* in the real projective plane is a *set* consisting of a Euclidean point **p** ≠ (0, 0, 0) and all the points in **V** identified with **p**. In other words, a projective *point* is a Euclidean *line* passing through the origin (with the origin removed). Similarly, a projective *line* is a Euclidean *plane* according to the following definition:

Definition *A **point** in the real projective plane* **P**$_2$ *is a line in three dimensional Euclidean space* **V** *passing through, but excluding, the origin. A **line** in the real projective plane is a plane passing through, but excluding, the origin.*

This definition is illustrated in Figure 13.2.   For clarity, we systematically use, throughout this book, italic letters (for example, $p, q, r, \ldots$) for points in plane geometries, and boldface letters (for example, $\mathbf{p}, \mathbf{q}, \mathbf{r}, \ldots$) for points in higher dimensional geometries.   In this chapter, therefore, the letter p represents a point (in the projective plane), which is actually a set of points $\mathbf{p}$ from $\mathbf{V}$ (a three-dimensional geometry).

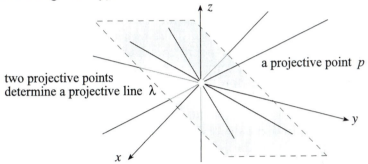

Figure 13.2 Projective points and lines

The following theorem summarizes the fundamental **incidence relationships** (meaning what lies on what) of projective geometry:

Theorem *In the real projective plane, two distinct points determine a unique line containing both points.   Two distinct lines determine a unique point contained in both lines.*

Proof: This follows right from fundamental properties of three-dimensional Euclidean lines and planes. The same arguments proved the same incidence relations for elliptic geometry in Chapter 11. ∎

A–Are there parallel lines in projective geometry?
B–Are projective lines reentrant?

Using Homogeneous Coordinates

Let p be a projective point.   Let $\mathbf{p} = (x, y, z)$ be one of the Euclidean points in p.   Then,

$$p = \{ \, k\mathbf{p} \mid k \text{ is a nonzero, real scalar} \, \}$$

The coordinates (x, y, z) are, therefore, not p's only coordinates: Any nonzero scalar multiple (kx, ky, kz) also serves as homogeneous coordinates for p.   Although p has infinitely many sets of coordinates, it is still possible to make good use of (x, y, z), as we soon show.

In projective geometry, lines have coordinates, too! Let λ be a projective line. Since λ is a Euclidean plane through the origin, there is a linear equation

$$ax + by + cz = 0 \qquad (+)$$

such that a Euclidean point $\mathbf{p} = (x, y, z)$ is on λ if, and only if, (x, y, z) satisfy (+). The coefficients of this equation, written $[a, b, c]$, are called **line coordinates** for λ. Square brackets are used to avoid confusion with point coordinates.

Like point coordinates, line coordinates are not unique. If scalar multiples $[ka, kb, kc]$ are plugged into equation (+), instead of $[a, b, c]$, the solutions of the equation are the same. Thus, coordinates $[ka, kb, kc]$ and $[a, b, c]$ determine the same projective line. In technical terms, line coordinates, like point coordinates, are **homogeneous**. The genesis of this terminology is that equation (+) is an example of a **homogeneous equation**, meaning an equation whose constant term is zero.

Two useful tools for calculation with point and line coordinates are the dot product and the cross product. Let $\mathbf{p} = (x, y, z)$ and $\mathbf{q} = (u, v, w)$ be two points in three-dimensional Euclidean space. The **dot product** of $\mathbf{p}$ and $\mathbf{q}$ is given by

$$\mathbf{p} \cdot \mathbf{q} = xu + yv + zw$$

In multivariable calculus, it is proved that $\mathbf{p} \cdot \mathbf{q}$ is zero if, and only if, the vectors $\mathbf{p}$ and $\mathbf{q}$ are perpendicular.

Let $\mathbf{L} = [a, b, c]$ be line coordinates for a projective line λ and $\mathbf{p} = (x, y, z)$ be point coordinates for a projective point p. Then, equation (+) can be rewritten

$$\mathbf{L} \cdot \mathbf{p} = 0 \qquad (+)$$

Now, this equation is satisfied if, and only if, the projective point p (represented by coordinates $\mathbf{p} = (x, y, z)$) is on the projective line λ (represented by coordinates $\mathbf{L} = [a, b, c]$). Therefore, the dot product can be used to calculate incidence relations. In other words, whether a given point is on a given line can be determined by calculating the dot product to see if it is zero. Note that, in Euclidean terms, equation (+) says that the line coordinates $\mathbf{L}$ represent a vector *perpendicular* to the vectors $\mathbf{p}$ in the plane λ.

Next, let $\mathbf{p} = (x, y, z)$ and $\mathbf{q} = (u, v, w)$ be two points in three-dimensional Euclidean space. The **cross product** is given by

$$\mathbf{p} \times \mathbf{q} = \begin{vmatrix} \mathbf{i} & \mathbf{j} & \mathbf{k} \\ x & y & z \\ u & v & w \end{vmatrix} = \left(yw - zv,\; zu - xw,\; xv - yu \right)$$

In multivariable calculus, it is proved that the cross product is a vector perpendicular to the plane through the origin containing the vectors $\mathbf{p}$ and $\mathbf{q}$.

Let p and q be projective points with coordinates $\mathbf{p} = (x, y, z)$ and $\mathbf{q} = (u, v, w)$. We know that $\mathbf{p}$ and $\mathbf{q}$ determine a projective line λ. What are homogeneous coordinates $\mathbf{L}$ for λ? Since the projective line determined by p and q is actually the Euclidean *plane* determined by the vectors $\mathbf{p}$ and $\mathbf{q}$, the line coordinates that we seek represent a vector perpendicular to $\mathbf{p}$ and $\mathbf{q}$. But this is exactly what the cross product provides. Therefore, $\mathbf{L} = \mathbf{p} \times \mathbf{q}$. In other words, the cross product calculates line coordinates for the line determined by two projective points. Similarly, if $\mathbf{L}$ and $\mathbf{M}$ are line coordinates for lines λ and μ, then $\mathbf{p} = \mathbf{L} \times \mathbf{M}$ are homogeneous coordinates for the projective point p that is the intersection of λ and μ.

C-Using three-dimensional geometric terms, explain why calculating $\mathbf{p} = \mathbf{L} \times \mathbf{M}$ gives coordinates for the intersection of the lines λ and μ.

Duality

That lines and points both have coordinates is one aspect of the

Principle of Duality *In every theorem about the real projective plane, the terms "point" and "line" may be interchanged and the resulting statement is also a theorem.*

Discussion: In the first place, every result of projective geometry that we have so far seen fulfills the principle of duality. The most important examples are the incidence relations: Two points determine a unique line, and, dually, two lines always determine (i.e., intersect in) a unique point. Figure 13.3 illustrates these facts: (a) shows two lines intersecting, and (b) shows two points determining a line. Pairs of figures like these, in which lines and points are interchanged while maintaining the same incidence relations are called **dual** figures.

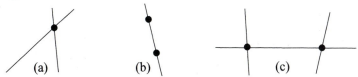

Figure 13.3 (a) and (b) are dual figures

Why should points and lines have interchangeable properties in projective geometry? From an analytic standpoint, coordinates are the key. In projective geometry, both lines and points have coordinates. Furthermore, line and point coordinates are both homogeneous. Thus, the set of all points and the set of all lines are algebraically identical: Each consists of all triples (x, y, z) of real numbers with scalar multiples identified. As far as the most fundamental geometric questions about lines and points are concerned everything comes down to considerations of incidence: what points are on what lines and what lines pass through what points. The algebraic expression of incidence is equation (+), in which point and line coordinates play *exactly the same role*. Hence, in the algebraic development of the projective plane, points and lines, so far, have exactly the same features. ■

Duality is a powerful idea. One consequence is that every proof in projective geometry proves two things at the same time: the given statement and its dual.

D–Draw the dual to Figure 13.3(c).

Other Models of the Projective Plane

The homogeneous model of the real projective plane is confusing: Points of the plane are (Euclidean) lines, lines in the plane are themselves (Euclidean) planes, and the projective plane itself incorporates (almost) all of (Euclidean) three-dimensional space! Before introducing projective transformations, and, thus, completing the description of projective geometry (according to the *Erlanger Programm*), let us clarify the nature of the projective plane by considering some alternative models.

The complexity of the homogeneous model lies in the ident-ifications used to construct it. We can obtain simpler models by taking a subset of space that includes only a few points from each ray emanating from the origin instead of the whole ray.

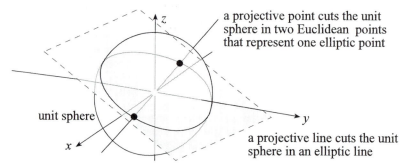

Figure 13.4 The elliptic plane models the projective plane.

A simple model along these lines is the unit sphere (double elliptic plane), which contains exactly two points from each (Euclidean) line through the origin. These diametrically opposite points remain to be identified to make the unit sphere into a model of the projective plane. Since this is precisely the identification performed in Chapter 11 to construct the single elliptic plane, we see that *the single elliptic plane is a model for the projective plane*. (See Figure 13.4.) In particular, as we have already pointed out, the incidence relations of projective geometry are the same as those of elliptic geometry. (The transformations of projective geometry, however, are much more complicated than those of elliptic geometry; see Chapter 14.)

For a still simpler model, we take a Euclidean plane *not* passing through the origin, for example, the plane $z = -1$, as depicted in Figure 13.5. Such a plane intersects (almost) every line through the origin in exactly one point, so that no identifications are necessary: A point in this model is just a point. Such a plane intersects (almost) every projective line in an ordinary line, so a line in this model is, likewise, just a line. So far, this model is great. It shows that, for some purposes, at least, the projective plane is like an ordinary plane.

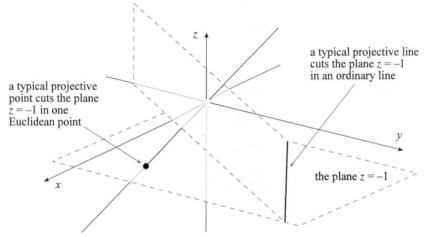

a typical projective line cuts the plane $z = -1$ in an ordinary line

a typical projective point cuts the plane $z = -1$ in one Euclidean point

the plane $z = -1$

Figure 13.5 An ordinary plane models the projective plane.

The catch is that there are points missing: The projective points in the homogeneous model that are (Euclidean) lines *parallel* to the chosen plane do not intersect the chosen plane in *any* point. Thus, our plane model for projective geometry is incomplete unless more points are added. We must add one point for every (Euclidean) line through the origin parallel to the plane model. Figure 13.6 illustrates this for the plane $z = -1$. The extra points added this way are called **ideal points**.

Actually, we have added a whole line of points, called the **ideal line**. This line corresponds to the one projective line in the homogeneous model which, because it is parallel to the chosen plane, does not intersect it. (In case the model is the plane $z = -1$, the ideal line is needed to represent the projective line that is the xy-plane.)

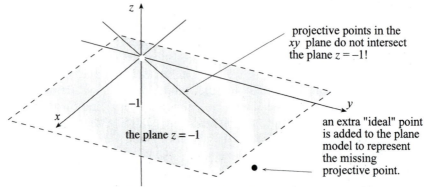

Figure 13.6 Adding ideal points

With the addition of an extra ideal line of "points-at-infinity," we *can* model the projective plane as an ordinary Euclidean plane. The extra line can be thought of as surrounding the plane. (See Figure 13.7.) In this view, the ideal line is an added *circle* with diagonally opposing points identified. Every ordinary line intersects the ideal line in just one point. Parallel ordinary lines share the same ideal point.

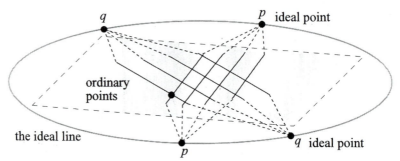

Figure 13.7 The projective plane as an ordinary plane with an added ideal line

---

E–Verify that two points determine a unique line and two lines intersect in one point when projective geometry is modeled as an ordinary Euclidean plane with an added ideal line. (Consider different cases: ideal points *vs.* ordinary points, and ideal lines *vs.* ordinary lines.)

Rules of Perspective

We can now describe the rules of perspective adopted by graphic artists over the course of years of experimentation. Consider once again, for this purpose, a single eye contemplating a scene. (See Figure 13.1 and also Figure 13.8.)

Three different models of the projective plane are relevant to the artistic depiction of this scene: the homogeneous model (which embodies our fundamental ideas about monocular vision and within which the scene is located) and two planar models: one associated with the **ground** plane of the scene and a second model associated with the plane of the artist's canvas, the **picture** plane.

The first rule of perspective is that lines that are parallel in three-dimensional space actually meet (in visual terms), since the eye sees projectively, and in projective geometry there are no parallel lines. Figure 13.8 illustrates this. The lines at the top and bottom of the house and along the sides of the roof must meet. (For the purpose of illustration, these meeting points have been brought further in than is strictly correct.) The points of intersection of parallel lines are called **vanishing points**.

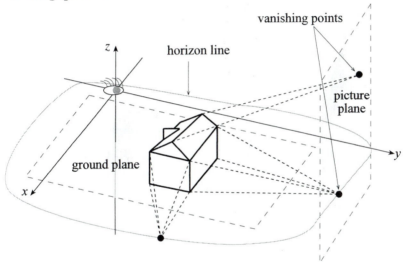

Figure 13.8 Perspective drawing uses multiple projective models.

Whether lines are depicted as intersecting *on canvas*, however, depends on the location of the vanishing point. If a vanishing point happens to be an *ideal* point (in the picture plane), then the lines meeting there are drawn genuinely parallel in the picture. On the other hand, if a vanishing point is an *ordinary* point in the picture plane, the lines must be drawn passing through that point. In Figure 13.8, all three of the vanishing points shown are ordinary points in the picture plane. (One of them doesn't look ordinary but is identified with an ordinary point in the picture plane.) Figure 13.9 shows the view from the origin projected onto the picture plane.

Of frequent occurrence, in many figures, are lines parallel to the *ground plane*. The vanishing point for a pair of such lines will be in the ideal line for the ground plane. This makes the ideal line for the ground plane a key line in any perspective picture. It is called the **horizon**. The horizon is *not* usually also the ideal line in the *picture plane* (unless the ground plane and the picture plane are the same, that is, the scene is being depicted from directly above!). Usually, the horizon is an ordinary line in the picture plane. It represents the eye level of the viewer (because in the homogeneous model it is the horizontal plane passing though the viewer's eye). The horizon contains the vanishing points of all lines parallel to the ground plane. Vanishing points of lines oblique to the ground plane (such as the roof lines in Figure 13.9) lie off the horizon.

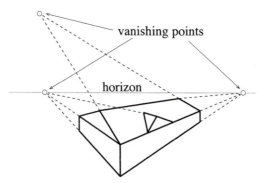

Figure 13.9 What the eye at the origin sees

The preceding is the essence of **geometric perspective**: the principles that dictate how objects are to be arranged in a picture if the goal is to render them as they are seen by a single eye. Codified as a set of rules, geometric perspective can be learned as a subject quite separate from projective geometry. In addition, tools were invented during the Renaissance to aid in the understanding of perspective. One of them is depicted by Albrecht Dürer in the print reproduced on page 136.

F–In Dürer's woodcut, find: a vanishing point, the horizon line, and two lines that are genuinely parallel.

G–Find the horizon line in the picture of the lute in Dürer's woodcut. Why is it so high?

Atmospheric perspective

Also important for the realistic depiction of three-dimensional scenes is **atmospheric perspective**: the visual rendering of the phenomenon that objects at a distance appear more obscure than objects close at hand. The most accurate representation of photographic reality in art relies on both techniques: geometric and atmospheric perspective. Of course, not every artist is interested in photographic realism; especially today. In fact, the invention of photography killed off, to a certain extent, the artistic pursuit of this kind of realism.

EXERCISES

Line and Point Coordinates

1. Find line coordinates for these projective lines:
    (a) the line determined by the xz-plane
    (b) the line determined by the points $(1, 2, 3)$ and $(-1, 2, 0)$
    (c) the line determined by the points $(0, 1, 4)$ and $(5, 0, 3)$
    (d) the ideal line in the ordinary plane model $z = -1$

2. True or False?
    (a) The point $(1, 2, 3)$ is on the line $[1, -2, 3]$.
    (b) The point $(2, 1, 4)$ is on the line determined by the points $(3, -2, 5)$ and $(4, -5, 6)$.
    (c) The line $[4, 3, 2]$ contains the point $(2, -2, 1)$.
    (d) The line $[1, 3, 2]$ contains the point of intersection of the lines $[0, 3, -1]$ and $[1, -3, 2]$.

3. Let lines λ and μ have coordinates **L** and **M**, respectively. Let p be the point of intersection of λ and μ. Prove that $\mathbf{L} \times \mathbf{M}$ gives a set of coordinates for p.

4. Let points p, q, and r in the projective plane have coordinates **p**, **q**, and **r**, respectively. Prove that $|\mathbf{p}\ \mathbf{q}\ \mathbf{r}| = 0$ if, and only if, the points p, q, and r are collinear.

*Hint: $|\mathbf{p}\ \mathbf{q}\ \mathbf{r}|$ is the determinant whose first column is **p**, second column is **q**, and third column is **r**. Recall that a determinant is zero if, and only if, one column can be written as a combination of the others.*

5. Let **L**, **M**, and **N** be line coordinates. Prove that $|\mathbf{L}\ \mathbf{M}\ \mathbf{N}| = 0$ if, and only if, the corresponding projective lines λ, μ, and ν are **concurrent**, meaning that they all pass through the same point.

6. When is the point (x, y, z) on the line $[x, y, z]$?

Perspective Drawing

7. Draw correct perspective renditions of the barn in Figure 13.1
    (a) from a viewpoint below the roof line
    (b) from a side
    (c) from directly above the barn

    Show clearly all vanishing points.

8. Analyze the use of perspective in the following works of art. Find the horizon line, and identify vanishing points that are explicitly indicated by lines in the woodcut. Comment on the location of the artist's point of view, as indicated by the location of the horizon line. Is it unusually high or low? Does the choice

of viewpoint contribute to the impact of the work of art as you understand it? Is any use made of vanishing points off the horizon line? If so, is this effective?

(a) Dürer's *Demonstration of Perspective Drawing of a Lute* (page 136)

(b) Bellotto's *Piazza San Marco, Venice* (page 136)

(c) Dali's *Corpus Hypercubicus* (page 192)

9. Choose a painting that uses perspective. (The period from 1550–1800 represents the high point of perspective art.) Analyze the use of perspective in the painting.

Duality

10. Explain how the results of Exercises 4 and 5 are evidence for the principle of duality.

11. Draw figures dual to those in Figure 13.10.

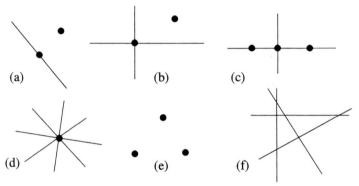

(a) (b) (c) (d) (e) (f)

Figure 13.10 Some projective configurations

usual rule for matrix multiplication. The matrix of T is calculated as follows: Let $\mathbf{e}_1 = (1, 0, 0)$, $\mathbf{e}_2 = (0, 1, 0)$, and $\mathbf{e}_3 = (0, 0, 1)$ be the standard basis vectors. Then $T\mathbf{e}_1$ is the first column of the matrix, $T\mathbf{e}_2$ is the second column, and $T\mathbf{e}_3$ is the third column. (See Exercise 4.)

Note: We make no geometric distinction between row and column vectors.

Inhomogeneous Coordinates

Matrix multiplication provides a means for calculating with projective transformations but does not lend itself readily to visualization of the geometric action of the transformations. In order to *see* projective transformations, a second coordinate system for the projective plane is useful.

For this purpose, we choose an ordinary plane model of the projective plane. The plane $z = 1$ is the standard choice. Coordinates for each projective point are obtained as follows: If (x, y, z) are homogeneous coordinates, we first divide by z getting $(x/z, y/z, 1)$. This is a point in the plane $z = 1$. Now, the third coordinate is no longer interesting, or informative, so it is dropped. The pair $(x/z, y/z)$ are **inhomogeneous coordinates** for the projective point. We will use capital letters (X, Y) for inhomogeneous coordinates to avoid confusion with homogeneous coordinates.

Inhomogeneous coordinates have two drawbacks: They do not exist for ideal points (for which $z = 0$) and they create more complicated formulae. For example, instead of the relatively simple matrix multiplication in equation (*), we have the following formula for the action of T on inhomogeneous coordinates:

$$Tp = T(X,Y) = \left(\frac{aX+bY+c}{gX+hY+i}, \frac{dX+eY+f}{gX+hY+i} \right) \quad (**)$$

While (**) is unwieldy in comparison with (*), it does allow the visualization of some projective transformations in conventional geometric terms. For example, if $b = d = g = h = 0$ and $a = e = i = 1$, then (**) takes the form

$$T(X,Y) = (X+c, Y+f)$$

which is an ordinary translation.

B–What other familiar transformations can you express as projective transformations using inhomogeneous coordinates?

The Fundamental Theorem

Projective transformations provide more flexibility of motion than any group we have yet studied, even more than Möbius geometry. Recall that with a Möbius transformation, any *three* points can be mapped to any other *three* points. With a projective transformation, in most circumstances any *four* points can be mapped to any other *four* points.

Theorem   *(Fundamental Theorem of the Projective Plane)   Let p_0, p_1, p_2, p_3 be four points in the projective plane such that no three are contained in the same projective line. Let q_0, q_1, q_2, q_3 be another four such points. Then, there is a projective transformation T such that $Tp_0 = q_0$, $Tp_1 = q_1$, $Tp_2 = q_2$, and $Tp_3 = q_3$.*

Proof: As in the proof of the fundamental theorem of Möbius geometry (in Chapter 5), it will suffice to find a transformation that maps four given points p_0, p_1, p_2, p_3 to four points in some standard position. For projective geometry, the conventional points are e_1, e_2, and e_3 whose coordinates are the standard basis vectors $\mathbf{e}_1 = (1, 0, 0)$, $\mathbf{e}_2 = (0, 1, 0)$, and $\mathbf{e}_3 = (0, 0, 1)$, plus the point e_0 with coordinates $\mathbf{e}_0 = (1, 1, 1)$. We therefore seek a projective transformation sending p_0 to e_0, p_1 to e_1, p_2 to e_2, and p_3 to e_3.

Actually, it is simpler to find the inverse transformation sending e_0 to p_0, e_1 to p_1, e_2 to p_2, and e_3 to p_3.

Let $\mathbf{p}_0$, $\mathbf{p}_1$, $\mathbf{p}_2$, and $\mathbf{p}_3$ be homogeneous coordinates for the projective points p_0, p_1, p_2, and p_3, respectively. By assumption, p_1 p_2, and p_3 do not lie on a single projective line. Therefore, the Euclidean points $\mathbf{p}_1$, $\mathbf{p}_2$, $\mathbf{p}_3$ do not lie in a single plane and so form a basis for three-dimensional Euclidean space. Therefore, we can express $\mathbf{p}_0$ as

$$\mathbf{p}_0 = k_1\mathbf{p}_1 + k_2\mathbf{p}_2 + k_3\mathbf{p}_3$$

for real constants k_1, k_2, and k_3. Furthermore, none of the constants k_1, k_2, or k_3 can be zero. If k_1 (for example) were zero, then p_0 would be on the projective line determined by p_2 and p_3, contrary to the hypothesis.

Let M be the matrix whose first column is $k_1\mathbf{p}_1$, whose second column is $k_2\mathbf{p}_2$, and whose third column is $k_3\mathbf{p}_3$. Let T be the linear transformation with this matrix. Clearly, $Te_1 = k_1\mathbf{p}_1$, $Te_2 = k_2\mathbf{p}_2$, $Te_3 = k_3\mathbf{p}_3$, and $Te_0 = \mathbf{p}_0$. Recall that in projective geometry $\mathbf{p}_1$ is identified with $k_1\mathbf{p}_1$, $\mathbf{p}_2$ identified with $k_2\mathbf{p}_2$, and $\mathbf{p}_3$ identified with $k_3\mathbf{p}_3$ (since all these constants are nonzero). Therefore, $Te_0 = p_0$, $Te_1 = p_1$, $Te_2 = p_2$, and $Te_3 = p_3$, as desired.   ∎

The Fundamental Theorem has this interesting consequence: It is impossible to define a nontrivial metric on the projective plane. In other words, projective geometry is not a metric geometry. Here's why. A metric is an invariant function $d(p, q)$ of two points. But in projective geometry, every pair of points is congruent to every other pair. Thus, the only possible distance functions in projective geometry are trivial ones that assign a constant as the distance between all distinct pairs of points.

C–Must an invariant function of three points be constant?
D–Must an invariant function of four points be constant?

Projective Configurations

In projective geometry, the simplest figures consist of points and lines only. Such figures are classified by incidence, that is, by counting how many points lie on what lines and how many lines pass through which points.

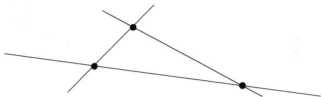

Figure 14.3 A projective triangle

One of the simplest figures, of course, is the **triangle** (Figure 14.3), which consists of three lines and three points: two points on each line and two lines through each point. Note that a projective triangle contains the whole of the lines determined by its vertexes. Projective geometry has no concept of a line segment because, projective lines being reentrant, there is no way to tell what points lie between two other points on a line.

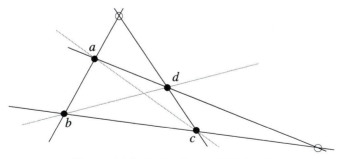

Figure 14.4 A simple quadrilateral

The notation (p_n, L_m) is used for any projective figure containing p points and L lines such that n lines pass through each point and m points lie on each line. Figures with this kind of regularity are called **configurations**. Thus, a triangle is a configuration of type $(3_2, 3_2)$, and a **simple quadrilateral** (Figure 14.4) is a configuration of type

$(4_2, 4_2)$. Note that the extra *points* (marked in the figure with circles) are not considered part of a *simple* quadrilateral. In the same spirit, the diagonal lines *ac* and *bd* (in gray in the figure) also do not belong to the configuration.

If the extra *intersections* are added to the simple quadrilateral, the figure, now called a **complete quadrilateral** [Figure 14.5a], is an example of a configuration of type $(6_2, 4_3)$. On the other hand, if the diagonal *lines* are added, the resulting figure [see Figure 14.5b], called the **complete quadrangle**, is of type $(4_3, 6_2)$. These are dual configurations.

There is more. In the complete quadrilateral, there are now *three* diagonal lines [in gray in Figure 14.5a] forming the **diagonal triangle**. In the dual configuration, there are three unnamed intersections [circled in Figure 14.5b], called **diagonal points**, which likewise form a **diagonal triangle**.

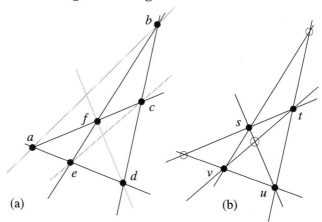

(a)   (b)

Figure 14.5 A complete quadrilateral (a) and a complete quadrangle (b)

One might hope, by drawing more diagonal points and lines, eventually to reach a figure that is "completely complete," meaning that no more diagonals (points or lines) can be drawn. However, it can be proved that, beginning with four points, no three of which are on a line, this process never ends.

E–Draw a configuration of type $(6_3, 9_2)$ and its dual.

Two Famous Theorems

We close the chapter with two famous theorems. To introduce them, some special terminology is needed.

Definition *Two triangles, $\triangle abc$ and $\triangle efg$, are **perspective from a point** if the lines determined by corresponding vertexes of the triangles (that is, the lines ae, bf, and cg) share a common point of intersection. Two triangles, $\triangle abc$ and $\triangle efg$, are **perspective from a line** if the points* *determined by the intersection of corresponding sides of the triangles (that is the points of intersection of ab and ef, bc and fg, and ca and ge) lie on a single line.*

Both kinds of perspective triangles are illustrated in Figure 14.6. In (a), the lines determined by corresponding vertexes are gray, and their common point *k* is the point from which Δ*abc* and Δ*efg* are perspective. In (b), *s*, *t*, and *u* are the points determined by the intersection of corresponding sides of the triangles, and the single line on which they lie; the line from which Δ*abc* and Δ*efg* are perspective, is dashed.

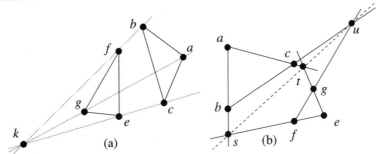

Figure 14.6 Triangles: Perspective from a point (a) and from a line (b)

The term "perspective from a point" is connected with the theory of perspective drawing. If Δ*abc* and Δ*efg* are perspective from a point, then an eye placed at the perspective point [*k* in Figure 14.6a] will see only one triangle, since the vertex pairs *ae*, *bf* and *cg* are on the same line of sight. The term "perspective from a line," however, has no particular justification from a visual point of view. Note that "perspective from a line" and "perspective from a point" are dual concepts.

The surprising result is

Desargue's Theorem *Two triangles are perspective from a point if, and only if, they are perspective from a line.*

Proof: Let Δ*abc* and Δ*efg* be perspective from the point *k* (see Figure 14.7). We will prove that they are perspective from a line; that is, the three points *s*, *t*, and *u* in Figure 14.7 are collinear.

To begin, we will apply a projective transformation to place the figure, consisting of the seven given points *a*, *b*, *c*, *e*, *f*, *g*, and *k* in a position chosen to simplify subsequent computation with their coordinates. According to the Fundamental Theorem, we can choose the location of any four points, as long as no three of them are collinear. For this purpose, we pick the points *a*, *b*, *c*, and *k*. Now *a*, *b*, and *c* are definitely not collinear as they form a triangle. However, *k* might be collinear with one of the sides of this triangle, that is, collinear with ab, bc, or ca. In that case, at least two of the lines in the figure are the same, and the proof of the theorem is actually easier. We leave to the reader the treatment of such "degenerate" cases and proceed under the assumption that no three of the four points a, b, c, and k are collinear.

By applying a projective transformation, we may assume that $\mathbf{a}$ = $(1, 0, 0)$, $\mathbf{b} = (0, 1, 0)$, $\mathbf{c} = (0, 0, 1)$, and $\mathbf{k} = (1, 1, 1)$ where we use boldface letters for homogeneous coordinates and ordinary letters for the points themselves. What, then, of $\mathbf{e}$, $\mathbf{f}$, and $\mathbf{g}$? Well, the point e, for example, lies along the line ka (since, by assumption, Δabc and Δefg are perspective from the point k). Therefore, the coordinates of e are a linear combination of the coordinates of k and a; that is,

$$\mathbf{e} \ = \ \alpha\mathbf{k} + \beta\mathbf{a} \ = \ (\alpha + \beta, \ \alpha, \ \alpha) \ = \ (x, 1, 1)$$

where $x = (\alpha + \beta)/\alpha$ and the last equality is justified because the coordinates are homogeneous. Similarly, we find that $\mathbf{f} = (1, y, 1)$ and $\mathbf{g} = (1, 1, z)$ for suitable real numbers y and z.

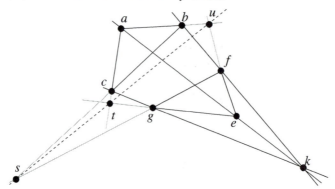

Figure 14.7 Proof of Desargues's theorem

The rest of the proof is mechanical computation. Homogeneous coordinates of the line ab, for example, are computed using a cross product,

$$ab = \mathbf{a} \times \mathbf{b} = [0, 0, 1]$$

Similarly,

$$
\begin{aligned}
bc &= [1, 0, 0] \\
ca &= [0, 1, 0] \\
ef &= [1 - x, \ 1 - y, \ xy - 1] \\
fg &= [yz - 1, \ 1 - z, \ 1 - y]
\end{aligned}
$$

and

$$ge \ = \ [1 - z, \ xz - 1, \ 1 - x]$$

Next to be determined are coordinates for s: the intersection of bc and fg. These coordinates are the cross product of the two line coordinates; that is,

$$\mathbf{s} = [1,0,0] \times [yz-1, 1-z, 1-y] = (0, y-1, 1-z)$$

Similarly, $\mathbf{t} = (1 - x, 0, c - 1)$ and $\mathbf{u} = (x - 1, 1 - y, 0)$.

To prove that s, t, and u are collinear, it suffices to check that the determinant

$$\begin{vmatrix} 0 & y-1 & 1-z \\ 1-x & 0 & z-1 \\ x-1 & 1-y & 0 \end{vmatrix}$$

is zero. This is left to the reader. (See Exercise 11.)

The foregoing actually proves only half of Desargues's theorem. We must also prove that two triangles perspective from a line are perspective from a point. This can be accomplished by an argument similar to that just used (see Exercise 13), or it suffices to note that this second part of the theorem is the dual of the first and so follows from the principle of duality. ∎

Our second famous theorem is:

Pappus's Theorem *Let points a, b, and c be colinear and likewise points a', b', and c'. Then, the three intersections of ab' with a'b, of ac' with a'c, and of bc' with b'c are collinear.*

Proof: The situation is depicted in Figure 14.8. A proof can be given using the same computational ideas used to prove Desargues's theorem. (See Exercise 15.)

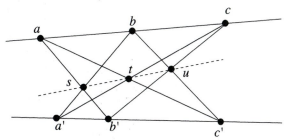

Figure 14.8 Proof of Pappus's Theorem

Conclusion

While the invention of projective geometry was motivated by problems of pictorial representation, as a mathematical subject it has a very different purpose. It functions as a kind of laboratory— providing a world in which geometrical ideas can be studied under certain ideal conditions, for example, where all pairs of lines intersect.

EXERCISES

Projective Transformations: Examples

1. A **shear** transformation in the plane has the form

$$T(x, y) = (x + ky, y)$$

where k is a constant.   Graph this transformation for $k = 1$ and for $k = 2$ using the style of graph introduced in Chapter 3.   What is the effect of different k's on the transformation?   Verify that T is a linear transformation.

2. A **strain** transformation in the plane has the form

$$T(x, y) = (kx, y)$$

where k is a constant.   Graph this transformation for $k = 1/2$ and for $k = 2$ using the style of graph introduced in Chapter 3.   What is the effect of different k's on the transformation?   Verify that T is a linear transformation.

3. Here are some projective transformations written in terms of *inhomogeneous* coordinates. Rewrite each in terms of *homogeneous* coordinates. Assume that a, b, k are real constants.
   (a) a translation: $T(X, Y) = (X + a, Y + b)$
   (b) a 90° rotation: $T(X, Y) = (-Y, X)$
   (c) a homothetic transformation: $T(X, Y) = (kX, kY)$
   (d) a strain transformation: $T(X, Y) = (kX, Y)$
   (d) a shear transformation: $T(X, Y) = (X + kY, Y)$
   (e) real inversion: $T(X, Y) = (X/(X + Y), (Y/(X + Y))$

Example: (a) To make the change from inhomogeneous to homogeneous coordinates, simply let $X = x/z$ and $Y = y/z$ and add a third coordinate of 1:

$$T(x, y, z) = T(X, Y) = (X + a, Y + b)$$

$$= \left(\frac{x}{z} + a, \frac{y}{z} + b, 1 \right) = (x + az, y + bz, z)$$

The last equation follows upon multiplying by z, permitted because the coordinates are homogeneous.

Projective Transformations: Theory

4. Let T be a linear transformation of V. Let $\mathbf{p}_1 = T\mathbf{e}_1$, $\mathbf{p}_2 = T\mathbf{e}_2$, and $\mathbf{p}_3 = T\mathbf{e}_3$, where $\mathbf{e}_1$, $\mathbf{e}_2$, and $\mathbf{e}_3$ are the standard basis vectors. Let M be the matrix whose first column is $\mathbf{p}_1$, second column is $\mathbf{p}_2$, and third column is $\mathbf{p}_3$. Prove equation (*); that is, for any vector P, $T\mathbf{p} = M\mathbf{p}$.

5. Derive equation (**) on page 153 from equation (*) on page 152.

Hint: If $\mathbf{p} = (x, y, z)$, then $\mathbf{p} = x\,\mathbf{e}_1 + y\,\mathbf{e}_2 + z\,\mathbf{e}_3$. *Use linearity.*

Fixed Points

6. Let T be a projective transformation. A fixed point p of T is a projective point whose coordinates $\mathbf{p} = (x, y, z)$ satisfy the algebraic equation $T\mathbf{p} = k\mathbf{p}$, where k is a real number. (In linear algebra, $\mathbf{p}$ is called an **eigenvector** of the transformation T, and k is called an **eigenvalue**.) Find the fixed points of these transformations:

   (a) $T\mathbf{p} = (2x, -y, 4z)$ (b) $T\mathbf{p} = (2y - z, 2z - x, 2x - y)$
   (c) $T\mathbf{p} = (y, z, x)$ (d) $T\mathbf{p} = (x + y, y + z, y)$
   (e) $T\mathbf{p} = (2x + 3y, -y, 4z)$

7. (for linear algebra students) Does every projective transformation have a fixed point?

Hint: The question of the existence of eigenvalues of a transformation is studied via the characteristic equation of a matrix (as described in any linear algebra text).

Configurations

8. Find configurations of the following types, or, if none exists, explain why.

   (a) $(4_1, 3_2)$ (b) $(2_3, 3_2)$
   (c) $(4_1, 2_2)$ (d) $(6_1, 3_2)$
   (e) $(5_3, 5_3)$ (f) $(4_1, 1_4)$

9. If a configuration of type (p_n, L_m) exists, show that $pn = LM$.

10. The most symmetric configurations are of type (p_n, p_n) abbreviated (p_n). Find examples of the following types, or, if none exists, explain why.

   (a) (3_2) (b) (4_2) (c) (7_2)
   (d) (2_2) (e) (3_3) (f) (4_3)

11. If a configuration is of type (p_n, L_m), what is the type of the dual configuration?

Theorems of Projective Geometry

12. Complete the proof of Desargues's theorem by checking that the determinant in the proof *is* zero.

13. Carry out the proof of Desargues's theorem in the degenerate case where the points a, b, and k are collinear.

14. Give a direct proof (without using duality) that triangles perspective from a line are also perspective from a point.

15. Prove Pappus's theorem.

16. The dual of Pappus's theorem is sometimes called Brianchon's theorem. State Brianchon's theorem and give a direct proof without using duality.

17. Prove that the diagonal points of a complete quadrangle are *not* collinear.

18. Let the vertexes of Δefg lie on the sides of Δabc. (One possibility is shown in Figure 14.9.) Suppose, in addition, that the lines ae, bf, and cg are concurrent. Let s be the intersection of ab and ef, t be the intersection of bc and fg, and u be the intersection of ca and ge. Prove that s, t, and u are collinear.

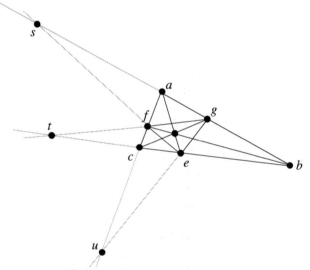

Figure 14.9 For Exercise 18

Hint: This can be proved by calculation or by application of a theorem already proven.

15 MULTIDIMENSIONAL PROJECTIVE GEOMETRY

This chapter is about projective geometries of all dimensions. The principal topics are incidence relations, the classification of conic sections, and duality. While multidimensional geometry can be a difficult subject, multidimensional *projective* geometry is relatively straightforward. This is because the algebra upon which it is based (linear algebra) is essentially the same in all dimensions. In geometry, the projective point of view, if applicable, is usually the simplest point of view. This is just one aspect of the *universality* of projective geometry explored in more detail in Chapter 16.

To begin, recall the definition of P_2, the real projective plane: The underlying set is Euclidean three-dimensional space (minus the origin), with the identification of all points that are scalar multiples of each other; the group of transformations is the group of all invertible, linear transformations of this space. What is odd about P_2 is that a *two*-dimensional geometry is modeled by a *three*-dimensional set. It is identification, of course, that cuts the dimension back from three to two.

The same ideas characterize *n*-dimensional projective geometry:

Definition Let n be a positive integer. Let P_n be the set of all points of $(n + 1)$-dimensional Euclidean space except the origin; that is,

$$P_n = \{(x_0, x_1, x_2, \ldots, x_n): x_0, x_1, x_2, \ldots, x_n \text{ are reals, not all zero}\}$$

Each point $p = (x_0, x_1, x_2, \ldots, x_n)$ of P_n is identified with all scalar multiples kp, where k is any nonzero real number. Let PG_n be the set of all invertible, linear transformations of P_n.
*The pair (P_n, PG_n) models **n-dimensional projective geometry**.*

The $(n + 1)$ numbers $x_0, x_1, x_2, \ldots, x_n$ defining a projective point are called **homogeneous coordinates** of the point. As with the projective plane, there are also **inhomogeneous coordinates**,

$$\left(X_1, X_2, \ldots, X_n\right) = \left(\frac{x_1}{x_0}, \frac{x_2}{x_0}, \frac{x_3}{x_0}, \ldots, \frac{x_n}{x_0}\right)$$

consisting of only n numbers, which seems more natural for an n-dimensional geometry, but the inhomogeneous coordinates do not exist at points where x_0 is zero.

A projective transformation T can be expressed using a matrix:

$$Tx = \begin{pmatrix} a & b & c & \dots \\ d & e & f & \dots \\ \dots & \dots & \dots & \ddots \end{pmatrix} \begin{pmatrix} x_0 \\ x_1 \\ x_2 \\ \vdots \end{pmatrix} = \begin{pmatrix} ax_0 + bx_1 + cx_2 + \dots \\ dx_0 + ex_1 + fx_2 + \dots \\ \dots \dots \end{pmatrix}$$

All this is a straightforward generalization of $\mathbf{P}_2$. *Note*: In this chapter, for clarity, dimension will always mean *projective* dimension (one less than Euclidean dimension).

Projective results are most easily formulated and proved using the terminology and results of linear algebra. We summarize the necessary background material without proof.

Linear Algebra Background *A set of points $\mathbf{p}_0, \mathbf{p}_1, \dots, \mathbf{p}_k$ from $\mathbf{P}_n$ is called* ***independent*** *if, whenever a linear combination*

$$\alpha_0\mathbf{p}_0 + \alpha_1\mathbf{p}_1 + \dots + \alpha_k\mathbf{p}_k$$

is zero, then all the coefficients $\alpha_0, \alpha_1, \dots, \alpha_k$ are also zero. The maximum size of an independent subset of $\mathbf{P}_n$ is $n + 1$ points.

A proper subset of $\mathbf{P}_n$ is a ***subspace*** *of $\mathbf{P}_n$, if it is closed under the operations of vector addition and scalar multiplication. Every subspace has a* ***dimension*** *k which is determined by the property that the maximum number of independent vectors in the subspace is $k + 1$.*

Let A be a subspace of $\mathbf{P}_n$ of dimension k. A set $\{\mathbf{p}_0, \mathbf{p}_1, \dots, \mathbf{p}_k\}$ of $k + 1$ independent points in A is called a ***basis*** *for A. Every basis* ***spans***, *meaning that every point $\mathbf{q}$ of A can be expressed as a linear combination of the basis points:*

$$\mathbf{q} = \alpha_0\mathbf{p}_0 + \alpha_1\mathbf{p}_1 + \dots + \alpha_k\mathbf{p}_k$$

if the coefficients $\alpha_0, \alpha_1, \dots, \alpha_k$ are chosen appropriately.

In geometric terminology, independent points are said to be **in general position**. This means that the points do not lie "unnecessarily" together on lines, or any other flat subsets of the geometry. (See Exercise 1.)

An important theorem based on independence is the Fundamental Theorem of Projective Geometry:

Fundamental Theorem of Projective Geometry *Let*

$$\mathbf{p}_0, \ \mathbf{p}_1, \ \mathbf{p}_2, \dots, \mathbf{p}_{n+1}$$

be $n + 2$ *points in projective space* $\mathbf{P}_n$ *and suppose that every subset of* $n + 1$ *of these points is in general position. Let*

$$\mathbf{q}_0, \ \mathbf{q}_1, \ \mathbf{q}_2, \ldots, \ \mathbf{q}_{n+1}$$

be another $n + 2$ *such points. Then there is a projective transformation* T *such that* $T\mathbf{p}_0 = \mathbf{q}_0$, $T\mathbf{p}_1 = \mathbf{q}_1$, $T\mathbf{p}_2 = \mathbf{q}_2, \ldots,$ *and* $T\mathbf{p}_{n+1} = \mathbf{q}_{n+1}$.

Proof: This is a straightforward generalization of the proof of the two-dimensional theorem in Chapter 14. (See Exercise 6.) ■

Flats

Flats are the most fundamental of all projective figures.

Definition *"Flat" is the geometric term for subspace. Thus, a **flat** is a proper subset of a projective geometry that is closed under vector addition and scalar multiplication. The **dimension** of a flat is its dimension as a subspace of* $\mathbf{P}_n$.

In the projective plane, $\mathbf{P}_2$, there are just two kinds of flats: points (zero-dimensional) and lines (one-dimensional). In projective solid geometry, $\mathbf{P}_3$, there are three kinds of flats: points (zero-dimensional), lines (one-dimensional), and planes (two-dimensional), just as in Euclidean solid geometry.

A–How many different kinds of flats are there in $\mathbf{P}_n$?

In general, the most important flats are **points** (zero-dimensional), **lines** (one-dimensional), and **hyperplanes** ($(n - 1)$-dimensional flats). A hyperplane, being one dimension smaller than the whole space, is the set of points $\mathbf{p} = (x_0, x_1, \ldots, x_n)$ satisfying a single linear equation

$$a_0 x_0 + a_1 x_1 + \ldots + a_n x_n = 0$$

where not all the coefficients $a_0, a_1, \ldots, a_n$ can be zero. Grouped together, the coefficients of this equation $[a_0, a_1, \ldots, a_n]$ are **homogeneous coordinates** of the hyperplane.

Incidence relationships among flats can be worked out from incidence relationships in vector spaces. Here is an example:

Theorem *(Fundamental Theorem on Incidence) Let A and B be flats of dimension m and s in* $\mathbf{P}_n$. *If* $m + s \geq n$, *then A and B intersect in at least a point.*

Proof: Since A is m-dimensional, A contains a basis of $m + 1$ independent points, say, $\mathbf{p}_0, \mathbf{p}_1, \ldots, \mathbf{p}_m$. Likewise, B contains a basis of $s + 1$ independent points: $\mathbf{q}_0, \mathbf{q}_1, \ldots, \mathbf{q}_s$. The union of these two collections of points has $m + s + 2$ points. By hypothesis, $m + s + 2 \geq$

$n + 2$. Since the maximum number of independent points in $\mathbf{P}_n$ is $n + 1$, the union of the $\mathbf{p}$'s and $\mathbf{q}$'s is not independent. Therefore, some linear combination

$$\alpha_0 \mathbf{p}_0 + \alpha_1 \mathbf{p}_1 + \ldots + \alpha_m \mathbf{p}_m + \beta_0 \mathbf{q}_0 + \beta_1 \mathbf{q}_1 + \ldots + \beta_s \mathbf{q}_s$$

equals zero, or, in other words,

$$\alpha_0 \mathbf{p}_0 + \alpha_1 \mathbf{p}_1 + \ldots + \alpha_m \mathbf{p}_m = -\beta_0 \mathbf{q}_0 - \beta_1 \mathbf{q}_1 - \ldots - \beta_s \mathbf{q}_s \ (= \mathbf{r})$$

where not all the coefficients (α's and β's) are zero. The two sides of this second equation represent a point, called $\mathbf{r}$, say. The point $\mathbf{r}$ is in A (since $\mathbf{r}$ is combination of the points $\mathbf{p}_0, \mathbf{p}_1, \ldots \mathbf{p}_m$) and in B (since $\mathbf{r}$ is also a combination of the points $\mathbf{q}_0, \mathbf{q}_1, \ldots, \mathbf{q}_s$). The point $\mathbf{r}$ cannot be zero, incidentally, because the vectors $\mathbf{p}_0, \mathbf{p}_1, \ldots, \mathbf{p}_m$ are independent *and* not all the coefficients are zero. Thus, A and B have at least the point $\mathbf{r}$ in common. ∎

Here are some examples of this theorem.

Example 1

There are no parallel lines in the projective plane $\mathbf{P}_2$. We already know this from Chapter 13, but our theorem gives us this result from a fresh perspective. Since lines have dimension 1, and $1 + 1 = 2$, and 2 is the dimension of $\mathbf{P}_2$, the theorem demands that two lines intersect.

Example 2

There *are* parallel lines in projective three-dimensional space $\mathbf{P}_3$. The possibility of parallel lines is admitted by the theorem, since $1 + 1$ is 2 (still), but 2 is not greater than or equal to 3 (the dimension of $\mathbf{P}_3$). The line joining the points (1, 0, 0, 0) and (0, 1, 0, 0), for example, is parallel to the line joining the points (0, 0, 1, 0) and (0, 0, 0, 1).

Example 3

There are no parallel *planes* in $\mathbf{P}_3$. This follows from our theorem, since planes have dimension 2, $2 + 2$ is 4, and 4 is greater than (or equal to) 3 (the dimension of $\mathbf{P}_3$).

B–Explain why the two lines in Example 2 are parallel.
C–There are parallel planes in $\mathbf{P}_n$ for what n?

Cross Ratio and the Projective Line

Even one-dimensional projective geometry has interesting features. In this section we introduce the cross ratio, an invariant on the projective line that has important applications in all dimensions.

The points of the projective line, $\mathbf{P}_1$, are represented by pairs of real numbers (x_0, x_1). According to the Fundamental Theorem, any *three* such points can be transformed to any other three points. (The independence condition of the Fundamental Theorem here demands only that the points be distinct.) In view of this result, there are no nontrivial invariants of one, two, or three points on the projective line. However, there is an important invariant of four points:

Definition *Let* $\mathbf{p}_0$, $\mathbf{p}_1$, $\mathbf{p}_2$, $\mathbf{p}_3$ *be four distinct points on a projective line. The* **cross ratio** *of these points is defined by*

$$CR[\mathbf{p}_0, \mathbf{p}_1, \mathbf{p}_2, \mathbf{p}_3] = \frac{|\mathbf{p}_0\ \mathbf{p}_2|\ |\mathbf{p}_1\ \mathbf{p}_3|}{|\mathbf{p}_1\ \mathbf{p}_2|\ |\mathbf{p}_0\ \mathbf{p}_3|}$$

where $|\mathbf{p}\ \mathbf{q}|$ *is the 2 by 2 determinant with first column the coordinates of the point* $\mathbf{p}$ *and second column the coordinates of* $\mathbf{q}$.

Theorem *The cross ratio is an invariant, that is if* $T: \mathbf{P}_1 \rightarrow \mathbf{P}_1$ *is a projective transformation, then*

$$CR[T\mathbf{p}_0, T\mathbf{p}_1, T\mathbf{p}_2, T\mathbf{p}_3] = CR[\mathbf{p}_0, \mathbf{p}_1, \mathbf{p}_2, \mathbf{p}_3].$$

Proof: This follows immediately from the formula

$$|T\mathbf{p}\ T\mathbf{q}| = |T|\ |\mathbf{p}\ \mathbf{q}|$$

where $|T|$ is the (2 by 2) determinant of T. (See exercise 7.) ∎

This cross ratio is directly related to the invariant of the same name used extensively in Möbius geometry. (See Chapter 5.) In fact, Möbius geometry is a form of projective geometry, as we will see in the next chapter.

D–Show that the formula for the cross ratio in this chapter is the same as the formula for cross ratio used in Chapter 5. (*Hint:* Use inhomogeneous coordinates.)

The cross ratio can be applied to four points from any projective line. To see how, let λ be a line in any projective geometry. Any two distinct points of λ, say, $\mathbf{s}$ and $\mathbf{t}$, are a basis for λ. This means that every point $\mathbf{p}$ of λ can be expressed as a combination of $\mathbf{s}$ and $\mathbf{t}$:

$$\mathbf{p} = \alpha\, \mathbf{s} + \beta\, \mathbf{t}.$$

The pair of coefficients (α, β) are called **relative coordinates** for $\mathbf{p}$ (meaning relative to the chosen basis of λ). Relative coordinates, being a pair of numbers, can be used in the 2 by 2 determinants of the cross ratio. It is easy to show that the value of the cross ratio, thus calculated, is independent of the choice of basis, that is, independent of which set of relative coordinates is used. (See Exercise 11.)

In projective geometry lines are reentrant. Given three points **p**, **q**, and **r**, on a reentrant line, it is not true that one of them is between the other two. (See Figure 15.1.) This is a major difference between reentrant and nonreentrant lines.

(a)	(b)
Three points on a Euclidean line: One is between the other two.	Three points on a reentrant line: None is between the other two.

Figure 15.1 There is no betweeness on reentrant lines.

On the other hand, given *four* points on a reentrant line, **p**, **q**, **r**, and **s**, there are two fundamentally different situations, depending on whether **p** and **q** separate, or do not separate, **r** and **s**. (See Figure 15.2.) These two situations are distinguished by the cross ratio. Thus, $CR[$**p**, **q**, **r**, **s**$]$ is negative if **p** and **q** separate **r** and **s**, and positive otherwise.

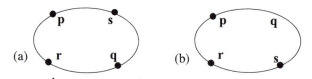

(a)	(b)
p and **q** separate **r** and **s**.	**p** and **q** do not separate **r** and **s**.

Figure 15.2 There is *separation* on re-entrant lines.

Quadrics

Flats are determined by linear equations. Next, in interest and importance, are figures determined by a quadratic equation.

Definition *A **quadric** in $\mathbf{P}_n$ is the set of points* **p** *in $\mathbf{P}_n$ whose coordinates* **p** $= (x_0, x_1, \dots , x_n)$ *satisfy a quadratic equation*

$$c_{0,0}x_0^2 + c_{0,1}x_0x_1 + \dots + c_{i,j}x_ix_j + \dots + c_{n,n}x_n^2 = 0 \quad (*)$$

whose coefficients $c_{i,j}$ are real numbers.
*In the projective plane, $\mathbf{P}_2$, a quadric is called a **conic**.*

Because the quadratic equation (*) is homogeneous, if a point **p** $= (x_0, \dots , x_n)$ satisfies (*), then so do all scalar multiples of **p**. This is essential, of course, in order that the equation define a projective figure.

In its most general form, (*) contains a term for every pair of variables from x_0, x_1, ... , x_n. In two dimensions, therefore, the complete quadratic is

$$ax_2^2 + bx_2x_1 + cx_1^2 + dx_2x_0 + ex_1x_0 + fx_0^2 = 0$$

If inhomogeneous coordinates are used, then we get the familiar general quadratic

$$aX_2^2 + bX_2X_1 + cX_1^2 + dX_2 + eX_1 + f = 0$$

whose graph in the (X_1, X_2) plane is either an ellipse, a parabola, or a hyperbola (or one of several degenerate forms). For example, the homogeneous equation

$$x_2^2 - 4x_1x_0 + 4x_0^2 = 0 \qquad (**)$$

corresponds to the inhomogeneous equation $X_2^2 - 4X_1 + 4 = 0$, which is the equation of a parabola. This connection with Euclidean conics is the justification for calling a quadric in the projective plane a conic.

E–Convert $x_2^2 - 4x_1^2 + 4x_0^2 = 0$ to an inhomogeneous form. What kind of Euclidean conic do you get?

The inhomogeneous equation, however, does not adequately represent the whole quadric, since it has fewer points than the homogeneous equation; that is, the points with $x_0 = 0$ are missing. In equation (**), for example, setting $x_0 = 0$ gives $x_2^2 = 0$, so the missing points, in this example, satisfy $x_2 = 0 = x_0$. The points satisfying these conditions have the form $(0, k, 0)$. They are all mutually identified, so actually just one projective point satisfies the homogeneous equation (**), but not the inhomogeneous equation. We describe this situation by saying that the parabola has one ideal point (or one point at infinity).

F–How many ideal points are on the conic $x_2^2 - 4x_1^2 + 4x_0^2 = 0$? Relate this to the Euclidean form of this conic.

In Euclidean geometry, conics are classified by applying a Euclidean transformation (rotation plus translation) to place them in standard position with their center at the origin. In the Euclidean plane, there are three important types of conics: ellipse, parabola, and hyperbola, plus a number of degenerate forms (including parallel lines, intersecting lines, a single point, and so forth). Now, there are more transformations in projective geometry than in Euclidean. Therefore, some conics not congruent in Euclidean geometry *are* congruent in projective geometry. This makes the projective classification of conics and quadrics (in dimensions other than 2) correspondingly simpler:

Theorem *(Projective classification of quadrics)* *In projective geometry, every quadric is congruent to one of the* **canonical forms***, namely, the quadrics with equations*

$$c_0 x_0^2 + c_1 x_1^2 + \ldots + c_n x_n^2 = 0$$

where each coefficient c_i is either 0, or 1, or −1.

Proof: Dimension is irrelevant to this argument, as we shall see, so we present it in the simplest case: the projective line. In $\mathbf{P}_1$, a quadric has the equation

$$a x_0^2 + b x_0 x_1 + c x_1^2 = 0$$

Completing the square, we get

$$a(x_0 + \alpha x_1)^2 + \beta x_1^2 = 0 \qquad (1)$$

where $\alpha = b/2a$ and $\beta = c - b^2/2a$ (except if $a = 0$; see Exercise 13). Next we apply a translation

$$(y_0, y_1) = T(x_0, x_1) = (x_0 + \alpha x_1, x_1)$$

If the point $\mathbf{p} = (x_0, x_1)$ satisfies equation (1), then $\mathbf{q} = T\mathbf{p} = (y_0, y_1)$ satisfies

$$a y_0^2 + \beta y_1^2 = 0 \qquad (2)$$

Thus, we have eliminated the middle term from (1), and already the resulting equation is close to the form predicted by the theorem. (In higher dimensions, we have to repeat this step for every nonzero middle term, but the process is exactly the same in each case.)

If a and β are positive, we next apply a homothetic transformation:

$$(z_0, z_1) = S(y_0, y_1) = (\sqrt{a}\, y_0, \sqrt{\beta}\, y_1)$$

If the point $\mathbf{q} = (y_0, y_1)$ satisfies equation (2), then $\mathbf{r} = S\mathbf{q} = (z_0, z_1)$ satisfies

$$z_0^2 + z_1^2 = 0 \qquad (3)$$

This is one of the canonical forms. If either a or β is negative, then we use either $\sqrt{|a|}$ or $\sqrt{|\beta|}$ accordingly, and equation (3) has one or more coefficients that are −1. Finally, if a or β is zero, then one or more of the coefficients in (3) is zero. ∎

G–Find the canonical form of the conic $x_0^2 - 4x_1 x_2 + 4x_2^2 = 0$.

Applying the Classification Theorem

In the real projective plane, P_2, the canonical form of a conic has three terms: $c_0x_0^2 + c_1x_1^2 + c_2x_2^2$. Here is a list of the forms this equation can take:

(1) $c_0 = c_1 = c_2 = 1$ (or -1). The equation is $x_0^2 + x_1^2 + x_2^2 = 0$, which is satisfied by *no* points. (Remember that $(0, 0, 0)$ is not in the projective plane.)   Conics that reduce to this case are called **imaginary**.

(2) $c_0 = c_1 = c_2 = 0$. In this case the conic is the whole projective plane! This form arises only if all the original coefficients are zero and is usually excluded from consideration.

(3) $c_0 = c_1 = 0$, $c_2 = \pm1$. The equation is $\pm x_2^2 = 0$. The points on the conic have the form $(u, v, 0)$ and form a projective space of dimension 1: a projective straight line.

(4) $c_0 = 0$, $c_1 = c_2 = \pm1$. The equation is $x_1^2 + x_2^2 = 0$. The points on the conic have the form $(u, 0, 0)$. This is a single point.

(5) $c_0 = 0$, $c_1 = +1$, $c_2 = -1$. The equation is $x_1^2 - x_2^2 = 0$. The points on the conic have the form $(u, \pm u, 0)$. This is a pair of points.

(6) $c_0 = c_1 = +1$, $c_2 = -1$. The equation is $x_0^2 + x_1^2 - x_2^2 = 0$. This is the only case that looks like what we usually call a conic.

Forms (2) to (5) are **degenerate** conics. They are not curved but rather consist of flats or unions of flats. For technical reasons, imaginary conics are considered nondegenerate, but only form (6) has the appearance of a "true" conic.

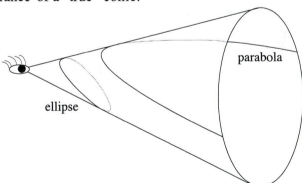

Figure 15.3 From the vertex of a cone, all conic sections look alike.

In inhomogeneous coordinates, form (6) is the equation of a hyperbola. But if, instead of dividing by x_0, we divide by x_2, we get the equation of a circle! In fact, the different Euclidean conics: ellipse, parabola and hyperbola, are all congruent in projective geometry! The reason for this is clear if we think of projective geometry as a model of monocular vision (recall Chapter 13). In Euclidean three-dimensional space, upon which the projective plane is based, conics are conic sections: plane sections of a cone. From the viewpoint of a single eye at the vertex of a cone, all conic sections look alike. (See Figure 15.3.) This is one aspect of the simplicity and beauty of projective geometry.

H–Consider the conic $x_2^2 - 4x_1x_2 + 4x_0^2 = 0$. Which Euclidean conic does this projective conic take the form of if $x_0 = 0$ is the ideal line? What if $x_1 = 0$ is the ideal line? If $x_2 = 0$ is the ideal line?

I–What 3-dimensional Euclidean surface is the graph of canonical form 6?

Duality

In the projective plane, both lines and points have coordinates. Furthermore, the set of all lines is algebraically identical to the set of points. This is the basis for the duality of points and lines described in Chapter 13, according to which every theorem of projective geometry remains true if the concepts of "point" and "line" are interchanged.

The situation in n-dimensions is only a bit more complicated. Here it is hyperplanes that have the same coordinate structure as points. Therefore, in n-dimensions, duality interchanges points with hyperplanes rather than lines. Duality also involves the other flats. For example, a line in $\mathbf{P}_n$ is determined (or spanned) by two points. Dually, two hyperplanes determine (by intersection) a flat of dimension $n - 2$. More generally, a flat of dimension k (the *span* of $k + 1$ points) is dual to a flat of dimension $n - k - 1$ (the *intersection* of $k + 1$ hyperplanes). This leads to

The Principle of Duality *Let* **T** *be a theorem of n-dimensional projective geometry. Then the statement* **T*** *that results from the systematic replacement in* **T** *of k-dimensional flats by (n − k − 1)-dimensional flats is also a theorem.*

Systematic replacement requires more than simple substitution of $(n - k - 1)$-dimensional flats for k-dimensional flats. For example, as the preceding paragraph suggests, the *intersection* of hyperplanes and the *spanning* of points are dual concepts. Therefore, the dualization of a theorem requires the interchange of these terms also.

The Dual Space

The duality of points and hyperplanes suggests that we consider a new kind of geometry: hyperplane geometry, a geometry in which hyperplanes, instead of points, are the fundamental building blocks. The main idea is expressed in the following definition:

Definition   *The **dual space** of $\mathbf{P}_n$, notated $\mathbf{P}_n{}^*$, is the set of all hyperplanes λ in $\mathbf{P}_n$ as represented by their homogeneous coordinates $\lambda = [a_0, a_1, a_2, \ldots, a_n]$. The connection between $\mathbf{P}_n$ and $\mathbf{P}_n{}^*$ is the equation*

$$\lambda \cdot \mathbf{p} \; = \; a_0 x_0 + a_1 x_1 + \ldots + a_n x_n \; = \; 0 \quad (+)$$

which, if true, means that the point $\mathbf{p}$ lies on the line λ.

Since hyperplane coordinates are homogeneous, the dual space $\mathbf{P}_n{}^*$ is also a projective space; its geometry is exactly the same as $\mathbf{P}_n$ itself. What is interesting is the correspondence between the two: Figures constructed out of *points* in $\mathbf{P}_n{}^*$ represent figures made out of *hyperplanes* in $\mathbf{P}_n$. Here are some examples:

Example 1: A Single Hyperplane

The simplest example is a single hyperplane. In $\mathbf{P}_n$ a hyperplane consists of many points; but in $\mathbf{P}_n{}^*$ it is a single point. Figure 15.4 illustrates this in two dimensions.

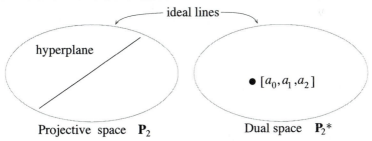

Projective space $\mathbf{P}_2$ Dual space $\mathbf{P}_2{}^*$

Figure 15.4 A hyperplane in a projective space corresponds to a point in the dual space.

Example 2: A Pencil of Hyperplanes

Let η and μ be two hyperplanes. In the dual space, they correspond to two points that span a one-dimensional subspace of $\mathbf{P}_n{}^*$, that is, a line. The dual figure in $\mathbf{P}_n$ is a one-dimensional family of hyperplanes, called a pencil. A **pencil** is a one-dimensional projective geometry (of hyperplanes), the dual of a line which is a one-dimensional geometry (of points). Figure 15.5 illustrates this concept (in two dimensions).

All the tools of one-dimensional projective geometry can be applied to pencils. We may, for example, take the cross ratio of four hyperplanes in a pencil! How? By using the usual formula for the cross ratio of points, but substituting line coordinates for point coordinates. This requires **relative line coordinates**, defined exactly as relative point coordinates were defined earlier: Choose any two lines in the pencil, for example η and μ, to serve as a basis. Then, the other lines λ in the pencil can be expressed using η and μ, namely, $\lambda = \alpha\eta + \beta\mu$.   The   pair $[\alpha, \beta]$ are   **relative   line coordinates** for λ.

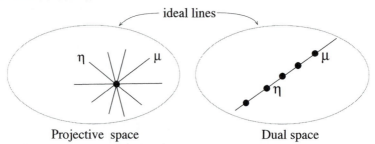

ideal lines

Projective space   Dual space

Figure 15.5 A line in the dual space corresponds to a pencil of hyperplanes in projective space.

Example 3: One More Point versus One More Hyperplane

Let P be a set of points, and H a set of hyperplanes The points span a subspace (flat); dually, the hyperplanes intersect in a subspace. If one more point, independent of the others, is added to P, then the new subspace spanned is of dimension one larger than before. If one more hyperplane is added to H, then the new subspace intersected is of dimension one *less* than before.

Algebraically, the reason for this difference is that the new hyperplane represents one more equation of the form (+) to be satisfied by the points in the intersection. One more equation, if independent of the other equations generated by the hyperplanes in H, reduces by one the dimension of the space of solutions.

J–Draw a pencil of hyperplanes in three dimensions.

K–The cross ratio of four hyperplanes $CR[\lambda_1, \lambda_2, \lambda_3, \lambda_4]$ in a pencil can be either positive or negative depending on whether the hyperplanes λ_1 and λ_2 separate the hyperplanes λ_3 and λ_4. Illustrate these two possibilities in two and three dimensions.

Example 4: Line Conics

There are also curves and surfaces in the dual space. For example, a quadric of hyperplanes is defined by an equation

$$c_{0,0}a_0^2 + c_{0,1}a_0a_1 + \ldots + c_{i,j}a_ia_j + \ldots + c_{n,n}a_n^2 = 0$$

where the a's are unknowns and the c's are constants. The solutions of this equation, $[a_0, a_1, a_2, \ldots, a_n]$, form an ordinary quadric in the dual plane that corresponds to a new figure, a **hyperplane quadric** in $\mathbf{P}_n$. In $\mathbf{P}_2$, this is called a **line conic**. (See Figure 15.6.)

L–In ordinary geometry, the tangent to a surface of points is a hyperplane that touches the surface at just one point. What is the dual concept?

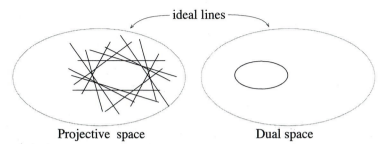

**Figure 15.6 A quadric in the dual space corresponds to a hyperplane
quadric in the original projective space.**

Summary

Projective geometry provides a laboratory for the study of figures
under certain ideal conditions. The special conditions occurring in
this laboratory are specially simple incidence relationships, specially
simple classification of quadrics, and the principle of duality, which
supplies an alternative vision of projective geometry as a geometry of
hyperplanes.

EXERCISES

Incidence

1. Let p_1, p_2, p_3, ... be points in general position in P_n. Explain
why no three of these points are collinear, and no four of them
lie on a plane.

2. Prove that the intersection of flats is either empty or a flat.

3. What possible dimension is the intersection of
   (a) two three-dimensional flats in P_5?
   (b) a two-dimensional flat and a three-dimensional flat in P_6?
   (c) a five-dimensional flat and a four-dimensional flat in P_7?

*Example: What possible dimension is the intersection of two two-dimensional
flats in P_3? The intersection can be two-dimensional, if the two flats are the
same. The intersection can be one-dimensional. But, these are the only
possibilities since, according to the Fundamental Theorem on Incidence, two
two-dimensional flats must intersect in P_3.*

4. What kind of flat can result from the intersection of a k-
dimensional flat and an m-dimensional flat in P_n?

5. In what spaces P_n do parallel k-dimensional flats exist?

6. Prove the Fundamental Theorem of Projective Geometry.

Hint: Imitate the proof of the Fundamental Theorem given in Chapter 14.

Cross Ratio

7. Prove that the cross-ratio is an invariant.

Hint: Use Exercise 9 of Chapter 5.

8. Compute these cross ratios in $\mathbf{P}_1$:

    (a) $CR[(1, 0), (3, 2), (2, 3), (3, 4)]$
    (b) $CR[(1, 1), (-1, 1), (3, -1), (0, 2)]$
    (c) $CR[(1, 2), (2, 1), (5, -1), (-4, 8)]$

9. On the real line plot the sets of points in Exercise 8 (each part (a)-(c) separately) using inhomogeneous coordinates. Determine, in each case, whether the first pair of points separates the second pair. Verify the suggestion made in the text that separation occurs when the cross ratio is negative.

10. Let **s** and **t** be distinct points in any projective space. Let **u** and **v** be a pair of distinct points on the line spanned by **s** and **t**. Let **s** and **t** be expressed in terms of **u** and **v**: $\mathbf{s} = a\mathbf{u} + b\mathbf{v}$; $\mathbf{t} = c\mathbf{u} + d\mathbf{v}$. Prove that $ad - bc \neq 0$.

11. Let **s** and **t** be distinct points in any projective space. Let $\mathbf{p}_0$, $\mathbf{p}_1$, $\mathbf{p}_2$, $\mathbf{p}_3$ be four points on the line spanned by **s** and **t**. Let $\mathbf{p}_0 = (\alpha_0, \beta_0)$, $\mathbf{p}_1 = (\alpha_1, \beta_1)$, $\mathbf{p}_2 = (\alpha_2, \beta_2)$, $\mathbf{p}_3 = (\alpha_3, \beta_3)$ be relative coordinates for $\mathbf{p}_0$, $\mathbf{p}_1$, $\mathbf{p}_2$, $\mathbf{p}_3$ based on **s** and **t**. Prove that the cross ratio of the points $\mathbf{p}_0$, $\mathbf{p}_1$, $\mathbf{p}_2$, $\mathbf{p}_3$ defined using these relative coordinates does not depend on the choice of points **s** and **t**.

*Hint: Let **u** and **v** be another pair of points which determine the same line as **s** and **t**. Let $\mathbf{p}_0 = (\sigma_0, \tau_0)$, $\mathbf{p}_1 = (\sigma_1, \tau_1)$, $\mathbf{p}_2 = (\sigma_2, \tau_2)$, $\mathbf{p}_3 = (\sigma_3, \tau_3)$ be relative coordinates for $\mathbf{p}_0$, $\mathbf{p}_1$, $\mathbf{p}_2$, $\mathbf{p}_3$ based on **u** and **v**.*
    *Now, **s** and **t** can be expressed in terms of **u** and **v**: $\mathbf{s} = a\mathbf{u} + b\mathbf{v}$; $\mathbf{t} = c\mathbf{u} + d\mathbf{v}$. Use these formulas to express the relative coordinates based on **u** and **v** in terms of the relative coordinates based on **s** and **t**. Then compute the cross ratio of $\mathbf{p}_0$, $\mathbf{p}_1$, $\mathbf{p}_2$, $\mathbf{p}_3$ using each set of relative coordinates to verify that they are the same.*

12. A **harmonic sequence** is a figure consisting of four points $\mathbf{p}_0$, $\mathbf{p}_1$, $\mathbf{p}_2$, $\mathbf{p}_3$ on a projective line such that $CR[\mathbf{p}_0, \mathbf{p}_1, \mathbf{p}_2, \mathbf{p}_3] = -1$. Prove that a harmonic sequence is an invariant figure in projective geometry.

Quadrics

13. Complete the proof of the classification of quadrics by considering the case where the coefficient a is zero.

14. Find the canonical forms of these quadrics:

    (a) $x_0^2 - 2x_0x_1 + 3x_1^2 = 0$
    (b) $x_0^2 + x_1^2 + 4x_2^2 - 2x_0x_1 - 4x_0x_2 + 4x_1x_2 = 0$
    (c) $9x_0^2 + x_1^2 + 2x_2^2 - 6x_0x_2 - 2x_1x_2 = 0$
    (d) $4x_0^2 + x_1^2 - 4x_1x_2 - 4x_2x_3 - x_3^2 = 0$
    (e) $9x_0^2 + 2x_2^2 - 6x_0x_2 - 2x_1x_2 = 0$
    (f) $x_0x_1 + x_1x_2 + x_2x_3 = 0$
    (g) $x_0x_1 + x_1x_2 + x_2x_3 + x_3x_4 = 0$

15. Which of the quadrics in Exercise 14 are conics? Classify these conics among the types (1) to (6) described in the text.

16. Describe the classification of quadrics in P_1 by listing all the possible canonical forms. How many are there? Which are degenerate?

17. In three dimensions, a quadratic equation describes a surface. Describe the projective classification of quadric surfaces in P_3 by listing all the possible canonical forms. How many are there? Which are degenerate?

18. Find the canonical form of these Euclidean surfaces as quadrics in P_3:

    (a) sphere   (b) ellipsoid

    (c) paraboloid   (d) cone

    (e) hyperboloid of one sheet   (f) cylinder

    (g) hyperboloid of two sheets (h) hyperbolic paraboloid

Hint: Standard equations for the conic surfaces are in many books on multi-variable calculus.

Example: The equation of a sphere is

$$x^2 + y^2 + z^2 = R^2$$

where R is the radius. To classify this quadric in projective terms, the Euclidean equation must be converted from inhomogeneous coordinates to homogeneous coordinates. Therefore, we set $x = x_1/x_0$, $y = x_2/x_0$, and $z = x_3/x_0$, and then clear fractions, getting

$$x_1^2 + x_2^2 + x_3^2 = R^2 x_0^2$$

This is almost in canonical form. Substituting x_0 for Rx_0, we get

$$-x_0^2 + x_1^2 + x_2^2 + x_3^2 = 0$$

19. Among the eight Euclidean surfaces listed in Exercise 18, how many are different viewed projectively?

Duality

20. What is the dual of

    (a) a line in P_2?

    (b) a line in P_5?

    (c) a three-dimensional subspace of P_5?

Example: What is the dual of a two-dimensional subspace of P_4? A two-dimensional subspace of P_4 is spanned by three points in general position. The dual figure is the intersection of three hyperplanes. In P_4, three hyperplanes in general position intersect in a subspace of dimension $4 - 3 = 1$, because each independent equation reduces the dimension by one. Thus, the dual of a two-dimensional subspace in P_4 is a line.

21. For each of these figures in P_3 describe the dual figure:

    (a) a line of points

(b) a plane of points

(c) the set of all planes passing through two given points

(d) the set of all lines passing through a single given point

(e) the pencil of all lines in a given plane passing through a given point in that plane

22. Discuss the nature of duality on the projective line P_1.

23. What figure in P_1 is dual to a harmonic sequence? In P_2? In P_3?

24. A k-dimensional flat in P_n^* is **self-dual** if the dual flat in P_n has the same dimension. For what dimensions k and n can this occur? What is the lowest dimensional example? What is the next lowest dimensional example?

25. Do two pencils always intersect? What is the nature of this intersection?

26. Are all pencils congruent?

27. Are all line conics congruent in P_2^*?

28. State and prove the dual of the Fundamental Theorem on Incidence.

16 UNIVERSAL PROJECTIVE GEOMETRY

The topics in this chapter are chosen to display the versatility of projective geometry: its flexibility as a mathematical tool and its connections with other branches of geometry and mathematics. Ultimately, as we shall see, all geometry is projective geometry.

Parameter Spaces

In studying a set of geometric objects, it is often useful to coordinatize them, meaning give them numerical names of some sort with which one can calculate. Descartes's coordinatization of the Euclidean plane is the original of this idea, and this whole book makes the case that coordinatization is a convenient way to deal with non-Euclidean geometries as well.

Coordinatization, however, is not limited to the *points* of a geometry; other objects can be coordinatized, too. If the objects being named are *not* points, the process is called **parametrization**, the resulting coordinates are called **parameters**, and the set consisting of the parameters of all the objects is called a **parameter space**.

For example, a Euclidean straight line is determined by its slope and intercept; therefore, the pair $(m, b) = $ (slope, intercept) parametrizes lines in Euclidean geometry. Both m and b can be any real number; therefore, the parameter space is a whole Cartesian plane. As the reader knows, this is a useful parametrization! From a purely geometric point of view, however, it is unsatisfactory because vertical lines are not parametrized.

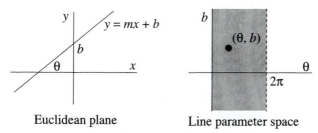

Euclidean plane   Line parameter space

Figure 16.1 A line in the Euclidean plane is a point in parameter space.

177

A more complete parametrization uses the angle θ between the line and the horizontal instead of the slope. (See Figure 16.1.) In the parametrization (θ, b), b can still be any real, but θ only ranges from 0 to 2π. Therefore, this parameter space is not a whole plane, but only an infinite strip.

A–The (θ, b) parametrization is further complicated by the identification of $\theta = 0$ with $\theta = 2\pi$. Why is this identification necessary?

Universal Parameter Space

One aspect of the universality of projective geometry is that parameter spaces for projective figures are frequently themselves projective geometries or subgeometries of projective geometries. Here are some examples:

Example 1. Hyperplanes

A hyperplane in n-dimensional projective space $\mathbf{P}_n$ is an $(n - 1)$-dimensional subspace of $\mathbf{P}_n$. The set of points on a hyperplane satisfies a single linear equation

$$a_0 x_0 + a_1 x_1 + \ldots + a_n x_n = 0$$

where not all the coefficients $a_0, a_1, \ldots, a_n$ can be zero. The numbers $[a_0, a_1, \ldots, a_n]$ are homogeneous coordinates for the hyperplane. As described in Chapter 15, the set of all these coordinates, $\mathbf{P}_n{}^*$, is called the **dual space** of $\mathbf{P}_n$. Thus, the parameter space of hyperplanes is a second projective space of dimension n. (See Chapter 15.)

Example 2. Quadrics

Curves and surfaces can also be parametrized. For example, consider all quadric surfaces in three-dimensional projective space $\mathbf{P}_3$. Each quadric satisfies a quadratic equation:

$$c_{0,0} x_0^2 + c_{1,1} x_1^2 + c_{2,2} x_2^2 + c_{3,3} x_3^2 + c_{0,1} x_0 x_1 + c_{0,2} x_0 x_2 + c_{0,3} x_0 x_3 +$$

$$c_{1,2} x_1 x_2 + c_{1,3} x_1 x_3 + c_{2,3} x_2 x_3 = 0 \quad (*)$$

The 10 coefficients of this equation

$$[c_{0,0}, c_{1,1}, c_{2,2}, c_{3,3}, c_{0,1}, c_{0,2}, c_{0,3}, c_{1,2}, c_{1,3}, c_{2,3}]$$

are **homogeneous coordinates** for the quadric: Homogeneous because the equation formed by replacing the coefficients in (*) with scalar multiples has the same solutions and hence determines the same quadric.

Thus, the set of all these coordinates, the parameter space of the quadrics in $\mathbf{P}_3$, is a projective space: $\mathbf{P}_9$. More generally, quadrics in any projective space $\mathbf{P}_n$ are parametrized by the points of a projective space of higher dimension.

Example 3. Other Flats

The k-dimensional flats in n-dimensional projective space are not normally parametrized by a whole projective space, except when $k = 0$ (points) or $k = n - 1$ (hyperplanes). Nonetheless, the parameter space for k-dimensional flats is an important and interesting subset of a projective space. The general situation is fairly complicated, so we present only a single example.

Consider the set of lines in $\mathbf{P}_3$. A line in $\mathbf{P}_3$ is determined not by a single equation, but by either two equations (i.e., as the intersection of two planes) or two points (i.e., as their span). Let us regard a line λ as determined by two distinct points, say, p and q, with coordinates $\mathbf{p} = (x_0, x_1, x_2, x_3)$ and $\mathbf{q} = (y_0, y_1, y_2, y_3)$. At first, one is tempted to put these together and use all eight coordinates as parameters for λ. The problem is that *any* two distinct points of λ span λ. Therefore, the parameter space for lines in $\mathbf{P}_3$ is not seven-dimensional (as might appear at first) but is a subset of a projective space of smaller dimension.

The trick to obtaining unique coordinates is to use combinations of the coordinates of p and q instead of the coordinates themselves. We begin by putting $\mathbf{p}$ and $\mathbf{q}$ in a matrix:

$$\begin{pmatrix} \mathbf{p} \\ \mathbf{q} \end{pmatrix} = \begin{pmatrix} x_0 & x_1 & x_2 & x_3 \\ y_0 & y_1 & y_2 & y_3 \end{pmatrix}$$

Now let $v[i, j]$ be the determinant of the ith and jth columns of this matrix. For example, $v[1, 3] = x_1 y_3 - x_3 y_1$. Let $\mathbf{v}$ be the vector consisting of these six determinants:

$$\mathbf{v} = (v[0, 1], v[0, 2], v[0, 3], v[1, 2], v[1, 3], v[2, 3])$$

For instance, if $\mathbf{p} = (1, 1, 3, 0)$ and $\mathbf{q} = (1, 2, 0, 1)$, then $\mathbf{v} = (1, -3, 1, -6, 1, 3)$. The numbers $v[i, j]$ are called **Plücker coordinates** of the line, and they are effective parameters for λ. Note that Plücker coordinates are homogeneous: If either $\mathbf{p}$ or $\mathbf{q}$ is replaced by $k\mathbf{p}$ or $k\mathbf{q}$, then $\mathbf{v}$ is multiplied by k, also.

B–Find the Plücker coordinates of the line determined by the points $(1, 2, 3, 4)$ and $(-2, 3, -1, 0)$.

To verify the effectiveness of Plücker coordinates, one must show (a) that except for homogeneity, the Plücker coordinates are *unique*, (that is, *different* choices of $\mathbf{p}$ and $\mathbf{q}$ on λ, produce the *same*

v, or a scalar multiple of **v**), and (b) that different lines have different Plücker coordinates. This is not difficult. (See Exercises 4 and 5.)

Thus, we conclude that lines in $\mathbf{P}_3$ are parametrized by points in the projective space $\mathbf{P}_5$. However, not all points in $\mathbf{P}_5$ are Plücker coordinates. The Plücker coordinates of a line in $\mathbf{P}_3$ satisfy a quadratic equation, namely,

$$v[0, 1]v[2, 3] - v[0, 2]v[1, 3] + v[0, 3]v[1, 2] \quad (+)$$

(See Exercise 6.) Moreover, it can be proven that *every* point in $\mathbf{P}_5$ satisfying (+) is the Plücker coordinate of a line in $\mathbf{P}_3$. Hence, the parameter space for lines in $\mathbf{P}_3$ is a quadric in $\mathbf{P}_5$!

This example is characteristic. In general, the k-dimensional flats in n-dimensional projective space can be parametrized by Plücker coordinates formed by taking $k + 1$ points spanning the flat, arranging these in rows in a matrix, and taking all $(k + 1)$ by $(k + 1)$ determinants of that matrix. These coordinates then satisfy one or more (usually more) quadratic equations, so the parameter space is a subset of a projective space formed by the intersection of quadrics. This space is called the **Grassmannian** of the flats.

C–Of what dimension is the projective space containing the Grassmannian of all lines in $\mathbf{P}_4$?

Application to Enumerative Geometry

Here is a famous problem of the sort that can be solved using parameter spaces. It was first stated by Jacob Steiner in 1848. *How many conics are tangent to five given conics in general position?*

The problem is illustrated in Figure 16.2 where the five given conics are drawn with thin lines (two ellipses, two circles, and a hyperbola). The problem is to count the number of conics tangent to all of them. (The thick parabola is an example.)

This is a problem in **enumerative geometry**, the branch of geometry that counts the number of occurrences of various phenomena, including tangency, incidence, and still more complicated types of coincidence.

Such questions are routinely addressed by examining a suitable parameter space. Obviously, the parameter space of conics in $\mathbf{P}_2$ is relevant to this problem. Although we will not give complete details of the solution, two features of this parameter space deserve mention. In the first place, the fact that it is a projective space gives meaning to the restrictive phrase "in general position." That the five given conics are *in general position* means that they are *independent* as a set of points in parameter space. In the second place, the parameter space of conics in $\mathbf{P}_2$ is five-dimensional. This is no coincidence. It is just this circumstance that makes this problem well posed: Throw in one more conic, and there will usually be *no* conics tangent to all six; take away a conic, and there are an *infinite* number of conics tangent to the remaining four.

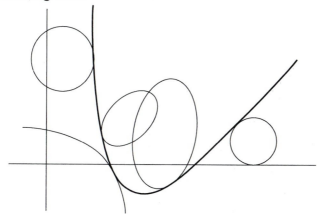

Figure 16.2 A conic tangent to five given conics

What is the answer to the question?   Steiner thought it was 7,776 but he was wrong.   There are actually only 3,264 different conics tangent to five given conics in general position!   The correct answer was found by Michel Chasles in 1864.

Universal Metric Geometry

A second universal feature of projective geometry is that the classical metric geometries—elliptic, parabolic and hyperbolic, their generalizations to higher dimensions, plus many other interesting and useful geometries that we haven't even mentioned—are all sub-geometries of projective geometry.

This works as follows: Begin with a fixed quadric

$$\mathbf{C}:\; a_{0,0}x_0{}^2 + a_{0,1}x_0x_1 + ... + a_{n,n}x_n{}^2 \; = \; 0$$

hereafter to be called the **absolute quadric.** For points **p** not on the absolute quadric, the quantity **C** (which is zero on the quadric) is either positive or negative. Points **p** for which **C** is negative will be called **ordinary points**, points for which **C** is positive will be called **ultra-ideal**, and points on the conic itself will simply be **ideal.** Note that multiplying by a scalar k (even a *negative* k) does not affect whether **p** is ordinary, ideal, or ultra-ideal. Thus, these are genuine projective concepts.

Let **G** be the subgroup of the projective group $\mathbf{P}_n$ consisting of the transformations that leave invariant the absolute quadric *and* the set **D** of ordinary points. Then (**D, G**) is a subgeometry of $\mathbf{P}_n$, denoted

$\mathbf{P}_n/\mathbf{C}$. This geometry takes on a variety of forms depending on the nature of the absolute conic.

Here is a planar example: Consider the geometry $\mathbf{H} = \mathbf{P}_2/\mathbf{C}$, where we choose the nondegenerate conic

$$\mathbf{C}: x_0{}^2 + x_1{}^2 - x_2{}^2 = 0$$

as absolute quadric. For straight lines in $\mathbf{H}$, we take the part of each line in $\mathbf{P}_2$ that is contained in $\mathbf{D}$. Two straight lines are **parallel** if the corresponding projective straight lines intersect on the absolute conic; two straight lines are **hyperparallel** if the corresponding projective straight lines intersect in an ultra-ideal point; otherwise the two straight lines intersect in $\mathbf{D}$. The distance function in $\mathbf{H}$ is

$$d(p, q) = \ln(CR[p, q, s, r])   (1)$$

where r and s are the ideal points on the straight line determined by p and q. These features of $\mathbf{H}$ are illustrated in Figure 16.3.

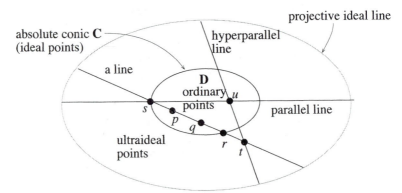

Figure 16.3 Hyperbolic geometry: The projective model

As the reader may have guessed, $\mathbf{H}$ is another model of hyperbolic geometry! Called the **projective model**, it was developed by Klein around 1871 and was the first model of the whole Lobatchevskian plane to be discovered. The circle model (see Chapters 7 to 10), which in this context is called the **conformal model**, was developed by Poincaré somewhat later (around 1882). As the terms suggest, the projective model is *not* conformal, so that the measure of an angle between two lines is not the same as the Euclidean measure of the angle between them but is given by a different formula.   In the projective model, the measure $\angle\lambda\mu$ of the acute angle between two lines λ and μ is

$$\angle\lambda\mu = -i/2 \ln(CR[\lambda, \mu, \alpha, \beta])   (2)$$

where α and β are the two tangents from the point u of intersection of λ and μ to the absolute conic and the cross ratio is a cross ratio of *lines*.

Equation (2) is a remarkable formula. Together with (1), it expresses a duality between point measurement (distance) and line measurement (angles). In general, the duality of points and lines, such a prominent feature of projective geometry, is *not* carried over to hyperbolic geometry or any of the other metric geometries. A hyperbolic line of points, for example, is *not* reentrant (although the dual object, a pencil of hyperbolic lines through a single point, *is* re-entrant in all three metric geometries). Formulas (1) and (2) therefore, represent a fragment of duality that somehow survives in the projective model of hyperbolic geometry. This is another aspect of the connection between linear and angular measure that is such a remarkable feature of hyperbolic geometry.

D–In the projective model of hyperbolic geometry, verify, by drawing a picture, that through a point not on a line there are exactly two distinct lines parallel to the given line.

Complex Projective Geometry

The alert reader may object at this point that the explanation we have given of formula (2) is nonsense, since there are *no* tangents to the absolute conic from a point *inside* the conic. (See the point u in Figure 16.3, for example.) The answer is that α and β are *imaginary* tangents, or more precisely, α and β are tangents to the absolute conic in the *complex* projective plane, which we now define.

Definition   Let n be a positive integer.   Let $\mathbf{C}_n$ be the set of all complex, $(n+1)$-dimensional, nonzero vectors; that is,

$$\mathbf{C}_n = \{(z_0, z_1, z_2, \ldots, z_n): z_0, z_1, z_2, \ldots, z_n \text{ are complex, not all zero}\}$$

Each point $\mathbf{p} = (z_0, z_1, z_2, \ldots, z_n)$ of $\mathbf{C}_n$ is identified with all scalar multiples $k\mathbf{p}$ where k is any non-zero complex number.   Let $\mathbf{GL}[\mathbf{C}, n]$ be the set of all invertible, complex, linear transformations of $\mathbf{C}_n$.

*The pair $\mathbf{P}_n[\mathbf{C}] = (\mathbf{C}_n, \mathbf{GL}[\mathbf{C}, n])$ models **n-dimensional complex projective geometry**.*

To understand how this definition works, note that the complex numbers have many important algebraic properties in common with the reals, namely, the **field axioms** (to be given shortly). As a consequence, all the projective geometric theory we have developed so far can be carried over intact to complex projective geometries. For example, the theorems cited earlier about linear algebra are true of complex projective spaces, as well as real spaces. In particular, the Fundamental Theorem of Projective Geometry is true of complex projective space, as well as the theory of flats and quadrics developed in Chapter 15.

A crucial difference between the real and the complex numbers is the Fundamental Theorem of Algebra, which states that every polynomial equation (quadratic and higher degree) has a solution in the complex numbers. As a result, complex projective geometry is often the preferred tool in advanced applications, since objects (such as tangents) exist more frequently there than in real projective geometry. One example is the tangents α and β in formula (2). A second example is in enumerative geometry: The 3,264 conics tangent to five given conics found by Chasles in 1864 are actually (it can now be revealed) complex conics.

Complex projective geometry has already played a major role in this book. Möbius geometry (Chapter 5) is actually $\mathbf{P}_1[\mathbf{C}]$, the complex projective *line* (one-dimensional, complex, projective geometry) expressed in inhomogeneous coordinates.

E–Write out the formula for a projective transformation in one dimension (using inhomogeneous coordinates), and compare it with the formula for a Möbius transformation.

Complex Geometry in Action

Let us calculate the angle between two lines in the hyperbolic geometry $\mathbf{H}$ using equation (2). For this purpose, let λ be the line connecting the points $(1, 0, 2)$ and $(2, 0, -3)$, and let μ be the line connecting the points $(0, 5, -2)$ and $(0, -1, 1)$.

F–Check that these points are in the hyperbolic plane $\mathbf{D}$.

We need the tangents α and β to the conic $x_0^2 + x_1^2 - x_2^2 = 0$ from the point of intersection of λ and μ. To find these tangents, we will work in the dual space, so we need line coordinates for λ and μ. These are obtained using cross products (Chapter 13). We find that $\lambda = [0, 1, 0]$ and $\mu = [1, 0, 0]$. Their point of intersection, $u = (0, 0, 1)$, is found by calculating another cross product.

The tangents that we seek, α and β, are in the pencil of all lines passing through u. Let $[a_0, a_1, a_2]$ be line coordinates. We must solve the equations

$$0 = u \cdot [a_0, a_1, a_2] = a_2$$

and

$$0 = a_0^2 + a_1^2 - a_2^2$$

simultaneously. The first equation says that the line $[a_0, a_1, a_2]$ passes through u; the second states that $[a_0, a_1, a_2]$ lies on the line conic $a_0^2 + a_1^2 - a_2^2 = 0$ (which consists of all tangents to the point conic $x_0^2 + x_1^2 - x_2^2 = 0$).

Since $a_2 = 0$, we have $[a_0, a_1, a_2] = [a_0, a_1, 0]$. Applying the second equation, we get $a_0^2 = -a_1^2$, or $a_0 = \pm ia_1$. Thus, the tangents α and β have line coordinates $[i, 1, 0]$ and $[-i, 1, 0]$. As we suspected, the two tangents do not exist in the real projective plane, and we have entered the realm of complex projective geometry.

To compute the angle between λ and μ requires the calculation of a cross ratio of lines, and that calculation, in turn, requires that we convert the line coordinates of lines, λ and μ, α and β, to relative line coordinates. A convenient basis for the pencil of lines through u consists of the two given lines λ and μ. In terms of these lines, we have

$$\alpha = \lambda + i\mu \quad \text{and} \quad \beta = \lambda - i\mu$$

Since $\lambda = \lambda$ and $\mu = \mu$, we now have these relative line-coordinates:

$$\lambda = [1, 0], \ \mu = [0, 1], \ \alpha = [1, i] \text{ and } \beta = [1, -i]$$

Thus, we can finally calculate $\angle \lambda\mu$:

$$< \lambda\mu = -\frac{i}{2}\ln(CR[\lambda, \mu, \alpha, \beta]) = -\frac{i}{2}\ln\left(\frac{\begin{vmatrix}1 & 0\\1 & i\end{vmatrix}\begin{vmatrix}0 & 1\\1 & -i\end{vmatrix}}{\begin{vmatrix}0 & 1\\1 & i\end{vmatrix}\begin{vmatrix}1 & 0\\1 & -i\end{vmatrix}}\right)$$

$$= -\frac{i}{2}\ln\left(\frac{-i}{i}\right) = -\frac{i}{2}\ln(-1)$$

To complete the calculation, we need to find $\ln(-1)$! This may seem impossible since there are no *real* logarithms of negative numbers. In the *complex* number system, however, they do exist. The logarithm function is the inverse of the exponential, and Euler's formula (page 18) gives $e^{\pi i} = -1$. It follows that $\ln(-1) = \pi i$, so that

$$\angle \lambda\mu = -\frac{i}{2}\pi i = \frac{\pi}{2}$$

Thus, λ and μ are perpendicular!

Finite Geometries

The existence of complex projective geometry suggests that other number systems might serve as coordinates of projective geometries. What properties must a system of numbers have in order to be used this way? The concept of a field, as defined below, includes more than enough properties to permit the development of projective geometry.

Definition *A field is a set F with two algebraic operations (written as addition and multiplication) that satisfy the following laws:*

(1) *Addition and multiplication are commutative and associative,*

(2) *F contains an additive identity (zero) and multiplicative identity (one),*

(3) *Every element of F has an additive inverse,*

(4) *Every non-zero element of F has a multiplicative inverse,*

(5) *Addition and multiplication satisfy the distributive law:*

$$a(b + c) = ab + ac$$

for all a, b and c in F.

Definition *Let F be a field and n a positive integer. Define* $\mathbf{F}_n$ *by*

$$\mathbf{F}_n = \{(x_0, x_1, x_2, \ldots, x_n): x_0, x_1, x_2, \ldots, x_n \text{ are in } F, \text{ not all zero}\}$$

Each point $\mathbf{p} = (x_0, x_1, x_2, \ldots, x_n)$ *of* $\mathbf{F}_n$ *is identified with all scalar multiples kp, where k is any nonzero element of F. Let* $\mathbf{GL}[F,n]$ *be the set of all invertible, linear transformations of* $\mathbf{F}_n$.

The pair, $\mathbf{P}_n[F] = (\mathbf{F}_n, \mathbf{GL}[F,n])$ *models* **n-dimensional projective geometry over F.**

Thus, for every field F, there are projective geometries $\mathbf{P}_n[F]$ of all dimensions. The whole theory of projective geometry, as developed so far, applies to these geometries: $\mathbf{P}_n[F]$ contains flats of all dimensions from 1 to $n - 1$, as well as other figures, curves, and surfaces, including quadrics. The Fundamental Theorem of Projective Geometry is valid in all these geometries. The proofs of all these results are the same, regardless of the field F.

Among the most interesting fields are the finite fields. The smallest field is $\mathbf{Z}/2$, with only two elements. By the definition of a field, these two elements must be 0 and 1, the additive and multiplicative identities.

G–Write out the addition and multiplication table for $\mathbf{Z}/2$.

Figure 16.4 shows the projective plane $\mathbf{P}_2[\mathbf{Z}/2]$ over $\mathbf{Z}/2$. Points in this geometry are triples (x, y, z) of elements from $\mathbf{Z}/2$. Since there are exactly two possibilities each for x, y, and z, there are eight such triples. But $(0, 0, 0)$ is not included, so there are only seven points in $\mathbf{P}_2[\mathbf{Z}/2]$. Because lines and points are dual in a projective plane, there are also exactly seven lines. These are represented as straight lines in the figure, except that a circular line is added to contain the three collinear points $(1,0,1)$, $(1,1,0)$, and $(0,1,1)$.

H–Why are $(1,0,1)$, $(1,1,0)$, and $(0,1,1)$ collinear?

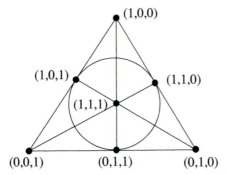

Figure 16.4 The seven-point plane $P_2[Z/2]$

Application to Combinatorial Design

Highly patterned finite sets of points appear in many fields. Crystallography, architecture, and solid-state physics are some examples. Combinatorial design is a less well-known, but nonetheless important, application.

Here is a typical problem: Our goal is to design an experiment to test the effect of different fertilizers on the growth of wheat. The wheat is to be planted in square plots in a rectangular pattern with different fertilizers applied to each plot. (See Figure 16.5.)

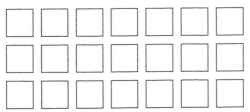

Figure 16.5 Layout of a possible experimental garden

Now, in any field, there are variations in sunlight, rainfall, and soil fertility that render certain parts of the field more advantageous than others for the growth of wheat. Our problem is to choose a pattern of application of the fertilizers so as to measure the contribution each fertilizer makes to the wheat harvest, *independent* of these other factors. Therefore, we decide that each fertilizer will be applied to plots in several columns, and, further, that every possible combination of two fertilizers will occur together in some column. The mathematics of this situation is described as follows:

***Definition** A **balanced, incomplete block design** (or **design**, for short) is an arrangement of v distinct objects into b blocks so that each block contains exactly k distinct objects, each object occurs in exactly r blocks, and every pair of distinct objects occurs together in exactly λ* *blocks. A design is **symmetric** if $v = b$ and $r = k$. The numbers v, b, r, k, and λ are called the **parameters** of the design.*

In our example, different fertilizers are the objects of a balanced, incomplete block design whose blocks are the columns of the experimental garden. Supposing that there are seven types of fertilizer, then we seek a symmetric design with parameters $v = b = 7$, $k = r = 3$, and $\lambda = 1$.

A solution to this particular problem is provided by the seven-point projective plane! The seven points can represent the fertilizers (objects) and the seven lines the columns (blocks). This produces the arrangement of fertilizers depicted in Figure 16.6 in which the seven fertilizers are given the labels p, q, r, s, t, u, and v.

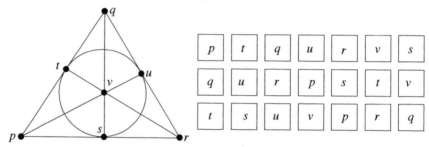

Figure 16.6 A symmetrical balanced incomplete block design

This example is a little artificial in that we have controlled the placement of the fertilizers in columns only, not rows. (Actually, note that every row contains exactly one application of each fertilizer.) Otherwise, this is a typical combinatorial design problem. Our solution is also a typical kind of design, as the following theorem explains:

Theorem *The hyperplanes of a projective geometry over a finite field F divide the points of the geometry into blocks of a symmetric balanced, incomplete block design with parameters*

$$b = v = \frac{q^{n+1} - 1}{q - 1} \qquad r = k = \frac{q^n - 1}{q - 1} \qquad \lambda = \frac{q^{n-1} - 1}{q - 1}$$

where q is the number of elements in the field F.

Proof: How many points are there, anyway, in a finite projective geometry? Let F be the field and q the number of elements in F. Each point p in $\mathbf{P}_n[F]$ is described by a vector $\mathbf{p} = (x_0, x_1, \ldots, x_n)$, an $(n + 1)$-tuple of field elements. There are q^{n+1} such vectors, but $(0, 0, \ldots, 0)$ is omitted, so only $q^{n+1} - 1$ vectors are used to form $\mathbf{P}_n[F]$. However, each vector $\mathbf{p}$ is identified with its nonzero scalar multiples. There are $(q - 1)$ nonzero scalars in F, so the $(q^{n+1} - 1)$ vectors are identified in sets of $(q - 1)$. Therefore, there are $(q^{n+1} - 1)/(q - 1)$ points in $P_n[F]$. Dually, this is also the number of hyperplanes. In the terminology of block designs, $b = v = (q^{n+1} - 1)/(q - 1)$.

All hyperplanes are congruent and hence contain the same number of points. Dually, every point is contained in the same number of hyperplanes. In block design notation, $k = r$.

How many points are there in a hyperplane? A hyperplane is itself a projective geometry: of dimension $(n - 1)$. Therefore, by the formula just derived, there are $(q^n - 1)/(q - 1)$ points in a projective space of this dimension. In the terminology of block designs, $r = k = (q^n - 1)/(q - 1)$.

A pair of points determines a line. The number of hyperplanes containing a given pair of points is the same for all lines, since all lines are congruent in projective geometry. Thus, the hyperplanes form a block design.

How many hyperplanes contain a particular line? A line is spanned by two points. Therefore, the number we seek is the number of *hyperplanes containing two given points*. Dually, this is the same as the number of *points* (dual to "hyperplanes") *contained in* (dual to "containing") *two given hyperplanes* (dual to "two given points"). But the intersection of two given hyperplanes is itself a projective geometry: of dimension $(n - 2)$. Therefore, there are $(q^{n-1} - 1)/(q - 1)$ hyperplanes containing a given pair of points. In the terminology of block designs, $\lambda = (q^{n-1} - 1)/(q - 1)$. ∎

Summary

In several ways, projective geometry is a universal geometrical tool. First, because the parameter spaces of important types of geometric objects tend to be projective spaces; second, because projective geometry contains many other geometries, including all the classical metric plane geometries; and third, because the coordinates of a projective geometry can be chosen from a variety of fields, so that projective geometry extends into areas of complex and discrete mathematics.

EXERCISES

Parameter Spaces

1. Invent parameter spaces for these families of Euclidean figures:

    (a) circles
    (b) ellipses
    (c) parabolas

Example: Circles have Euclidean equations of the form $(x - h)^2 + (y - k)^2 = R^2$, where (h, k) is the center of the circle and R the radius. Clearly, for geometric as well as algebraic reasons, it is sensible to take (h, k, R) as a *parametrization of the circles in Euclidean geometry. The parameters h and k can be any real number, while R must be nonnegative. Therefore, the parameter space (h, k, R) is half of a Euclidean three-dimensional space.*

2. What is the dimension of the projective space of all conics in $\mathbf{P}_4$? In $\mathbf{P}_5$? In $\mathbf{P}_n$?

3. Let λ be a line through the two points p and q in $\mathbf{P}_2$. We have two ways to compute line coordinates for λ: using the cross product (see Chapter 13) and using Plücker coordinates. Verify that these give the same result (if Plücker coordinates are chosen properly).

4. Let λ be a line in $\mathbf{P}_3$. Let $\mathbf{p}$ and $\mathbf{q}$ be two distinct points on λ. Let $\mathbf{s}$ and $\mathbf{t}$ be a second pair of points on λ. Show that the Plücker coordinates generated by $\mathbf{s}$ and $\mathbf{t}$ are the same as those generated by $\mathbf{p}$ and $\mathbf{q}$.

Hint: Express $\mathbf{s}$ and $\mathbf{t}$ in terms of $\mathbf{p}$ and $\mathbf{q}$.

5. Let λ and μ be two distinct lines in $\mathbf{P}_3$. Let $\mathbf{p}$ and $\mathbf{q}$ be two distinct points on λ. Let $\mathbf{s}$ and $\mathbf{t}$ be two distinct points on μ. Show that the Plücker coordinates generated by $\mathbf{s}$ and $\mathbf{t}$ are different from those generated by $\mathbf{p}$ and $\mathbf{q}$.

Hint: Use an indirect proof. Assume that the Plücker coordinates are the same. Then, show that $\mathbf{s}$ and $\mathbf{t}$ can be expressed in terms of $\mathbf{p}$ and $\mathbf{q}$; hence, the lines λ and μ are the same.

6. Let $\mathbf{v} = (v[0, 1], v[0, 2], v[0, 3], v[1, 2], v[1, 3], v[2, 3])$ be the Plücker coordinates of a line in $\mathbf{P}_3$. Show that $\mathbf{v}$ satisfies the quadratic equation (+) on page 182.

Hint: A computer algebra system may be useful.

7. What is the dimension of the projective space containing the Grassmannian of

   (a) lines in $\mathbf{P}_4$?
   (b) planes in $\mathbf{P}_4$?
   (c) three-dimensional flats in $\mathbf{P}_5$?
   (d) three-dimensional flats in $\mathbf{P}_6$?
   (e) four-dimensional flats in $\mathbf{P}_6$?

8. Let $G(k, n)$ be the Grassmannian of k-dimensional flats in n-dimensional projective space. $G(k, n)$ is a subset of a projective space $\mathbf{P}_N$. Find a formula for N in terms of n and k.

9. Find the canonical form of the Grassmannian of lines in $\mathbf{P}_3$ considered as a quadric in $\mathbf{P}_5$.

Hint: Apply the classification theorem (from Chapter 15) to the quadric (+) on page 182.

Subgeometries

10. Find the angle between the lines $\lambda = [1, 0, 0]$ and $\mu = [1, 1, 0]$ in **H**.

11. Prove that two lines λ and μ are perpendicular in **H** if, and only if, the sequence of lines λ, μ, α, β in formula (2) on page 184 is a harmonic sequence in $\mathbf{P}_2{}^*$.

12. What is the numerical relationship between $CR[\lambda, \mu, \alpha, \beta]$ and $CR[\lambda, \mu, \beta, \alpha]$? What is the geometric relationship between the angles measured by $-i \ln(CR[\lambda, \mu, \alpha, \beta])/2$ and $-i \ln(CR[\lambda, \mu, \beta, \alpha])/2$?

13. Explain why the four points p, q, s, r in formula (1) on page 184 are never a harmonic sequence.

14. Show that $\mathbf{P}_2/\mathbf{C}$ is elliptic geometry if $\mathbf{C}$: $x_0{}^2 + x_1{}^2 + x_2{}^2 = 0$.

Hint: Use inhomogeneous coordinates.

15. $\mathbf{P}_2/\mathbf{C}$ is called **affine geometry** if $\mathbf{C}$: $x_2 = 0$. Show that the transformation group of affine geometry contains all the transformations listed in Exercise 3 of Chapter 14. What is the one exception? Show that affine geometry contains Euclidean geometry as a subgeometry.

16. What other plane geometries can be modeled as $\mathbf{P}_2/\mathbf{C}$ for a suitable choice of absolute conic **C**? What can you find out about them?

Solid Subgeometries

The next five exercises assume knowledge of solid geometries. (See Chapters 18 and 19.)

17. Prove that the geometry **H** is hyperbolic geometry by showing that it is the same as hyperboloidal geometry (Chapter 18).

18. Find a conic **C** so that $\mathbf{P}_3/\mathbf{C}$ is solid elliptic geometry.

18. Find a conic **C** so that $\mathbf{P}_3/\mathbf{C}$ is solid hyperbolic geometry.

20. Find a conic **C** so that $\mathbf{P}_3/\mathbf{C}$ is pseudo-Euclidean geometry.

21. What other solid geometries can be modeled by $\mathbf{P}_3/\mathbf{C}$ for a suitable choice of absolute conic **C**? What can you find out about these geometries?

Complex Geometry

22. Prove that a conic and a line in the complex projective plane always intersect in at least one point. What is the maximum number of points of intersection?

23. The classification of quadrics in complex projective geometry is like the real classification of quadrics, but simpler, because the

coefficients in the canonical forms can be limited to 1 and 0. Explain why.

24. Describe the classification of conics in C_2. How many different types of conics are there? Which of these are degenerate?

25. Describe the classification of quadrics in C_3. How many different types of conics are there? Which of these are degenerate?

Finite Geometries

26. $Z/3$ is a field of only three elements. Let them be represented by 0, 1, and -1. Complete addition and multiplication tables for $Z/3$. (See Table 16.1.)

+	0	1	−1
0	0	1	−1
1	1	−1	
−1	−1		

*	0	1	−1
0	0	0	0
1	0	1	
−1	0		

Table 16.1 The finite field $Z/3$

27. Draw the plane $P_2[Z/3]$, showing all points and lines.

28. $GF(4)$ is a field of four elements. Let them be represented by 0, 1, a, and b. Complete addition and multiplication tables for $GF(4)$. (See Table 16.2.)

+	0	1	a	b
0	0	1		b
1		0	b	a
a			0	
b				0

*	0	1	a	b
0	0	0	0	0
1	0			
a	0			1
b	0		1	

Table 16.2 The finite field $GF(4)$

29. How many points are in the projective space $P_1[GF(4)]$ (the projective line with coordinates from $GF(4)$)? How many points are **1**.

30. Draw the plane $P_2[GF(4)]$, showing all points and lines.

Combinatorial Designs

31. How many lines are in the projective space $\mathbf{P}_n[F]$, where F has q elements?

32. Prove that the set of all lines in a finite projective geometry forms a balanced, incomplete block design.

33. Find the parameters of the block design of all lines in a finite projective space.

34. Prove that the set of all planes in a finite projective geometry forms a balanced, incomplete block design.

35. Find the parameters of the block design of all planes in a finite projective space.

PART IV

SOLID GEOMETRY

GEOMETRY AND ART: PLATE IV

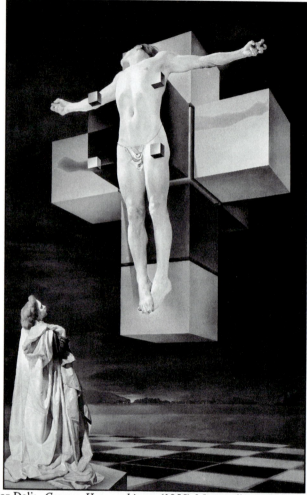

Salvador Dali: *Corpus Hypercubicus* (1955) Metropolitan Museum of Art.

This work by the twentieth-century Spanish artist Salvador Dali uses perspective to create a careful, even clinical, portrayal of the crucifixion in a fantastic landscape. Dali, moreover, goes beyond three-dimensional reality by using a four-dimensional cube (a **hypercube**) unfolded into three dimensions as the cross. This specifically mathematical image supports Dali's religious message by suggesting that, just as the hypercube transcends ordinary three-dimensional space, so Christ transcends the ordinary physical world.

KORBOWSKI *(held by the detectives)* It doesn't matter. There's a revolution going on. We'll meet again. This won't last long. Perhaps today we'll all be free. Alice, I loved you and you only even in the thick of crimes so monstrous as to be four-dimensional and non-Euclidean in their swinishness.

–from *The Water Hen* (1921) by Stanislaw Witkiewicz (reprinted in *The Madman and the Nun and Other Plays*, translated by Daniel Gerould and C. S. Durer, Applause Theater Book Publishers, New York)

The next three chapters extend non-Euclidean geometrical ideas into the third (and fourth) dimension. In addition to generalizations of both hyperbolic and elliptic geometries to three dimensions, there are other interesting higher-dimensional geometries to consider, in particular, pseudo-Euclidean geometry, which is the geometry of the special theory of relativity.

The first step is to consider extending the complex numbers to a higher-dimensional number system. This is necessary if we want to continue to use the analytic treatment of geometry (based on the algebra of the complex numbers) upon which we have depended so far.

Unfortunately, this extension is *not* possible to three dimensions. That is, there is no way to define a multiplication of points in three dimensions that is commutative, associative, and distributive, and that satisfies the cancellation laws: In a word, that possesses all the algebraic properties the real and complex numbers possess.

However, there is a *four-dimensional* multiplication that satisfies most of these properties. (The only property lost is the commutative law of multiplication.) This multiplication leads to the **quaternions**, discovered by William Hamilton in 1843 in the course of efforts to understand the nature of three-dimensional space.

In Chapter 17, therefore, we introduce the four-dimensional geometry and algebra of the quaternions, which is applied subsequently to various three- and four-dimensional geometries.

Chapter 18 prepares for the study of non-Euclidean geometries in higher dimensions by setting forth the basics of Euclidean geometry in three dimensions from the point of view of the *Erlanger Programm*. Pseudo-Euclidean geometry is also presented here.

Chapter 19 treats non-Euclidean geometries, elliptic and hyperbolic, in three dimensions.

17 QUATERNIONS

The complex numbers were constructed by adding i, a single square root of -1, to the real numbers. The quaternions are obtained by adding *three* square roots of -1!

*Definition A **quaternion** is a number of the form*

$$a + bi + cj + dk$$

*where a, b, c, and d are real numbers, **i**, **j**, and **k** are square roots of -1, and, in addition, **ijk** $= -1$.*

The quaternions, including **i**, **j**, and **k**, will be written in boldface to distinguish them from complex numbers. Thus, **i** is a quaternion, while i continues to be the complex unit.

Quaternion Algebra

To add quaternions, just combine separately the coefficients of **i**, **j**, and **k**. For example,

$$(1 + 2i + 3j + 4k) + (2 - 3i + 4j - 5k) = 3 - i + 7j - k$$

This is simply four-dimensional vector addition.

To multiply quaternions, all that is needed is the distributive law plus the formula

$$i^2 = j^2 = k^2 = ijk = -1 \quad (*)$$

For example, to find **ij**, we calculate as follows:

$$ij = -ij(-1) = -ijk^2 = -(ijk)k = -(-1)k = k$$

A similar computation yields **ji** $= -k$. Note that we are assuming that real numbers (like -1) commute with quaternions. However, these examples show that the multiplication of non-real quaternions is *not* commutative.

A–Using (*), deduce that **ij** $= -$**ji** $= $**k**, **jk** $= -$**kj** $= $**i**, and that **ki** $= -$**ik** $= $**j**.

B–What is **kji**?

Although, quaternion multiplication is not commutative, all the other familiar algebraic properties of the complex and real numbers are satisfied by quaternions. They are listed in the following theorem:

Theorem *Quaternion multiplication has the following properties:*
(a) *Associativity:* $\mathbf{q}\ (\mathbf{r}\ \mathbf{s})\ =\ (\mathbf{q}\ \mathbf{r})\ \mathbf{s}$
(b) *Distributivity:* $\mathbf{q}\ (\mathbf{r} + \mathbf{s})\ =\ \mathbf{q}\,\mathbf{r} + \mathbf{q}\,\mathbf{s}$
(c) *Inverses: For every quaternion* $\mathbf{q} \neq 0$, *there is a quaternion* $\mathbf{r}$
such that $\mathbf{q}\ \mathbf{r} = 1$.
(d) *Cancellation: If* $\mathbf{q}\ \mathbf{r} = \mathbf{q}\ \mathbf{s}$, *then* $\mathbf{r} = \mathbf{s}$.

Proof: The simplest proof is to point out that the quaternions

$$\mathbf{q}\ =\ t + x\mathbf{i} + y\mathbf{j} + z\mathbf{k}$$

can be modeled by the 4 x 4 matrixes:

$$\begin{bmatrix} t & y & x & -z \\ -y & t & z & x \\ -x & -z & t & -y \\ z & -x & y & t \end{bmatrix}$$

in other words, by taking

$$\mathbf{i} = \begin{bmatrix} 0 & 0 & 1 & 0 \\ 0 & 0 & 0 & 1 \\ -1 & 0 & 0 & 0 \\ 0 & -1 & 0 & 0 \end{bmatrix} \quad \mathbf{j} = \begin{bmatrix} 0 & 1 & 0 & 0 \\ -1 & 0 & 0 & 0 \\ 0 & 0 & 0 & -1 \\ 0 & 0 & 1 & 0 \end{bmatrix} \quad \text{and } \mathbf{k} = \begin{bmatrix} 0 & 0 & 0 & -1 \\ 0 & 0 & 1 & 0 \\ 0 & -1 & 0 & 0 \\ 1 & 0 & 0 & 0 \end{bmatrix}$$

All we have to do is verify that this choice gives **i**, **j**, and **k** all the properties we have assumed of them. (See Exercise 3.) Then, the conclusion of the theorem follows because matrix multiplication has the desired algebraic properties. ■

C–Multiply out: $(2\mathbf{i} + \mathbf{j})(\mathbf{j} + \mathbf{k})$ and $(2 + 3\mathbf{j})(\mathbf{i} + \mathbf{j} - \mathbf{k})$.

Cartesian Form

Written

$$\mathbf{q}\ =\ t + x\mathbf{i} + y\mathbf{j} + z\mathbf{k}$$

a quaternion is said to be in **Cartesian form**. This is analogous to the form $a + bi$ of a complex number.

Building on this analogy, the **scalar part** of **q** is defined by

$$S\mathbf{q} = t$$

and the **vector part** by

$$Vq = x\mathbf{i} + y\mathbf{j} + z\mathbf{k}.$$

These are like the real and imaginary parts of a complex number. Note that Sq is a real number, but Vq is a quaternion. We also have the **conjugate** of q,

$$q* = Sq - Vq = t - x\mathbf{i} - y\mathbf{j} - z\mathbf{k}$$

and the **modulus**,

$$|q| = (t^2 + x^2 + y^2 + z^2)^{1/2}$$

If $|q| = 1$, then q is called a **unit** quaternion. If $Sq = 0$, then q is called a **pure** quaternion.

D–Every pure, unit quaternion is a square root of -1. Check it!

Pure Quaternions

As in the study of complex numbers, we are interested primarily in *geometric* properties of the quaternions. In this section, we consider, in particular, the pure quaternions $q = x\mathbf{i} + y\mathbf{j} + z\mathbf{k}$, because they represent points in three-dimensional space. It is this that makes the quaternions useful for studying solid geometry.

Pure quaternions are intimately connected with the three-dimensional vectors familiar from multivariable calculus. This is clear as far as addition and scalar multiplication are concerned, which we have seen are the same for quaternions and vectors. In addition, pure quaternion multiplication, as we now show, is directly connected with the two vector products: the dot product and the cross product.

If $q = x_1\mathbf{i} + y_1\mathbf{j} + z_1\mathbf{k}$ and $r = x_2\mathbf{i} + y_2\mathbf{j} + z_2\mathbf{k}$ are pure quaternions, then

$$-S(qr) = x_1x_2 + y_1y_2 + z_1z_2 \qquad (**)$$

which is the **dot product** of q and r. According to the usual geometric interpretation of the dot product,

$$-S(qr) = q \cdot r = |q||r|\cos(\phi)$$

where ϕ is the angle between the vectors q and r (in Euclidean three-dimensional space).

Next, observe that

$$V(qr) = (y_1z_2 - y_2z_1)\mathbf{i} - (x_1z_2 - x_2z_1)\mathbf{j} + (x_1y_2 - x_2y_1)\mathbf{k} \qquad (***)$$

This is the **cross product** of q and r, which has the geometric interpretation

$$V(qr) = (q \times r) = |q||r|\sin(\phi)\,\mathbf{u}$$

where ϕ is as before, and **u** is a pure unit quaternion representing a unit vector in three-dimensional space perpendicular to the plane of the two three-dimensional vectors represented by **q** and **r** and forming a right-handed system with them as shown in Figure 17.1.

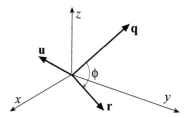

Figure 17.1 Locating the cross product in space

Putting these two results together:

scalar part ⟶   ⟵ vector part

$$\mathbf{qr} = -(\mathbf{q}\cdot\mathbf{r})+(\mathbf{q}\times\mathbf{r}) \qquad (+)$$

This gives a geometric picture of the product of pure quaternions.

E–Check (**) and (***).

Transformations

Our reason for studying quaternions is to be able to express important transformations of three- and four-dimensional space, just as we used complex numbers to express important transformations of the plane. As in the theory of the complex numbers, this depends on the *polar form* of quaternions which we now describe.

To begin, recall the polar form of a complex number:

$$z =|z|(\cos(\theta) + i\sin(\theta))$$

For quaternions, we have instead,

$$\mathbf{q} \;=\; |\mathbf{q}| \,(\cos(\theta) + \mathbf{u}\,\sin(\theta))$$

Note that, where complex polar form uses two coordinates, $|z|$ and θ, quaternion polar form uses three: $|\mathbf{q}|$, θ, and **u**. The additional coordinate, **u**, is not a number, but a pure, unit quaternion. It stands in the place of the i in complex polar form, and, like i, it is a square root of -1. Its appearance can be explained as follows. The complex square roots of -1 are $\pm i$, which, as vectors, point in opposite directions along a *line*. Either i or $-i$ could be used in the complex polar form; it makes no difference. In contrast, there are infinitely many quaternion square roots of -1. They are the pure, unit quaternions which, as vectors, point in all possible directions in three-dimensional *space*. This, in a nutshell, is the difference between the complex numbers and the quaternions: The roots of –1 are one-dimensional in the former, three-dimensional in the latter.

Theorem *Every quaternion can be represented in the form:*

$$\mathbf{q} = |\mathbf{q}| \ (\cos(\theta) + \mathbf{u} \ \sin(\theta))$$

where **u** *is a pure, unit quaternion.*

Proof: See Exercises 9 and 10.

We now have the tools to express a variety of transformations in three dimensions. The transformations we are interested in are the usual suspects already studied in the plane: translation, rotation, and homothetic transformation.   For translation and homothetic transformation, the formulas are nearly the same in two and three dimensions. (See Exercise 11.) Rotation, however, is different. The algebra of rotations is much more complicated in three dimensions than in two, but can be neatly expressed using quaternion multiplication as follows:

Theorem *Let* **r** *be a unit quaternion.   Let R be the transformation defined by*

$$R\mathbf{q} = \mathbf{r} \, \mathbf{q} \, \mathbf{r}^*$$

where **q** *is a pure quaternion.   Then R is a rotation of the three-dimensional space of pure quaternions about an axis passing through the origin. Specifically, if the polar form of* **r** *is*

$$\mathbf{r} = \cos(\theta) + \mathbf{u} \sin(\theta)$$

where **u** *is a pure, unit quaternion, then Rq is the pure quaternion obtained by rotating* **q** *about* **u** *by the angle* 2θ. *Every rotation of three-dimensional space (about an axis passing through the origin) can be expressed in this way.*

Proof: We examine *R*q for some particular quaternions **q** and then put these special cases together.

Case 1: **q** = **u**. Then

$$R\mathbf{u} = \mathbf{r} \, \mathbf{u} \, \mathbf{r}^* = (\cos(\theta) + \mathbf{u} \sin(\theta)) \ \mathbf{u} \ (\cos(\theta) - \mathbf{u} \sin(\theta))$$

Expanding this product gives

$$R\mathbf{u} = \cos(\theta)^2 \mathbf{u} - \sin(\theta)^2 \mathbf{u}^3 = \cos(\theta)^2 \mathbf{u} - \sin(\theta)^2 (-\mathbf{u}) = \mathbf{u}$$

since $\mathbf{u}^2 = -1$.   Thus, **u** is a fixed point of *R*. Of course, this is no more than we should expect if *R***u** *is* the rotation of **u** about the axis

u! Thus, the conclusion of the theorem is confirmed in this simple case.

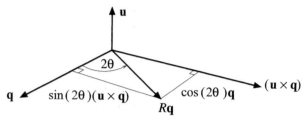

Figure 17.2 Rotation in space

Case 2: **q** is perpendicular to **u**. Such a quaternion **q** is illustrated in Figure 17.2. Then,

$$R\mathbf{q} = \mathbf{r}\,\mathbf{q}\,\mathbf{r}^* = (\cos(\theta) + \mathbf{u}\sin(\theta))\,\mathbf{q}\,(\cos(\theta) - \mathbf{u}\sin(\theta))$$
$$= \cos(\theta)^2\,\mathbf{q} + \mathbf{u}\,\mathbf{q}\,\cos(\theta)\sin(\theta) - \mathbf{q}\,\mathbf{u}\,\cos(\theta)\sin(\theta) - \mathbf{u}\,\mathbf{q}\,\mathbf{u}\,\sin(\theta)^2$$

Now, using equation (+), we get

$$\mathbf{u}\,\mathbf{q} = -(\mathbf{u}\cdot\mathbf{q}) + (\mathbf{u}\times\mathbf{q}) = (\mathbf{u}\times\mathbf{q})$$

because **u** and **q** are perpendicular. Similarly,

$$\mathbf{q}\,\mathbf{u} = (\mathbf{q}\times\mathbf{u}) = -(\mathbf{u}\times\mathbf{q})$$

so that

$$\mathbf{u}\,\mathbf{q}\,\mathbf{u} = (\mathbf{u}\cdot(\mathbf{u}\times\mathbf{q})) - (\mathbf{u}\times(\mathbf{u}\times\mathbf{q})) = -(\mathbf{u}\times(\mathbf{u}\times\mathbf{q}))$$

since $(\mathbf{u}\times\mathbf{q})$ is also perpendicular to **u**. Finally, using the right-hand rule, we obtain

$$\mathbf{u}\,\mathbf{q}\,\mathbf{u} = \mathbf{q}$$

Putting all this together gives

$$R\mathbf{q} = \cos(\theta)^2\,\mathbf{q} + 2\cos(\theta)\sin(\theta)(\mathbf{u}\times\mathbf{q}) - \sin(\theta)^2\,\mathbf{q}$$
$$= \cos(2\theta)\,\mathbf{q} + \sin(2\theta)(\mathbf{u}\times\mathbf{q})$$

This shows that, indeed, $R\mathbf{q}$ is the rotation of **q** through an angle of 2θ about the axis of **u**.

Case 3: **q** is *any* pure quaternion. First note that R is a linear transformation of three-dimensional space; that is, R satisfies

$$R(\mathbf{s} + \mathbf{t}) = R(\mathbf{s}) + R(\mathbf{t})$$

and

$$R(\alpha\mathbf{s}) = \alpha\,R(\mathbf{s})$$

where **s** and **t** are three-dimensional vectors and α is a real number. (See Exercise 12.) Now, rotation of three-dimensional space is *also* a linear transformation. (See Exercise 13.) The importance of these facts is that linear transformations are determined by their action on a set of basis vectors. Therefore, to prove that two linear transformations are equal, it suffices to prove that they have the same effect on the vectors from some basis of three-dimensional space.

The present theorem concerns *two* linear transformations: on the one hand, R, and, on the other hand, rotation about the axis **u** by the angle 2θ. So far we have shown that these two transformations are equal in two cases: when **q** = **u**, and when **q** is perpendicular to **u**. However, a basis for three-dimensional space can be found consisting of the vector **u** plus two orthogonal vectors. Therefore, since we have already shown that R acts like a rotation for these particular vectors, it follows (by the argument in the preceding paragraph) that R acts like a rotation on *all* three-dimensional vectors. •

Summary

The quaternions are a four-dimensional number system satisfying all the algebraic laws of the real and complex number systems except the commutative law of multiplication. Quaternions play a role in three- and four-dimensional analytic geometry comparable to the role the complex numbers play in one- and two-dimensional analytic geometry. Thus, various algebraic ideas defined for quaternions have geometric interpretations in Euclidean geometry. In particular, *length* is represented algebraically by the *modulus* of a quaternion, and *directions* (in three dimensions) are represented by the *pure, unit quaternions*. The two concepts, length and direction, are united in the polar form of a quaternion.

Also, the quaternion operations of addition and multiplication have geometric interpretations as transformations of three-dimensional space:

$$T\mathbf{q} = \mathbf{q} + \mathbf{b} \quad \text{(translation)}$$

and

$$R\mathbf{q} = \mathbf{r}\,\mathbf{q}\,\mathbf{r}^* \quad \text{(rotation)}$$

This is the essence of analytic geometry: Algebraic operations have a geometric interpretation, and geometric operations can be expressed algebraically.

EXERCISES

Quaternion Arithmetic

1. Calculate
   (a) $(1 - 5\mathbf{i} - 4\mathbf{j} + 25\mathbf{k}) - (-3 - \sqrt{2}\mathbf{i} + 3\mathbf{j})$
   (b) $(1 - \mathbf{i} - \mathbf{j})(2\mathbf{k} - 3\mathbf{i})$ (c) $(\mathbf{i} - \mathbf{j} + \mathbf{k})(2 - 4\mathbf{j})$

2. Find the Cartesian form of these quaternions:

$$\frac{1}{1+i}, \qquad \frac{1}{i+j+2k}, \qquad \frac{1}{2-i+j}$$

Hint: Recall Exercise 10 in Chapter 2.

Basic Properties of Quaternions

3. Complete the proof of the algebraic properties of quaternions by showing that the 4 x 4 matrices chosen for **i**, **j**, and **k** have the properties attributed to them in the definition of quaternions. Also check that the reals (in the form of the diagonal matrices) commute with all other 4x4 matrices.

4. Prove these properties of the conjugate and modulus:
   (a) $(\mathbf{q}\,\mathbf{r})^* = \mathbf{q}^*\mathbf{r}^*$ (b) $|\mathbf{q} + \mathbf{r}| \leq |\mathbf{q}| + |\mathbf{r}|$
   (c) $|\mathbf{q}|^2 = \mathbf{q}\,\mathbf{q}^* = \mathbf{q}^*\,\mathbf{q}$ (d) $S(\mathbf{q}) = (\mathbf{q} + \mathbf{q}^*)/2$
   (e) $|\mathbf{q}\,\mathbf{r}| = |\mathbf{r}\,\mathbf{q}|$ (f) $V(\mathbf{q}) = (\mathbf{q} - \mathbf{q}^*)/2$

5. Prove that every nonzero quaternion **q** has a multiplicative inverse $\mathbf{q}^{-1} = \mathbf{q}^*/|\mathbf{q}|^2$.

Pure Quaternions

6. Show that the reciprocal of a pure quaternion is pure.

7. Verify that all pure, unit quaternions are square roots of -1. Prove that these are the only quaternion square roots of -1.

8. Prove that two three-dimensional vectors **q** and **r** (represented by pure quaternions) are orthogonal if, and only if, $\mathbf{q}\,\mathbf{r} = -\mathbf{r}\,\mathbf{q}$. (Then **q** and **r** are called **anticommutative**.)

Polar Form

9. Prove the existence of the polar form of a quaternion.
*Hint: **u** is the normalized vector part of **q**.*

10. Is the polar form of a quaternion unique?

Quaternions and Transformations

11. Find formulas for translation and homothetic transformation (stretching or shrinking from the origin) in three dimensions using quaternions.

12. Prove that $R(\mathbf{q}) = \mathbf{r}\,\mathbf{q}\,\mathbf{r}^*$ is a linear transformation where **r** is any quaternion.

13. Prove that a rotation of three-dimensional space is a linear transformation.
Hint: Give a geometric proof. (Use no formulas!)

14. Let r_1 and r_2 be quaternions. Let $R_1q = r_1 \, q \, r_1{}^*$ and $R_2q = r_2 \, q \, r_2{}^*$ be the corresponding rotations. Show that the composition $R_1R_2(q)$ is also a rotation.

Remark: It's somewhat surprising that the composition of two rotations in three dimensions is only a rotation rather than some more complicated transformation.

15. Exercise 14 shows that the set of all rotations $Rq = r \, q \, r^*$ form a transformation group of the space of pure quaternions, and, therefore, define a geometry. What is that geometry?

16. Let R_1 be a rotation of 90 degrees about an axis e_1. Let R_2 be a second rotation of 90 degrees about an axis e_2 perpendicular to e_1. Using quaternions, find a formula for the transformation R of three-dimensional space that is the result of composing R_1 and R_2. R is a rotation: Of what angle and about what axis?

Remark: This exercise points out the complex nature of the rules governing the combination of rotations about a common center in three dimensions. In contrast: What simple rule governs the combination of rotations about a common center in two dimensions?

17. Using quaternions, find a formula for a rotation of three-dimensional space about an axis *not* passing through the origin.

18. Let u be a pure, unit quaternion. Let ρ be the plane in three-dimensional space perpendicular to u and passing through the origin. Prove that the transformation

$$Tq = u \, q \, u$$

reflects the vector q across the plane ρ.

19. Investigate what happens when two reflections are composed. In particular, let R_1 be a reflection about a plane perpendicular to the axis e_1. Let R_2 be a second reflection about a plane perpendicular to the axis e_2 perpendicular to e_1. Using quaternions, find a formula for the transformation R of three-dimensional space that is the result of composing R_1 and R_2. What sort of transformation is R?

18 EUCLIDEAN AND PSEUDO-EUCLIDEAN SOLID GEOMETRY

This chapter and the next present several three- and four-dimensional geometries. The underlying set for each geometry is a set of quaternions, and the transformation group of each geometry is based on quaternion operations. A truly bizarre feature of this chapter is the dramatic appearance of *complex* quaternions.

Euclidean Solid Geometry

For the purpose of comparing different solid geometries, we start with Euclidean solid geometry. Here is a formal definition:

Definition *Let* **V** *be the set of pure quaternions; that is,*

$$\mathbf{V} = \{\mathbf{v} = x\mathbf{i} + y\mathbf{j} + z\mathbf{k} : x, y, z \text{ real}\}$$

Let **R** *be the set of transformations T of* **V** *described by the formula*

$$T\mathbf{v} = \mathbf{r}\,\mathbf{v}\,\mathbf{r}^* + \mathbf{b}$$

where **v**, **r**, *and* **b** *are quaternions:* **b** *is pure, and* **r** *is a unit. The pair* (**V**, **R**) *models* **Euclidean solid geometry**.

From Chapter 17, recall that $\mathbf{r}\,\mathbf{v}\,\mathbf{r}^*$ is a rotation of **V**. Thus, in both two and three dimensions, a Euclidean transformation consists of a rotation (about an axis passing through the origin) followed by a translation.

Screw Motions

These transformations help us to understand the nature of three-dimensional Euclidean geometry by describing its transformations in simple physical terms. Thus, let $T\mathbf{v} = \mathbf{r}\,\mathbf{v}\,\mathbf{r}^* + \mathbf{b}$ be a three-dimensional Euclidean transformation where $\mathbf{r} = \cos(\theta) + \mathbf{u}\sin(\theta)$. If **u** (the axis of the rotation part of *T*) and **b** (the translation part) are *parallel*, then *T* is called a **screw motion**. (See Figure 18.1.) Thus, a screw motion rotates about an axis while translating in the same direction.

A–Make a screw motion with your hand.

A screw motion is the most general type of Euclidean transformation in three dimensions (if we agree that plain rotations and translations are particularly simple examples of screw motion). In other words:

Theorem *Every Euclidean transformation is a screw motion.*

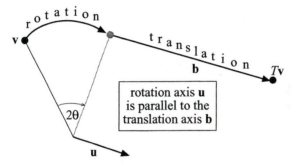

Figure 18.1 Screw motion = rotation + translation

This theorem deserves to be called the fundamental theorem of Euclidean solid geometry. It implies, for example, that, in a Euclidean world, the motion of a rigid body, unaffected by any external force, must follow a helix, the path traced out by a point under the repeated action of a screw motion. (See Figure 18.2.)

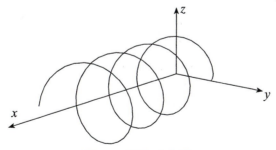

Figure 18.2 A helix

The proof of the theorem depends on two lemmas.

Lemma 1 *Every Euclidean transformation with a fixed point is a rotation.*

Proof: Outlined in Exercise 4. ∎

Lemma 2 *Let* $Tv =$ **r** **v** **r*** $+$ **b** *be a Euclidean transformation. Let* **r** $= \cos(\theta) +$ **u** $\sin(\theta)$. *If* **u** *and* **b** *are perpendicular, then* T *is a rotation about an axis parallel to* **u**.

Proof of the theorem: Suppose for a moment that we have proved lemma 2. Then the proof of the theorem is completed as follows: Let $Tv =$ **r** **v** **r*** $+$ **b** be a Euclidean transformation. Now **b** can be split into two parts, **b** $=$ **b**$_1$ $+$ **b**$_2$, in such a way that **b**$_1$ is perpendicular to **u**, and **b**$_2$ is parallel to **u**. Then T becomes

$$Tv = \mathbf{r}\,\mathbf{v}\,\mathbf{r}^* + \mathbf{b} = \mathbf{r}\,\mathbf{v}\,\mathbf{r}^* + \mathbf{b}_1 + \mathbf{b}_2 = (\mathbf{r}\,\mathbf{v}\,\mathbf{r}^* + \mathbf{b}_1) + \mathbf{b}_2$$

Now, the transformation in parentheses has its translation axis **b**$_1$ perpendicular to its rotation axis **u**, so that, according to lemma 2, this transformation is a simple rotation whose axis is parallel to **u**. If we now consider the whole transformation, we see that T is the combination of a rotation (with axis parallel to **u**) plus translation by **b**$_2$ (which is parallel to **u**). Thus, T is a screw motion. ■

Proof of lemma 2: Let $Tv =$ **r** **v** **r*** $+$ **b**, and suppose that **b** is perpendicular to **u**. We will prove that T has a fixed point. Then lemma 1 implies T is a rotation. Let

$$\mathbf{v} = \frac{1}{2\sin(\theta)}\mathbf{r}^*\mathbf{u}\mathbf{b}$$

We will prove that **v** is a fixed point of T. To calculate Tv, however, we need these relations: **b** **u** $= -$**u** **b**, **u** **r** $=$ **r** **u**, and **b** **r*** $=$ **r** **b**. The first follows because **b** and **u** are perpendicular, hence anticommute. (See Exercise 8 in Chapter 17.) Using this and the polar form of **r**, the other two formulas follow easily. (See Exercise 5 in this chapter.)

We first calculate the rotation portion of T applied to **v**:

$$\mathbf{r}\,\mathbf{v}\,\mathbf{r}^* = \mathbf{r}\left(\frac{1}{2\sin(\theta)}\mathbf{r}^*\mathbf{u}\mathbf{b}\right)\mathbf{r}^*$$

$$= \frac{1}{2\sin(\theta)}\mathbf{u}\mathbf{b}\mathbf{r}^* = \frac{1}{2\sin(\theta)}\mathbf{r}\mathbf{u}\mathbf{b}$$

Then,

$$Tv = \mathbf{r}\,\mathbf{v}\,\mathbf{r}^* + \mathbf{b} = \frac{1}{2\sin(\theta)}\mathbf{r}\mathbf{u}\mathbf{b} + \mathbf{b}$$

$$= \frac{1}{2\sin(\theta)}(\mathbf{r}\mathbf{u} + 2\sin(\theta))\mathbf{b} = \frac{1}{2\sin(\theta)}\mathbf{r}^*\mathbf{u}\mathbf{b} = \mathbf{v}$$

since (as is easily verified) **r** **u** $+ 2\sin(\theta) =$ **r*** **u**. ■

Complex Quaternions

In order to develop pseudo-Euclidean geometry, a significant alternative solid geometry, we extend the concept of quaternion as follows:

Definition *A complex quaternion is a quantity of the form*

$$\mathbf{q} = t + ix\mathbf{i} + iy\mathbf{j} + iz\mathbf{k}$$

where i is the ordinary complex unit.

In other words, a complex quaternion is a quaternion with three *imaginary* coefficients and one *real* coefficient. As with ordinary quaternions the **scalar part** of **q** is defined by

$$S\mathbf{q} = t$$

the **vector part** is defined by

$$V\mathbf{q} = ix\mathbf{i} + iy\mathbf{j} + iz\mathbf{k}$$

the **conjugate** of **q** is

$$\mathbf{q}^* = S\mathbf{q} - V\mathbf{q} = t - ix\mathbf{i} - iy\mathbf{j} - iz\mathbf{k}$$

and the (**Minkowski**) **norm** of **q** is

$$\| \mathbf{q} \| = |t^2 - x^2 - y^2 - z^2|^{1/2}$$

Note that absolute values are needed to define the norm, since the quantity $t^2 - x^2 - y^2 - z^2$ can be negative.

B–Find a nonzero complex quaternion for which ‖q‖ is zero. Can such a quaternion have a multiplicative inverse?

The algebra and geometry of complex quaternions present many contrasts with the algebra and geometry of ordinary quaternions. These are explored in Exercises 7 to 10. The most important result is the existence of a **polar form** for complex quaternions, analogous to polar form for ordinary quaternions: Every complex quaternion can be written in the form

$$\mathbf{q} = \| \mathbf{q} \| (\cosh(\tau) + i\mathbf{u} \sinh(\tau))$$

where **u** is a pure, unit quaternion, cosh and sinh are the hyperbolic functions (introduced in Exercise 21 of Chapter 2), and τ is real. To prove this, we simply imitate the derivation of polar form for ordinary quaternions, substituting hyperbolic for trigonometric functions. (See Exercise 12.) Clearly complex quaternions are a "hyperbolic" version of quaternions.

Pseudo-Euclidean Geometry

Pseudo-Euclidean geometry is an important non-Euclidean solid geometry. It is defined using complex quaternions but otherwise is quite analogous to three-dimensional solid Euclidean geometry.

Definition *Let* **W** *be the set of complex quaternions with zero* **k** *term; that is,*

$$\mathbf{W} = \{\mathbf{w} = t + ix\mathbf{i} + iy\mathbf{j} : t, x, y \text{ real}\}$$

Let **PE** *be the set of transformations T of* **W** *described by the formula*

$$T\mathbf{w} = \mathbf{r}\,\mathbf{w}\,\mathbf{r} + \mathbf{b}$$

where **w**, **r**, *and* **b** *are in* **W** *and* $\|\mathbf{r}\| = 1$. *The pair* (**W**, **PE**) *models* **three-dimensional pseudo-Euclidean geometry.**

The transformation $L\mathbf{w} = \mathbf{r}\,\mathbf{w}\,\mathbf{r}$ is called a **Lorentz transformation**. Lorentz transformations are the rotations of pseudo-Euclidean geometry. To get the feel of them, let us compute one. Let **r** be the particular unit given by

$$\mathbf{r} = (\cosh(\tau) + i\mathbf{j}\sinh(\tau))$$

where, for simplicity, we choose **r** with only two nonzero components. Let $\mathbf{w} = t + ix\mathbf{i} + iy\mathbf{j}$ be a typical element of **W**. Then,

$$L\mathbf{w} = \mathbf{r}\,\mathbf{w}\,\mathbf{r}$$
$$= (\cosh(\tau) + i\mathbf{j}\sinh(\tau))(t + ix\mathbf{i} + iy\mathbf{j})(\cosh(\tau) + i\mathbf{j}\sinh(\tau))$$

Multiplying out, we obtain

$$L\mathbf{w} = [(\cosh(\tau)^2 + \sinh(\tau)^2)\,t + 2\cosh(\tau)\sinh(\tau)\,y]$$
$$+ i\,(\cosh(\tau)^2 - \sinh(\tau)^2)\,x\,\mathbf{i}$$
$$+ i\,[2\cosh(\tau)\sinh(\tau)\,t + (\cosh(\tau)^2 + \sinh(\tau)^2)\,y]\,\mathbf{j}$$

Then, using the hyperbolic identities:

$$\cosh(\tau)^2 - \sinh(\tau)^2 = 1$$
$$\cosh(2\tau) = \cosh(\tau)^2 + \sinh(\tau)^2$$
$$\sinh(2\tau) = 2\cosh(\tau)\sinh(\tau)$$

we get

$$L\mathbf{w} = [\cosh(2\tau)t + \sinh(2\tau)y] + ix\mathbf{i} + i[\sinh(2\tau)t + \cosh(2\tau)y]\mathbf{j}$$

Note: This formula is closely analogous to the formula for a Euclidean rotation (from Chapter 17):

$$R\mathbf{v} = [\cos(2\theta)x + \sin(2\theta)z]\mathbf{i} + y\mathbf{j} + [-\sin(2\theta)x + \cos(2\theta)z]\mathbf{k}$$

C–Give a detailed derivation of the preceding formula for $L\mathbf{w}$.
D–Verify the foregoing hyperbolic identities.

The Special Theory of Relativity

Einstein's special theory of relativity describes physical events as seen by an ideal observer occupying a special position in space–what is called an *inertial reference frame*–which is, essentially, a position free of gravitational influence. Einstein maintained that under these special conditions (whence the term *special* relativity), the laws of physics would appear the same to all observers.

According to the special theory of relativity, physical events relative to an inertial frame of reference are described by four coordinates: three space coordinates (x, y, and z), plus time (t). The resulting four-dimensional coordinate system is called **space-time**.

The coordinates of this system correspond exactly to the four components of a complex quaternion. Thus, let

$$\mathbf{w}_1 = t_1 + ix_1\mathbf{i} + iy_1\mathbf{j} + iz_1\mathbf{k}$$

and

$$\mathbf{w}_2 = t_2 + ix_2\mathbf{i} + iy_2\mathbf{j} + iz_2\mathbf{k}$$

be two space-time events as seen by an observer in an inertial frame of reference. In classical physics, there are two ways to measure the coordinate distance between events: temporal distance (separation in time) and spatial distance (separation in space). In quaternion terms, these are

$$\text{time separation} = |S(\mathbf{w}_1) - S(\mathbf{w}_2)| = |t_1 - t_2| = ((t_1 - t_2)^2)^{1/2}$$

and

$$\text{space separation} = |V(\mathbf{w}_1) - V(\mathbf{w}_2)| = ((x_1 - x_2)^2 + (y_1 - y_2)^2 + (z_1 - z_2)^2)^{1/2}$$

In classical physics, observers of the events $\mathbf{w}_1$ and $\mathbf{w}_2$ would measure the same time separation and space separation.

In relativistic physics, however, observers of $\mathbf{w}_1$ and $\mathbf{w}_2$ would *not* necessarily observe the same time and space separations, although, they will observe the same **relativistic** or **Minkowski separation**:

$$\left\| \mathbf{w}_1 - \mathbf{w}_2 \right\| = \left| (t_1 - t_2)^2 - (x_1 - x_2)^2 - (y_1 - y_2)^2 - (z_1 - z_2)^2 \right|^{1/2}$$

In other words, the Minkowski separation is an invariant of relativistic space-time geometry, but space separation and time separation alone are not.

Now, the quantity

$$M = (t_1 - t_2)^2 - (x_1 - x_2)^2 - (y_1 - y_2)^2 - (z_1 - z_2)^2$$

can be negative as well as positive or zero. This gives rise to three different types of intervals between events. If M is positive, then the time separation predominates, and the interval between $\mathbf{w}_1$ and $\mathbf{w}_2$ is called **timelike**. If M is negative, then the space separation predominates, and the interval is called **spacelike**. (See Figure 18.3.)

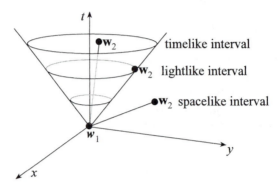

Figure 18.3 Regions in pseudo-Euclidean geometry

If M is zero, the interval is called **lightlike**. (Units are chosen so that the speed of light is 1.) Lightlike intervals are observed when it is possible for a ray of light to travel between the two events. Any other particle travels at a speed less than that of light, that is, less than 1. Therefore, in the interval separating two events along the path followed by nonlight particles, *time separation predominates over space separation*. In other words, particles other than rays of light always travel so as to create timelike intervals between events along their path through space-time.

So far it appears that the geometry of space time is like a *four-dimensional*, pseudo-Euclidean geometry. But what about transformations? Are any transformations relevant to the theory of relativity?

Consider two observers in inertial frames of reference. Suppose that the reference frame $\mathbf{w}_2$ of observer 2 is traveling at a constant speed σ in relation to the frame $\mathbf{w}_1$ of observer 1. For simplicity, suppose further that motion of observer 2 in relation to observer 1 is in the y-direction. Then there *is* a transformation connecting $\mathbf{w}_1$ and $\mathbf{w}_2$, namely,

$$
\begin{aligned}
t_1 &= t_2\,(1 - \sigma^2)^{-1/2} + y_2\,\sigma\,(1 - \sigma^2)^{-1/2} \\
x_1 &= x_2 \\
y_1 &= y_2\,\sigma\,(1 - \sigma^2)^{-1/2} + y_2\,(1 - \sigma^2)^{-1/2} \\
z_1 &= z_2
\end{aligned}
$$

(A physical explanation of this transformation is given, for example, in Taylor and Wheeler [F5].) For our purposes, the important thing is to notice that the coefficients $a = (1 - \sigma^2)^{-1/2}$ and $b = \sigma\,(1 - \sigma^2)^{-1/2}$ satisfy $a^2 - b^2 = 1$. Therefore, according to Exercise 22 of Chapter

2, there is a number τ such that $a = \cosh(\tau)$ and $b = \sinh(\tau)$. When written using τ, it becomes apparent that the preceding transformation is the Lorentz transformation

$$\mathbf{w}_1 = L\mathbf{w}_2 = \mathbf{r}\,\mathbf{w}_2\,\mathbf{r}$$

where $\mathbf{r} = a + ib\,\mathbf{j} = \cosh(\tau) + i\sinh(\tau)\,\mathbf{j}$. The quantity τ is called the **velocity parameter** of the transformation. Thus, a Lorentz transformation connects the viewpoints of two observers in relativistic space-time.

In the late nineteenth century, a number of discoveries in physics suggested the importance of Lorentz transformations. Einstein took the step of postulating Lorentzian invariance, and was led thereby to the special theory of relativity. From our viewpoint, the heart of this theory is that the relativistic geometry of space-time *is* a pseudo-Euclidean geometry of dimension four. The mathematical background of Einstein's work includes, therefore, both non-Euclidean geometry and the *Erlanger Programm*.

Hyperboloidal Geometry

Consider the "unit sphere" **H** in **W** consisting of the points such that

$$\|\mathbf{r}\|^2 = t^2 - x^2 - y^2 = 1$$

Of course the surface **H** is not a sphere but is actually a **hyperboloid of two sheets**. (See Figure 18.4.) Because its equation is similar to that of the sphere, the hyperboloid is sometimes referred to as a "sphere with imaginary radius."

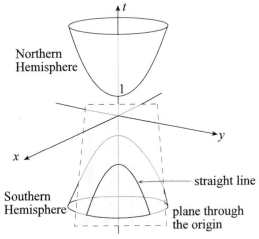

Figure 18.4 The unit hyperboloid

On the surface **H** we can build a geometry, called **hyperboloidal geometry**. The construction of this geometry, inside three-dimensional pseudo-Euclidean geometry (**W**, **PE**), proceeds in a manner analogous to the construction, inside ordinary three-dimensional Euclidean geometry, of elliptic geometry on the surface of the unit sphere **S**.

For example, in elliptic geometry, straight lines are the great circles on **S**, that is, circles obtained by intersecting **S** with a plane passing through the origin. These same planes also intersect **H**. We take the resulting curves as the straight lines in hyperboloidal geometry. Angles between straight lines in elliptic geometry are defined as the angles between the Euclidean tangents to great circles. The same definition can be used for angles on **H**.

One complication is the fact that the hyperboloid has two **sheets** (i.e., is in two pieces) whereas the sphere has only one piece. We concentrate here on the "Southern Hemisphere" of the hyperboloid (where $t < 0$).

E–Do two points determine a line in hyperboloidal geometry?

The Return of Stereographic Projection

The reader may suspect that hyperboloidal geometry is not a new geometry at all but is really just another model of hyperbolic geometry. That this is so can be proved by applying to **H** the same stereographic projection applied earlier to **S**!

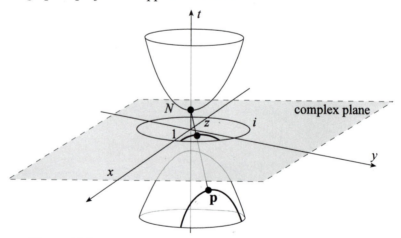

Figure 18.5 Stereographic projection of the unit hyperboloid

Therefore, let **C** be the xy (i.e., complex) plane. Stereographic projection of the hyperboloid **H** onto **C** proceeds as follows: Each point **p** on the hyperboloid is connected by a straight line with the "North Pole" N on the "Northern Hemisphere" of **H**. This line passes through the plane **C** at some point z. The point z is the stereographic projection of **p**. (See Figure 18.5.)

The exercises that follow explore some of the properties of this mapping. In particular, they show how the "straight lines" we have defined on **H** are transformed by stereographic projection into the familiar "straight lines" of hyperbolic geometry in the unit disk model.

Summary

All the geometries introduced in this chapter are subgeometries of four-dimensional space-time, defined as follows:

Definition *Let* $\mathbf{Q_C}$ *be the set of all (four-dimensional) complex quaternions*

$$\mathbf{q} = t + ix\,\mathbf{i} + iy\,\mathbf{j} + iz\,\mathbf{k}$$

Let **L** *be the set of all transformations* T *of* $\mathbf{Q_C}$ *for which the Minkowski separation is an invariant, that is, such that*

$$\| T\mathbf{q} - T\mathbf{r} \| = \| \mathbf{q} - \mathbf{r} \|$$

L *is called the* ***full Lorentz group***. *Then* $(\mathbf{Q_C}, \mathbf{L})$ *is the geometry of (four-dimensional)* ***relativistic space-time***.

The relationships among all the geometries mentioned in this chapter are displayed in this table:

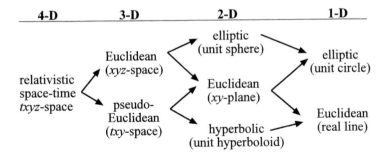

4-D	3-D	2-D	1-D

Here, the geometry at the end of each arrow is a subgeometry of the geometry at the beginning of each arrow.

EXERCISES

Euclidean Geometry

1. The Euclidean transformation, $T\mathbf{v} = \mathbf{r}\,\mathbf{v}\,\mathbf{r}^* + \mathbf{b}$, consists of a rotation followed by a translation. Show that the action of T can

also be accomplished by a translation followed by a rotation (as in Euclidean plane geometry).

2. Show that the set **R** of Euclidean transformations is indeed a transformation group.

3. Show that the Euclidean distance between two points,

$$d(\mathbf{u}, \mathbf{v}) \ = \ |\mathbf{u} - \mathbf{v}|$$

is an invariant of Euclidean solid geometry.

4. Prove lemma 1.

*Hint: Let T be a Euclidean transformation with a fixed point. If the fixed point is the origin, the proof of the lemma is simple. (Use the formula for T.) If the fixed point is **a** (not the origin), then consider the transformation STS⁻¹, where S is the translation from **a** to the origin.*

5. Complete the proof of lemma 2 by showing that if $\mathbf{r} = \cos(\theta) + \mathbf{u} \sin(\theta)$ is a unit quaternion, and **b** is a pure quaternion perpendicular to **u**, then

    (a) **b u** = –**u b**   (b) **u r** = **r u**
    (c) **b r*** = **r b**   (d) **r u** + 2 sin(θ) = **r* u**

Hint: Remember that **r r*** = 1.

6. Verify that the fixed point $\mathbf{v} = \mathbf{r}^* \mathbf{u}\, \mathbf{b}/(2\sin(\theta))$ in the proof of lemma 2 is a pure quaternion. Why is this important?

7. If **v** is a fixed point of $T\mathbf{v} = \mathbf{r}\,\mathbf{v}\,\mathbf{r}^* + \mathbf{b}$, where $\mathbf{r} = \cos(\theta) + \mathbf{u} \sin(\theta)$, show that $\mathbf{v} + k\mathbf{u}$ is also a fixed point (where k is real). Explain how this proves that the rotation axis of T is parallel to **u**.

8. Show that, in Euclidean solid geometry, the angle θ between two line segments **q r** and **q s** (where **q, r,** and **s** are in **V**) can be calculated from the formula

$$\theta = \arccos\!\left(\frac{S((\mathbf{r} - \mathbf{q})(\mathbf{s} - \mathbf{q})^*)}{|\mathbf{r} - \mathbf{q}||\mathbf{s} - \mathbf{q}|} \right)$$

Complex Quaternions

9. Prove that complex quaternions, like ordinary quaternions, obey the associative and distributive laws.

10. Prove these properties of the conjugate and norm for complex quaternions:

    (a) $(\mathbf{q}\,\mathbf{r})^* = \mathbf{q}^* \, \mathbf{r}^*$
    (b) $\| \mathbf{q} \|^2 = |\mathbf{q}\,\mathbf{q}^*| = |\mathbf{q}^* \, \mathbf{q}|$
    (c) $\| \mathbf{q}\,\mathbf{r} \| = \| \mathbf{r}\,\mathbf{q} \|$

11. Verify that *not* every nonzero complex quaternion has an inverse and that complex quaternions do *not* satisfy the cancellation law.

12. **Polar Form** Prove that a complex quaternion of nonzero Minkowski norm can be written in the form

$$\mathbf{q} \;=\; \pm \| \mathbf{q} \| \; (\cosh(\tau) + \mathbf{u} \sinh(\tau))$$

where **u** is a pure, unit complex quaternion, and τ is real.

Hint: Use properties of the hyperbolic functions cosh and sinh developed in the exercises of Chapter 2.

Pseudo-Euclidean Geometry: Transformations

13. The transformation $T\mathbf{w} = \mathbf{r} \, \mathbf{w} \, \mathbf{r} + \mathbf{b}$ consists of a Lorentz transformation followed by a translation. Show that the action of T can also be accomplished by a translation followed by a Lorentz transformation.

14. Show that the set **PE** is a transformation group.

15. Show that the **Minkowski separation** between two points

$$M(\mathbf{w}, \mathbf{v}) \;=\; \| \mathbf{w} - \mathbf{v} \|$$

is an invariant of pseudo-Euclidean geometry.

Note: The Minkowski separation, however, is not a metric. (Why not?)

16. Imitating Exercise 6, write out a formula to define the measure of the angle between line segments **qr** and **qs** in pseudo-Euclidean geometry (where **q**, **r**, and **s** are elements of **W**). Use this formula to show that it is possible for a line segment to be perpendicular to itself in pseudo-Euclidean geometry.

Hyperboloidal Geometry

17. Prove that two points determine a line in hyperboloidal geometry.

Hint: Two points on the hyperboloid plus the origin determine a . . . ?

18. A straight line in hyperboloidal geometry is actually what Euclidean curve?

19. Do parallel lines occur in hyperboloidal geometry? If so, then what types of parallelism can occur?

Hint: Questions of parallelism in hyperboloidal geometry can be settled by considering the planes determining the straight lines.

20. Discuss cycles in hyperboloidal geometry. What Euclidean curves are they? What different types of cycles arise?

Hint: In spherical geometry, circles are plane sections of the sphere. Why should hyperbolic geometry be any different?

21. Elliptic geometry, hyperboloidal geometry, and projective geometry are three plane geometries modeled by subsets of three-dimensional Euclidean space. In all three geometries, points are determined by identifying Euclidean points with their scalar multiples. In all three geometries, lines are the intersections of the underlying set with planes passing through the origin.

Elliptic geometry and projective geometry have the same incidence properties: Two points determine a line and two lines determine a point. Why does hyperboloidal geometry *not* have these incidence properties?

Stereographic Projection

22. Show that the stereographic projection of the Southern Hemisphere of **H** is the inside of the unit circle on **C,** and stereographic projection of the Northern Hemisphere is the outside of the unit circle.

23. Let **p** have coordinates (x, y, z). Find the coordinates of the point z, the stereographic projection of **p** from **H** to the complex plane, in terms of the coordinates of **p**. Conversely, find formulas for the coordinates of **p** in terms of those of z.

24. Two points on **H** are called **diametrically opposite** if the Euclidean straight line connecting them passes through the origin. Prove that if p_1 and p_2 are diametrically opposite, then their stereographic projections z_1 and z_2 are inverse with respect to the unit circle, that is, $z_1 = 1/z_2$.

25. Prove that the stereographic projections of "straight lines" on **H** are circles on **C,** unless the line passes through a "pole," in which case the projection is a Euclidean straight line.

26. Prove that the stereographic projection of a straight line on **H** is an arc of a circle perpendicular to the unit circle.

Hint: Prove and use the fact that these projections must be self-inverse; that is, they are identical to their image under inversion in the unit circle. (See Exercise 24.)

19 HYPERBOLIC AND ELLIPTIC SOLID GEOMETRY

Chapter 18 presented three-dimensional geometries that are subgeometries of the *complex* quaternions, that is, subgeometries of four-dimensional pseudo-Euclidean geometry (or relativistic space-time). In this chapter we turn to subgeometries of the *ordinary* quaternions, that is, subgeometries of ordinary four-dimensional Euclidean geometry. These include three-dimensional versions of the non-Euclidean plane geometries: hyperbolic geometry and elliptic geometry.

A Spherical Model

Just as models for plane hyperbolic geometry are found within plane Euclidean geometry, so models for solid hyperbolic geometry can be constructed inside three-dimensional Euclidean geometry. The first model we consider is the analog of the disk model. In this model, hyperbolic space is the solid inside of a sphere. (See Figure 19.1.)

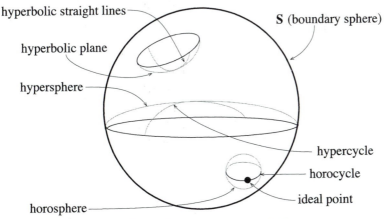

Figure 19.1 Three-dimensional hyperbolic space

In plane hyperbolic geometry, straight lines are the arcs of circles that are perpendicular to the boundary sphere. In solid hyperbolic geometry, we adopt the same definition. All the familiar phenomena associated with parallelism in hyperbolic geometry appear in solid

221

hyperbolic geometry: parallel lines, hyperparallel lines, horocycles, and hypercycles.

In addition to these one-dimensional geometric objects, there are surfaces of various kinds. In the first place, there are **planes**, defined as the portions of Euclidean spheres lying inside hyperbolic space and intersecting the boundary sphere **S** at right angles. Other Euclidean spheres also play a role in solid hyperbolic geometry. A Euclidean sphere entirely contained in hyperbolic space is called a (hyperbolic) **sphere**, a Euclidean sphere that is tangent to **S** is a **horosphere**, and a Euclidean sphere which meets **S** in an angle different from 90° is a **hypersphere**. All these are illustrated in Figure 19.1. Figure 19.2 is a close-up of a hyperbolic dodecahedron: a solid formed by the intersection of 12 planes in hyperbolic space.

Figure 19.2 A hyperbolic dodecahedron

A–In hyperbolic space, what different possibilities for parallelism (or incidence) arise between two planes?

The Half-Space Model

For a formal definition of solid hyperbolic geometry, we use a half-space instead of a sphere, because the formulae are simpler.

Definition *Let* **U** *be the set of ordinary quaternions of the form*

$$\mathbf{q} = t + x\mathbf{i} + y\mathbf{j}$$

where t, x, and y are real numbers and y > 0. **U** *is called* **upper half-space.** *Let* **M** *be the* **full Möbius group,** *the set of transformations of* **U** *defined by*

$$Tq = (aq + b)(cq + d)^{-1}$$

where a, b, c, and d are complex constants (of the form u + vi, u and v real) such that ad − bc = 1. The pair (**U**, **M**) *models* **three-dimensional hyperbolic geometry.**

Note that quaternion **i**, in this chapter, is treated as the same as the complex number *i*. Also, the half-space **U** is half of the *three-dimensional* space of quaternions of the form $t + x\mathbf{i} + y\mathbf{j}$ (no **k**). Of course, it must be proved that the Möbius transformations form a transformation group for **U**. (See Exercises 1 and 2.) This shouldn't be too surprising, however, since, with real coefficients, these transformations map the upper half plane into itself. What we have done is add one more coordinate to the points of the half plane (getting upper half-space) and to the coefficients of the transformations (getting the full Möbius group). This is summarized in the following table:

	hyperbolic plane geometry	hyperbolic solid geometry
points	$x + iy$, $y > 0$ (the upper half plane)	$t + \mathbf{i}\,x + \mathbf{j}\,y$, $y > 0$ (upper half-space)
group	Möbius transformations (real coefficients)	Möbius transformations (complex coefficients)

The new feature is the noncommutative nature of the quaternions.

B–Draw examples of parallel and hyperparallel hyperbolic planes in the half-space model.

We will point out only a few features of solid hyperbolic geometry, since much of the development of this geometry is a natural extension of plane hyperbolic geometry.

Ideal Elements

The Euclidean plane consisting of quaternions of the form $t + x\mathbf{i}$ (the complex plane) functions as the **plane at infinity.** The points in this plane (including the point at infinity) are called **ideal points.**

Planes and Lines

A **hyperbolic straight line** is a half circle or Euclidean straight line in **U** perpendicular to the plane at infinity. A **hyperbolic plane** is a Euclidean hemisphere or half plane perpendicular to the ideal plane. The intersection of a hyperbolic plane with the plane at infinity is called the **horizon** of the plane.

Parallelism

If two hyperbolic planes intersect, the intersection must be a hyperbolic line. (See Exercise 7.) Otherwise, two hyperbolic planes are either **parallel** (meaning that their horizons are tangent) or **hyperparallel**, in which case neither they nor their horizons intersect. Two hyperbolic lines are either **coplanar** (a self-explanatory term) or **skew**. Two coplanar lines are parallel, hyperparallel, or intersect.

Cycles and Spheres

A **cycle** is any Euclidean circle or straight line (or portion thereof) in **U** that is not perpendicular to the plane at infinity. As in the hyperbolic plane there are hyperbolic circles, horocycles, and hypercycles. Similarly, **spheres**, **horospheres**, and **hyperspheres** are portions of Euclidean spheres and planes that are not perpendicular to the plane at infinity.

Arc Length

The differential of arc length is

$$ds^2 = \frac{dt^2 + dx^2 + dy^2}{y^2}$$

C–Two horospheres intersect in what kind of curve(s)?

D–Using two known formulae—one for the area of a hyperbolic circle in hyperbolic plane geometry (see Chapter 10), and the other for the volume of a sphere in Euclidean geometry—guess a formula for the volume of a sphere in hyperbolic geometry.

Two-dimensional Subgeometries

A resident of solid hyperbolic space studying mathematics will naturally study the geometry of various two-dimensional hyperbolic surfaces. The first such surface geometry to be studied would, of course, be hyperbolic plane geometry, the natural plane subgeometry of solid hyperbolic space. This geometry is, of course, the plane hyperbolic geometry we have already studied. (See Exercise 19.) Also of importance to an inhabitant of hyperbolic space is the natural geometry of a sphere in solid hyperbolic space. This is the same as the Euclidean spherical geometry we studied in Chapter 11. (See Exercise 22.)

In addition, however, a hyperbolic mathematics student would study **horospherical geometry**, the natural geometry of a horosphere. In this geometry, a horocycle is the shortest path between two points, so horocycles play the role of straight lines. It is amazing, but true: horospherical geometry is the same as Euclidean geometry! (See Exercise 20.) Therefore, inhabitants of solid hyperbolic space would study Euclidean geometry as a *natural* geometry, whereas the poor inhabitants of a Euclidean world only study hyperbolic geometry unnaturally!

Volume

The differential for arc length leads to a differential for volume

$$dV = \frac{dt\ dx\ dy}{y^3}$$

As an example of the use of this formula, we will find the volume of a **horospherical cylinder**: a solid whose top and bottom are regions enclosed by circles on parallel horospheres and whose sides are formed by diameters of the horospheres connecting the two circles. (See Figure 19.3.)

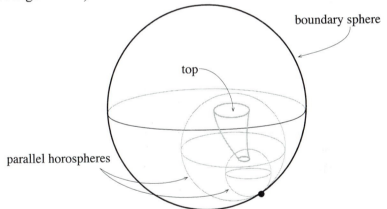

Figure 19.3 A horospherical cylinder

To calculate this volume, we will use the half-space model. Here the most useful horocycles have their ideal point at infinity, and hence, are Euclidean planes parallel to the *tx*-plane. This is depicted in Figure 19.4, in which y_1, y_2, and r are *Euclidean* measurements, but R_1, and R_2 are the *hyperbolic* radii of the top and bottom of the cylinder, and H its *hyperbolic* height. In this position, the horospherical cylinder is a Euclidean cylinder. Using cylindrical coordinates (see Exercise 14), we get

$$V = \int_{y_1}^{y_2} \int_0^{2\pi} \int_0^r \frac{r dr\ d\theta\ dy}{y^3} = \frac{\pi}{2} r^2 \left(\frac{1}{y_1^2} - \frac{1}{y_2^2} \right)$$

This gives V in Euclidean terms. However, using arc length integrals we find that $R_1 = r/y_1$ and $R_2 = r/y_2$; therefore,

$$V = \frac{1}{2}\left(\pi R_1^2 - \pi R_2^2\right)$$

Notice that the height H does not appear explicitly in this formula, suggesting that H has no effect on the volume. This is absurd. What is happening is that the three measurements R_1, R_2, and H are not independent. They are linked, that is, there is a formula connecting these three dimensions of the horospherical cylinder from which an alternative formula for V can be derived that does use H. (See Exercise 15.)

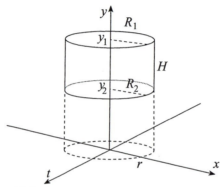

Figure 19.4 A horospherical cylinder in half-space

E–What other kinds of cylinders exist in hyperbolic space?
F–What kinds of cones exist in hyperbolic space?

Solid Elliptic Geometry

Recall that plane elliptic geometry arises "naturally" as spherical geometry, the natural geometry of the **2-sphere**, that is, the set of points $\mathbf{r} = x\mathbf{i} + y\mathbf{j} + z\mathbf{k}$ in three-dimensional Euclidean space such that

$$|\mathbf{r}| = (x^2 + y^2 + z^2)^{1/2} = 1$$

Although the 2-sphere is defined in three-dimensional Euclidean space, it is intrinsically two-dimensional (a surface), hence the term 2-sphere.

A generalization of this geometry to three dimensions is the "natural" geometry of the **3-sphere**.

Definition *The **3-sphere** $\mathbf{S}^3$ is the set of points $\mathbf{r} = t + x\mathbf{i} + y\mathbf{j} + z\mathbf{k}$ in four-dimensional Euclidean space such that*

$$|\mathbf{r}| = (t^2 + x^2 + y^2 + z^2)^{1/2} = 1$$

Let **O** *be the group of transformations of* S^3 *given by*

$$T\mathbf{q} = \mathbf{a}\,\mathbf{q}\,\mathbf{b}$$

where **a** *and* **b** *are* unit *quaternions.   The geometry* $(S^3, \mathbf{O})$ *models* **three-dimensional elliptic geometry.**

We do not pursue solid elliptic geometry any further but allow the reader this opportunity in Exercises 24–27.

Summary

The geometries discussed in this chapter can be displayed as follows:

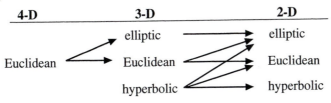

4-D	3-D	2-D
	elliptic	elliptic
Euclidean	Euclidean	Euclidean
	hyperbolic	hyperbolic

EXERCISES

Möbius Transformations

1. Prove that the transformation $T(\mathbf{q}) = (a\mathbf{q} + b)(c\mathbf{q} + d)^{-1}$ maps **U** onto itself.

Hint: Rationalize the denominator again!

2. Show that **M** (the full Möbius group) is a transformation group.

3. Express the following transformations of **U** as Möbius transformations (assuming that a and b are complex):

   a) $T\mathbf{q} = \mathbf{q} + b$
   b) $T\mathbf{q} = 1/\mathbf{q}$
   c) $T\mathbf{q} = a\,\mathbf{q}\,a$

4. Show that the transformation $T\mathbf{q} = a\,\mathbf{q}\,a$ (where a is complex) consists of a homothetic transformation plus a rotation of **U**.

Hint: Use the polar form of a.

5. Show that every Möbius transformation can be expressed as a sequence of operations of the three types in Exercise 3.

6. Prove that the Möbius transformations **M** of U are conformal.

Hint: Apply Exercise 5.

Theorems of Solid Hyperbolic Geometry

7. Prove that the intersection of two hyperbolic planes is either a hyperbolic line, an ideal point, or the empty set.

8. Prove that two points determine a line in solid hyperbolic geometry and that three points determine a plane.

9. Prove that, given a plane and a point not on it, there exists a unique perpendicular line from the point to the plane.

10. Prove that two nonintersecting planes have a unique common perpendicular plane.

11. Prove that two planes are perpendicular if, and only if, their horizons are perpendicular.

12. Give a formal definition of **spheres**, **horospheres**, and **hyperspheres** in solid hyperbolic geometry.

Measurement in Solid Hyperbolic Geometry

13. Derive the formula for the volume differential from the given arc length differential.

14. Justify these additional volume differential formulas:

   a) cylindrical coordinates:

$$dV = \frac{rdr\, d\theta\, dy}{y^3}$$

   b) spherical coordinates:

$$dV = \frac{d\rho\, d\theta\, d\phi}{\rho^3 \cos(\phi)^3}$$

15. Derive this formula for the volume of a horospherical cylinder:

$$V = \frac{\pi}{2} R_2^2 \left(e^{2H} - 1\right)$$

Hint: Integrate along the edge of the cylinder to get a formula connecting H with R_1 and R_2.

16. A **cone** (with an ideal vertex) is formed by rotating a right-angled triangle with one ideal vertex about the side opposite the nonzero acute angle. Draw this solid in both spherical and half-space models of hyperbolic solid geometry. Find a formula for its volume.

Hint: One side of the right triangle can be a Euclidean line.

17. An **equidistant cylinder** is a solid bounded on the side by a family of equidistant curves, all the same distance from a common axis, and at the top and bottom by two hyperbolic circles centered on this axis and contained in planes perpendicular to the axis. Draw this solid in both spherical and half-space models of hyperbolic solid geometry. Find a formula for its volume.

Hint: The axis can be a Euclidean line. Use spherical coordinates.

18. Show that a plane and a horosphere intersect in a circle. Find a formula for the area on the horosphere enclosed by this circle in terms of its hyperbolic radius.

Hint: In the half-space model, choose a horosphere that is a Euclidean plane.

Subgeometries

19. Show that the geometry of a plane in solid hyperbolic space is the same as hyperbolic plane geometry.

Hint: In the half-space model, choose a hyperbolic plane that is a Euclidean plane.

20. Show that horospherical geometry (with horocycles as the straight lines) is the same as Euclidean geometry.

Hint: In the half-space model, choose a horosphere that is a Euclidean plane.

21. Show that a horocycle is the shortest distance between two points on a horosphere.

22. In addition to horospherical geometry, inhabitants of solid hyperbolic space would study **spherical geometry** and **hyperspherical geometry**. Describe these geometries. What curves might play the role of straight lines? What are the incidence relations in these geometries?

23. Find a four-dimensional geometry with all three, three-dimensional geometries, elliptic, parabolic, and hyperbolic, among its subgeometries.

Solid Elliptic Geometry

24. Prove that **O** is a transformation group.

25. What are the straight lines on S^3? What are the planes?

Hint: In spherical geometry, straight lines are the great circles. What is a "great sphere" on the 3-sphere?

26. Following are four different types of transformation of the 3-sphere onto itself. Which types form subgroups of **O**? What is the overlap among these different types of transformation? Can you characterize what makes these transformations different geometrically (as opposed to algebraically)?

   (a) $Tq = \mathbf{a}\,\mathbf{q}$ where $|\mathbf{a}| = 1$
   (b) $Tq = \mathbf{q}\,\mathbf{b}$ where $|\mathbf{b}| = 1$
   (c) $Tq = \mathbf{a}\,\mathbf{q}\,\mathbf{a}$ where $|\mathbf{a}| = 1$
   (d) $Tq = \mathbf{a}\,\mathbf{q}\,\mathbf{a}^{-1}$ where $\mathbf{a} \neq 0$

Hint: Consider the action of each type of transformation on "coordinate" great spheres or great circles, that is, where one or two coordinates are zero.

27. Stereographic projection maps the 2-sphere onto the Euclidean plane with one extra point added (at ∞). Find an analogous

transformation from the 3-sphere onto three-dimensional Euclidean space. What point or points need to be added to three-dimensional Euclidean space in order that this transformation be one to one?

Product Geometries

28. Let S^2 stand for two-dimensional elliptic geometry. Using the disk model of S^2, we can regard the product geometry $S^2 \times E^1$ as a sub*set* of E^3 (three-dimensional Euclidean geometry). Sketch this subset. Describe some of the transformations of $S^2 \times E^1$. Is $S^2 \times E^1$ a sub*geometry* of E^3?

29. Let H^2 stand for two-dimensional hyperbolic geometry. Using the disk model of H^2, the product geometry $H^2 \times E^1$ can be regarded as a sub*set* of E^3 (three-dimensional Euclidean geometry). Sketch this subset. Describe some of the transformations of $H^2 \times E^1$. Is $H^2 \times E^1$ a sub*geometry* of E^3?

30. Find an example of a three-dimensional geometry besides S^3 that is finite.

PART V

DISCRETE GEOMETRY

Pablo Picasso: *Girl Before a Mirror* Oil on Canvas (1932) 64 x 51 1/2" The Museum of Modern Art, New York. Gift of Mrs. Simon Guggenheim

By the twentieth century, artists had largely lost interest in photographic realism. Pablo Picasso (among others) played with perspective: showing multiple views of objects and dissecting them into pieces of raw geometry. The woman in this picture is portrayed head on, in profile, and reflected in a mirror. Repeating discrete decorative patterns bring out clothing and background dynamically. Her body is partially formed from geometric figures such as circles and straight lines. Note the dramatic psychological contrast between the woman and her reflected image. Mirror reflection, both in art and in mathematics, is considered mysterious, problematical and a bit sinister.

THOMASINA ... Septimus, if there is an equation for a curve like a bell, there must be an equation for one like a bluebell, and if a bluebell, why not a rose? Do we believe nature is written in numbers?
SEPTIMUS We do.
THOMASINA Then why do your equations only describe the shapes of manufacture?
SEPTIMUS I do not know.
THOMASINA Armed thus, God could only make a cabinet.
SEPTIMUS He has mastery of equations which lead into infinities where we cannot follow.
THOMASINA What a faint heart! We must work outward from the middle of the maze.
–from *Arcadia* by Tom Stoppard (Faber and Faber, London, 1993)

The geometries presented in this book so far have been continuous geometries, "continuous", in this context, meaning "infinitely divisible." Euclidean geometry, for example, is a continuous geometry because Euclidean objects—triangles, circles, and the like—may be divided into smaller and smaller parts ad infinitum. A square of side k, for instance, is the union of four squares of side $k/2$, nine squares of side $k/3$, and so on. There is no smallest square, so this process can go on forever.

These geometries have actually been continuous in two ways: in their underlying set, and their transformation group. Thus, not only are Euclidean *objects* infinitely divisible, but so are Euclidean *transformations*. A rotation by the angle α, for instance, can be broken down into two rotations by the angle $\alpha/2$, or three rotations by the angle $\alpha/3$, and so on. There is no smallest rotation, so this process can go on forever.

In this part, we present **discrete** geometries: geometries where either the underlying set or its transformations are *not* infinitely divisible. Instead, there are **atoms**: objects or transformations that cannot be broken down into smaller parts. We will see both kinds of discrete geometry: geometries with discrete underlying sets, and geometries with discrete sets of transformations. We are able, however, to present only a small sample of the great variety of discrete geometries that are known.

Discrete geometries have become very important in recent decades because they lend themselves to computer implementation. Many computational problems, and the programs or algorithms that solve them, involve discrete geometry.

Chapter 20 introduces **matroids**, a class of discrete geometries with a discrete underlying set. Matroids are connected with graphs, another kind of discrete geometry, and with linear algebra. An important algorithm, the greedy algorithm, is characterized by this theory.

Chapter 21 introduces **reflections**, an intrinsically discrete transformation. Reflections are atomic; there is no such thing as a partial reflection. Reflections are therefore more fundamental than the transformations we have examined so far: rotations, translations, homothetic transformations, and inversion. The study of reflections gives a deeper understanding of the continuous plane geometries, elliptic, parabolic, and hyperbolic, as well as providing an example of a discrete transformation.

Chapters 22 and 23 build the theory of reflections into a general theory of **discrete symmetry**. The symmetry of a figure is measured by the group of transformations that leave the figure invariant. Symmetry groups are often discrete groups, and they have important applications to scientific fields such as solid-state physics and crystallography, and to artistic fields such as architecture, visual design, and music. Chapter 22 treats discrete Euclidean symmetry; Chapter 23 treats non-Euclidean symmetry.

Note: The finite projective geometries discussed in Chapter 16 are further examples of discrete geometries, as are some of Bachmann's geometries described in Chapter 25.

20 Graphs and Matroids

Graphs

In a geometry with a discrete underlying set, some aspect of the geometry is atomic, that is, cannot be subdivided. Here is an example:

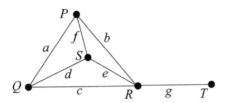

Figure 20.1 A graph

Imagine that Figure 20.1 depicts a whole geometry. It has only the five labeled points and seven labeled lines. Points, of course, are always atomic, but here the lines are also. These lines, or *edges* as they are usually called in this kind of geometry, do not contain an infinite number of points; they contain exactly 2 points apiece. This type of discrete geometry is called a graph. Here is the definition:

Definition *A **graph** consists of two finite sets V and E. The elements of V are called **points** or **vertices**. The elements of E are called **edges**. Edges are simply pairs of distinct points. If a = {P, Q} is an edge, then the vertices P and Q are the **endpoints** of a.*

For example, the graph in Figure 20.1 consists of the two sets:

$$V = \{P, Q, R, S, T\}$$

and

$$E = \{a, b, c, d, e, f, g\}$$

where

$$a = \{P, Q\}$$
$$b = \{P, R\}$$
$$c = \{Q, R\}$$
$$d = \{Q, S\}$$
$$e = \{R, S\}$$
$$f = \{P, S\}$$
$$g = \{R, T\}$$

235

In a graph, points are abstract objects; their position in a plane or in space does not matter. What does matter is that there are certain things called points, that certain pairs of points are connected by edges, and that other pairs are not. To study graphs, it is not necessary to draw pictures of them (like Figure 20.1), but pictures are extremely helpful.

Graphs have many applications: to transportation networks (street maps, airline routes, rail and subway systems), utility systems (water, natural gas, electricity, telecommunications, sewage), organizational charts (corporation hierarchies, family trees), electrical wiring diagrams, and floor plans. In studying these systems, the physical nature of the points and edges is unimportant, at least to the solution of some problems; all that matters is what points are connected to what other points.

In comparison with the other geometries in this book, it may seem that graphs are not geometries at all. But the questions asked about graphs—questions about paths and connectedness—are geometric questions and the whole subject has a distinctly geometric flavor.

The following definition gives the basic terminology used to pose questions about connectivity and routing.

Definition A ***path*** in a graph G is a nonempty, finite set of edges that can be arranged in a sequence $\{P_0, P_1\}$, $\{P_1, P_2\}$, $\{P_2, P_3\}$, . . ., $\{P_{m-1}, P_m\}$ where the vertices $P_1, \ldots P_{m-1}$ are distinct. The ***length*** of a path is the number m of edges, and the path is said to ***connect*** P_0, and P_m. If $P_0 = P_m$, then the path is a ***cycle***.

If all pairs of vertices in a graph are connected by paths, the graph is called ***connected***.

For example, the edges $\{Q, P\}$, $\{P, R\}$, $\{R, T\}$ in the graph of Figure 20.1, that is, the edges a, b, and g form a path from Q to T. The edges a, b, c form a cycle. Figure 20.1 is a connected graph because there is a path from every vertex to every other vertex.

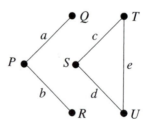

Figure 20.2 Another graph

If a graph is not connected, it, nonetheless, always consists of a finite number of connected pieces, called **components**, which are the maximal connected pieces of the graph. For example, the graph in Figure 20.2 has two components.

A-Is the point T on any cycle in Figure 20.1?   What about in Figure 20.2?

B- How many cycles are there in Figure 20.1?

Graphs can be much more interesting than the two examples we have given so far. Here is a fascinating one, the Peterson graph, about which a whole book has been written [H1]:

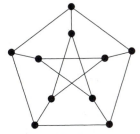

Figure 20.3 The Peterson graph

The Peterson graph is composed of two interlinked cycles of length five. It is best known for things that it isn't. For example, it isn't *planar*, that is, it cannot be drawn in a plane without 'extra' intersections: intersections of edges at points that are not vertices of the graph. It is also not *Hamiltonian*, meaning that there is no cycle that passes through every vertex. (Try to find one!)

Cycles

The oldest questions in graph theory concern cycles, and cycles remain a central object of study. Matroids generalize the set of cycles of a graph. To describe them, we need the following property:

Theorem *(Cycle elimination) If C_1 and C_2 are distinct cycles in a graph G and x is an edge in the intersection $C_1 \cap C_2$, then there is a cycle C_3 such that $C_3 \subseteq C_1 \cup C_2 - \{x\}$.*

Proof: This is called the cycle elimination property because it gives a condition under which there is a cycle C_3 which doesn't contain (i.e., eliminates) a given edge x of C_1. The condition is the existence of a second cycle C_2 containing x.

Our proof uses this lemma:

Lemma *If S is a nonempty set of edges from a graph G' such that every vertex of G' is on an even number of edges of S, then S contains a cycle.*

Proof: Take any edge $\{P, Q\}$ of S. Because the number of edges of S meeting at Q is even, there must be a second edge of S with the vertex Q. Let that edge be $\{Q, R\}$ and list this edge next. Similarly there must be another edge of S containing R. List that edge next. We are always able to continue this process, adding a new edge at every vertex we visit, until we complete a cycle by returning to a vertex already visited. ∎

Returning to the proof of the theorem, consider the graph G' whose edges are the edges in the symmetric difference set:

$$C_1 \oplus C_2 = (C_1 \cup C_2) - (C_1 \cap C_2) = (C_1 - C_2) \cup (C_2 - C_1)$$

One easily verifies that every vertex of G is on an even number of edges (possibly zero) of $C_1 \oplus C_2$ (see Exercise 5). Furthermore, $C_1 \oplus C_2$ is non-empty because C_1 and C_2 are distinct. By the lemma, $C_1 \oplus C_2$ contains a cycle. That cycle is contained in $C_1 \cup C_2 - \{x\}$. ∎

Cut Sets

Cycles connect. Cut sets disconnect. Here is the definition:

Definition *A set D of edges of a graph G is a **cut set** if the graph obtained by removing the edges of D from G has one more component than G, and D is a minimal set with this property.*

For example, $\{a, b, f\}$ is a cut set for the graph in Figure 20.1. That graph falls apart into two components, if these edges are removed. At the same time, all three edges are necessary to accomplish this. Removing only two of these edges does not disconnect the graph. Thus, $\{a, b\}$ is *not* a cut set. The edges $\{a, b, c, d\}$ are also *not* a cut set because a *subset* of these edges suffice to disconnect the graph.

C-Is $\{g\}$ a cut set for Figure 20.1?

D-Find a one-edge cut set in Figure 20.2. Is there a two-edge cut set?

Cut sets and cycles are completely different in appearance and purpose, yet they have many similar properties. For example, the elimination property holds for cut sets as well as cycles:

Theorem *(Cut set elimination) If D_1 and D_2 are distinct cut sets and x is an edge in the intersection $D_1 \cap D_2$, then there is a cut set D_3 such that $D_3 \subseteq D_1 \cup D_2 - \{x\}$.*

Our proof uses this lemma:

Lemma *Let V be a subset of the vertices of a graph G. Let **D**(V) be the set of all edges of G connecting a vertex in V with a vertex not in V. Every cut set is of the form **D**(V) for some subset V.*

Proof: Let D be a cut set of G. Because D is a minimal set of edges disconnecting G, the edges of D must be contained in a single connected component of G.   Thus, we can assume that G is connected.

Removing the edges of D disconnects G into several components. However, again because D is minimal, removing the edges of D must divide G into at most two components.   Let V be the vertices of one of these components.   Then V is a subset of the vertices of G and clearly $D = \boldsymbol{D}(V)$. ∎

For example, the cut set $\{a, b, f\}$ in the graph in Figure 20.1 is in fact $\boldsymbol{D}(\{P\})$, because it consists of all edges connecting P with the other vertices of this graph.

Turning to the proof of the elimination property, the cut sets D_1 and D_2 are associated, according to the lemma, with vertex sets V_1 and V_2 in the same component of G such that $D_1 = \boldsymbol{D}(V_1)$ and $D_2 = \boldsymbol{D}(V_2)$.

Consider the symmetric difference,

$$V = V_1 \oplus V_2 = (V_1 - V_2) \cup (V_2 - V_1)$$

Then

$$\boldsymbol{D}(V) \subseteq D_1 \cup D_2$$

and

$$x \notin \boldsymbol{D}(V)$$

These conclusions are depicted in Figure 20.4.   The graph G is arranged so that V_1 is the set of vertices to the left of a vertical divider and V_2 is the set of vertices above a horizontal divider. Therefore, D_1 consists of the edges that cross the vertical divider; D_2 is the set of edges that cross the horizontal divider. The set V consists

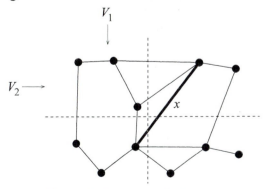

Figure 20.4 Cut set elimination

of the vertices in the upper right and lower left quadrants. Any edge in $D(V)$ must cross one of the dividers, hence is in either D_1 or D_2. On the other hand the edge x, crosses both dividers, hence cannot be in $D(V)$.

Finally, $D(V)$ disconnects the graph G but may not be a cut set (because it is not minimal). However, because it disconnects G, $D(V)$ contains a cut set which is contained in $D_1 \cup D_2 - \{x\}$. ∎

Definition and Examples of Matroids

Matroids model the common properties of cycles and cut sets. This makes them a very general kind of discrete geometry. Here is the definition.

*Definition A **matroid** M consists of an underlying set S and a set **C** of subsets of S. The members of **C** are called **circuits** and have the following properties:*
   (C1) *The empty set is not a circuit,*
   (C2) *No circuit is a proper subset of another circuit,*
   (C3) *(elimination property) If C_1 and C_2 are distinct circuits and x is an element of S such that $x \in C_1 \cap C_2$, then there is a circuit C_3 such that $C_3 \subseteq C_1 \cup C_2 - \{x\}$.*

The next theorem is no surprise.

*Theorem Let G be a graph with edge set S. Let **C** be the set of all cycles of G, and let **D** be the set of all cut sets of G. Then S is a matroid if either the set **C** or the set **D** is chosen as the set of circuits. The first is called the **cycle matroid** of G; the second is called the **cut set matroid of G**.*

Proof: We know that cut sets and cycles satisfy (C3). Proving (C1) and (C2) is fairly easy. (See Exercises 6 and 11.) ∎

Thus, every graph is associated with two matroids: a cycle matroid and a cut set matroid. A matroid that is the cycle matroid of a graph is called **graphic**; a matroid that is the cut set matroid of a graph is called **co-graphic**. These are very important types of matroids, but there are many others. Here are two examples:

*Definition   Let S be a matroid and let **C** be its set of circuits. If **C** is empty, then S is called a **free matroid**. If **C** consists of all subsets of S with k elements (k, a fixed integer), then S is called a **uniform matroid**.*

That free matroids and uniform matroids are actually matroids is left as an exercise (see Exercises 16 and 18). These two simple types are often used to test conjectures about matroids.

Independence

More kinds of matroids arise in other geometric contexts, for example, in the geometry of linear algebra. To introduce them we need to learn more about matroids.

*Definition A subset of the underlying set of a matroid is **independent** if it contains no circuit.*

In a graphic matroid, for example, an independent set is a set of edges that contains no cycle. Such a set of edges is called a **forest**. If it is connected, it is called a **tree**. The edges $\{a, e, g\}$ are a forest in Figure 20.1; the edges $\{b, e, g\}$ are a tree.

In a co-graphic matroid, an independent set is a set of edges that contains no cut set and, therefore, does *not* disconnect the graph. We call this a **co-forest**.

E-Find a forest in Figure 20.1 that has as many edges as possible.

F-Find a co-forest in Figure 20.1 that has as many edges as possible.

Here are some properties of independent sets.

Theorem Independent sets of a matroid G have these properties:
(I1) *The empty set is independent,*
(I2) *Any subset of an independent set is independent,*
(I3) *If I_1 and I_2 are independent sets and $|I_1| < |I_2|$, then there is an element $x \in I_2 - I_1$ such that $I_1 \cup \{x\}$ is independent.*

Proof: (I1) and (I2) are obvious. To prove (I3), consider all sets J with these three properties: (i) J is independent, (ii) J is contained in $I_1 \cup I_2$, and (iii) $|J| > |I_1|$. There is at least one such set, namely I_2. Among all of these sets, let J be one that contains as many elements of I_1 as is possible. We hope that this J contains all of I_1. If so, the theorem is proven.

Suppose, contrary to what we want to prove, that $I_1 - J$ is non-empty, and let $x \in I_1 - J$. By construction, $J - I_1$ is also non-empty. For every $y \in J - I_1$, let

$$J_y = J \cup \{x\} - \{y\}$$

Then J_y is contained in $I_1 \cup I_2$, and $|J_y| > |I_1|$. These are properties (ii) and (iii). However, J_y also contains one more element of I_1 than J does. Therefore, J_y cannot be independent [property (i)], because by assumption J contains the *maximum* number of points of I_1 among the sets satisfying all *three* properties (i), (ii), and (iii). Thus J_y contains a circuit. Choose a circuit in J_y and call it C_y. Note that y is *not* in C_y, but x is. (Otherwise, C_y is contained in J—but J is independent!)

Again, let $y \in J - I_1$. If $C_y \cap (J - I_1)$ is empty, then C_y is contained in I_1 which is impossible because I_1 is independent. Therefore, there is a element $z \in C_y \cap (J - I_1)$.

Now C_y and C_z are distinct circuits, because z is in one but not the other. Further, x is in both C_y and C_z. Therefore, by the circuit elimination property, there is a circuit C such that

$$C \subseteq (C_y \cup C_z) - \{x\} \subseteq J$$

This contradicts the independence of J. ∎

This proof is based on one in Oxley [H3]. It uses a trick, often found in matroid theory: Choose a set with a maximal property (the set J above) and draw conclusions based on this property.

Property (I3) says that elements can be added to an independent set, creating a larger independent set, as long as there is *some* independent set with more elements. If this is done to all independent sets at the same time, eventually they will all have the same number of elements. This number is the **rank** of the matroid, and the maximal independent sets are called **bases**. In the cycle matroid of a connected graph a basis is called a **spanning tree**. The rank of the cycle matroid of a graph is called its **cycle rank**; the rank of the cut set matroid of a graph is called its **co-cycle rank**.

For example, edges $\{b, e, d, g\}$ are a spanning tree in Figure 20.1. This graph, therefore, has cycle rank 4.

G-Find a spanning co-forest in Figure 20.1. What is the cut set rank of this graph?

H-Find the cycle rank and the cut set rank of the graph in Figure 20.2.

Cryptomorphic Matroid Theory

The language of independent sets and bases comes from linear algebra. Rank is also a linear algebra concept, although it is usually called **dimension**. The interesting fact is that these linear algebra concepts, in particular the idea of independence, completely characterize matroids. The following theorem explains how.

Theorem *Let S be a set and* **I** *be a collection of subsets of S satisfying properties* (I1), (I2), *and* (I3). *Let* **C** *be the collection of those subsets of S that are not in* **I**, *but minimally so, meaning that removing any one point from such a set produces a set that is in* **I**. *Then the sets* **C** *satisfy the circuit axioms* (C1), (C2), *and* (C3). *In the resulting matroid, the independent sets are exactly the sets* **I**.

Proof: Let's call the sets in **I** "independent", using quotes until we show that they are actually the independent sets of a matroid.

Likewise, we call the sets in **C** "circuits" until we prove that they satisfy the circuit axioms.

By definition a "circuit" is a set that becomes "independent" when just one element is removed. It follows that the empty set is not a "circuit" because you can't remove anything from it. This is (C1). Further, a "circuit" cannot be a proper subset of another "circuit" because all proper subsets of a "circuit" are "independent", but a "circuit", by definition, is not "independent". This is (C2).

To prove (C3), let $x \in C_1 \cap C_2$ where $C_1 \cap C_2$ are distinct "circuits". We want to prove that $C_1 \cup C_2 - \{x\}$ contains a "circuit". If not, then $C_1 \cup C_2 - \{x\}$ is "independent". We will find that this leads to a contradiction, by choosing a suitable set with a maximal property.

Let y be an element of $C_2 - C_1$. Note that x and y are different because x is in C_1 and y is not. Because C_2 is a "circuit", $C_2 - \{y\}$ is "independent". Let J be the largest "independent" subset of $C_1 \cup C_2$ containing $C_2 - \{y\}$. Then because C_1 is a "circuit", there is some element z of C_1 not in J. Thus,

$$|J| \leq |C_1 \cup C_2 - \{y, z\}| = |C_1 \cup C_2| - 2 < |C_1 \cup C_2 - \{x\}|$$

Both J and $C_1 \cup C_2 - \{x\}$ are "independent". Therefore, by (I3), J can be made into a larger "independent" set by adding a suitable element from $C_1 \cup C_2 - \{x\}$. This contradicts the assumption that J is the largest "independent" set containing $C_2 - \{y\}$. This proves (C3).

We have now proved that the "circuits" **C** are the circuits (no quotes) of a matroid on the set S. That the independent sets of that matroid are the "independent" sets **I** is proved by the following chain of reasoning. A set J is independent if and only if J contains no circuit. This is the definition of independent set. But J contains no circuit if, and only if, J contains no "circuit", since the two concepts are the same. This, in turn, means that J contains no set that is minimally not "independent", since this is the definition of "circuit", and this is true if, and only if, J is "independent" itself. Thus a set is independent if and only if it is "independent". ∎

A matroid, therefore, can be specified by giving its circuits, which must satisfy (C1)–(C3), or, alternatively, its independent sets, which must satisfy (I1)–(I3). The language of circuits and the language of independent sets are equivalent; they describe the same subject from different points of view. This is sometimes summarized by saying that matroid theory has different **cryptomorphic** forms: one using circuits and one using independence. Among mathematical subjects, matroid theory is particularly rich in cryptomorphic forms. Each of the concepts: rank, basis, spanning set, and hyperplane has associated properties that are the basis for a cryptomorphic version of matroid theory. (Some are explored in the exercises.)

Representable Matroids

We see that matroid theory is a geometric theory uniting the language and ideas of graph theory and linear algebra. So far we have seen concrete examples of matroids derived from graphs: cycle and cut set matroids. Examples of matroids can also arise, according to the previous theorem, using independent sets instead of circuits.

Let S be a finite set of vectors from any vector space. If we let **I** be the subsets of S that are independent according to linear algebra, then properties (I1), (I2) and (I3) follow immediately from theorems of linear algebra. Therefore S is a matroid. This kind of matroid is called a **representable** matroid.

Representable matroids are important because they can be studied using linear algebra. Great efforts have been made to discover exactly which matroids are representable and which are not.

As an example, let S consist of the vectors:

$$\mathbf{a} = (1, 0, -1, 0) \qquad \mathbf{b} = (0, -1, 1, 0) \qquad \mathbf{c} = (-1, 1, 0, 0)$$
$$\mathbf{d} = (1, 0, 0, 0) \qquad \mathbf{e} = (0, 1, 0, 0) \qquad \mathbf{f} = (0, 0, 1, 0)$$

and $\mathbf{g} = (0, 0, 0, 1)$. In this matroid, $\{\mathbf{d}, \mathbf{e}, \mathbf{f}, \mathbf{g}\}$ is clearly independent. On the other hand, $\{\mathbf{a}, \mathbf{b}, \mathbf{c}\}$ is dependent, but minimally so, so it is a circuit. The matroid S could be presented by arranging these vectors as the columns of a matrix as in Figure 20.5. It was by considering this kind of example, that Hassler Whitney was led to coin the term matroid (1935).

Careful comparison of the matroid S and the cycle matroid of the graph in Figure 20.1 will reveal that they are the same matroid. Thus it is possible for a matroid to be both graphic and representable. Notice that the independence of the tail of the graph in Figure 20.1 which consists of the edge g and the vertex T, is brought out by this representation of the matroid: The vector $\mathbf{g}$ is the only one with a nonzero entry in the last column.

$$\begin{pmatrix} 1 & 0 & -1 & 1 & 0 & 0 & 0 \\ 0 & -1 & 1 & 0 & 1 & 0 & 0 \\ -1 & 1 & 0 & 0 & 0 & 1 & 0 \\ 0 & 0 & 0 & 0 & 0 & 0 & 1 \end{pmatrix}$$

Figure 20.5

I-Find a representation for the cycle matroid of Figure 20.2.

Note that there are matroids that are not graphic and matroids that are not representable Some examples are given in the exercises.

The Greedy Algorithm

We now consider some combinatorial applications of matroids. First among these is an important graph theory problem: finding a minimum weight spanning tree. As an example, consider Figure 20.6. This is the Peterson graph with numerical weights attached to each edge.

Think of the vertices as locations that we genuinely wish to connect, nodes in a computer network, for example, and the weights as the cost of building the edges. The problem is to choose a set of edges connecting all the vertices of the graph with least cost. Thus we want the sum of the weights of the edges chosen to be minimal. A spanning *tree* connects all the vertices using the fewest edges because it has no cycles. Therefore, the problem is to find a minimum cost spanning tree.

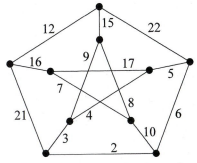

Figure 20.6 A weighted graph

The following procedure solves the problem:

Kruskal's Algorithm *Let G be a graph with weights assigned to each edge. To find a tree T of minimal weight spanning G:*
(a) Start with the empty set $T = \varnothing$.
(b) Among the edges currently not in T, pick one of least weight that does not create a cycle with the edges already in T. Add this edge to T.
(c) Repeat step (b) until every remaining edge creates a cycle with the edges already in T. Then T is a minimum weight spanning tree.

This is called a **greedy** algorithm because at each step the cheapest feasible edge is chosen. For the graph in Figure 20.6, edges are chosen in the order 2, 3, 4, 5, 7, 8, 9, 12, 15. These edges form the spanning tree shown in Figure 20.7. Note that the edge with

weight 6 is not chosen because it makes a cycle with previously
chosen edges. So does the edge with weight 10.

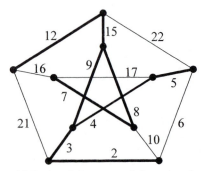

Figure 20.7 A minimum weight spanning tree

Clearly Kruskal's algorithm produces a spanning tree (or forest, if
the graph is not connected), but is it of minimal cost? We will prove
the algorithm is correct but in the more general context of matroids.
Let the elements of a matroid S be given weights. Instead of trees, S
has independent sets; instead of spanning trees, bases. The minimum
cost, spanning tree problem becomes the minimum cost, basis
problem. Thus, the algorithm reads:

Matroid Greedy Algorithm *Let S be a matroid with weights assigned
to each element. To find a basis B for S of minimum total weigh,*
    (a) Let $B = \varnothing$.
    *(b) Among the elements x of S not in B, pick the one of least
weight such that $B \cup \{x\}$ is independent. Add this edge to B.*
    *(c) Repeat step (b) until every edge not in B creates a circuit
with the edges already in B. Then B is a minimum weight basis.*

Proof: Let $B = \{x_1, x_2, \ldots, x_r\}$ be the set chosen by this algorithm,
where r is the rank of the matroid S. Then B is clearly a basis for S,
but is it minimal? Let $F = \{y_1, y_2, \ldots, y_r\}$ be another basis, and
suppose that the elements of both B and F are listed in order of
increasing weight. Let ω be the weighting function. We will show
that $\omega(x_i) \leq \omega(y_i)$ for every i, proving that B is minimal.

Suppose, contrary to what we want to prove, that $\omega(x_i) > \omega(y_i)$ for
some i. Let i be the first such i, and set $I_1 = \{x_1, x_2, \ldots, x_{i-1}\}$ and $I_2 =
\{y_1, y_2, \ldots, y_i\}$. Then $|I_1| < |I_2|$, so by (13), there is an element y_k in I_2
$- I_1$ such that $I_1 \cup \{y_k\}$ is independent. Now

$$\omega(x_i) > \omega(y_i) \geq \omega(y_k)$$

Therefore, when the greedy algorithm choose the ith element of B, y_k
should have been chosen, not x_i. This is the desired contradiction. ∎

Why stop with matroids? Given a set S whose elements have weights ω and any collection $\mathbf{I}$ of subsets of S, we can ask: How can we find a maximal subset from $\mathbf{I}$ with the least total weight? The greedy algorithm in this case reads as follows:

General Greedy Algorithm *Let S be a set with weights assigned to each element, and let $\mathbf{I}$ be any collection of subsets of S. To find a maximal element of $\mathbf{I}$ of minimum total weigh,*
  (a) Let $B = \varnothing$.
  (b) Among the elements x of S not in B, pick the one of least weight such that $B \cup \{x\}$ is in $\mathbf{I}$. Add this edge to B.
  (c) Repeat step (b) until every remaining edge creates a set not in $\mathbf{I}$ when added to B. Then stop.

However, this algorithm only works when S is a matroid!

In the first place, we must assume that $\mathbf{I}$ satisfies (I1) and (I2), the first two axioms of independence, or else it makes no sense to build a maximal subset from $\mathbf{I}$, element by element, starting with the empty set. The following theorem then proves that S must be a matroid:

Theorem *Let S be a set. Let $\mathbf{I}$ be a collection of subsets of S satisfying (I1) and (I2). Then if the general greedy algorithm produces a maximal set from $\mathbf{I}$ with minimal weight for all choices of weights, then $\mathbf{I}$ satisfies (I3).*

Proof: The proof again is by contradiction. Let I_1 and I_2 be sets from $\mathbf{I}$ such that $|I_1| < |I_2|$ but suppose that there is no element x of I_2 such that $I_1 \cup \{x\}$ is in $\mathbf{I}$. Under these circumstances (see Exercise 38),

$$0 < |I_1 - I_2| < |I_2 - I_1| < |S| \qquad (*)$$

Now choose weights according to this formula:

$$\omega(x) = \begin{cases} -|S| & \text{if } x \in I_1 \cap I_2 \\ -|I_1 - I_2| & \text{if } x \in I_1 - I_2 \\ -|I_2 - I_1| & \text{if } x \in I_2 - I_1 \\ 0 & \text{otherwise} \end{cases}$$

Then the greedy algorithm will first pick all the elements of $I_1 \cap I_2$, and then all the elements of $I_1 - I_2$. At this point all of I_1 has been chosen. No element of $I_2 - I_1$ can be chosen from now on because, by assumption, they all produce a set not in $\mathbf{I}$ when added to I_1. If B is the set eventually chosen by the greedy algorithm, then B consists of I_1 plus possibly further elements of weight 0. The total weight of B is

$$\omega(B) = -|S||I_1 \cap I_2| - |I_1 - I_2|^2$$

On the other hand, I_2 is contained in some maximal element of **I**, say I_3, whose total weight is *less* than (or equal to) the total weight in I_2 (because the weights are negative or zero). Hence

$$\omega(I_3) \le \omega(I_2) = -|S||I_1 \cap I_2| - |I_2 - I_1|^2 < \omega(B)$$

where the last inequality follows from (*). This contradicts the assumption that the general greedy algorithm produces a set of least weight. ∎

Thus matroids embody the idea of the greedy algorithm in an essential way. Let (G) be the statement that "the greedy algorithm finds a maximal element from **I** for all possible weights." Then given (I1) and (I2), (G) is equivalent to (I3). Thus the greedy algorithm furnishes another cryptomorphic description of matroids.

Matching and Transversals

Our second application of matroids is another famous graph theory problem. Suppose the vertices of a graph G divide into *two* sets, S and T and all the edges connect a vertex of S with one of T. This kind of graph is called **bipartite**. An example is in Figure 20.8.

T:

S:

Figure 20.8 A bipartite graph G links two sets

Choosing a set of edges with no common endpoints sets up a one-to-one correspondence between a subset of S and a subset of T. This is called a **matching**. An example is given in Figure 20.9. Here all four points of T are matched with S creating a **complete matching**.

T:

S:

Figure 20.9 A complete matching on G

These terms are formally defined as follows:

Definition   *A graph G is **bipartite** if the vertices of G can be partitioned into two sets S and T and all edges of G have one vertex in S and one it T. We say that the sets S and T are **linked** by G.*

*A **matching** is a set M of edges in G that have no endpoints in common. If, as we shall always assume, $|S| \geq |T|$, then the matching is **complete** if every vertex of T lies on an edge of M.*

Finding a complete matching is called the **marriage problem** because S and T can be interpreted as sets of women and men, while the edges represent potential marriages (presumably acceptable to both partners). A complete matching sets up the largest possible number of simultaneous, compatible marriages. There are many other scenarios. For example, let S be a group of people and T a set of tasks. Edges link each person with the tasks they are able to do. A matching is an assignment of people to appropriate tasks. A complete matching accomplishes as many tasks simultaneously as possible. Or S may be a group of people who are on several committees while T is the set of committees. Edges join each committee with its members. A complete matching appoints a council with one representative from each committee. Many such scheduling and arrangement questions lead to complete matchings.

A last application may seem frivolous but there is a serious idea behind it. It begins with a game board, like the one shown in Figure 20.10. The problem is to place dominos on this board, as many as possible. A domino covers a pair of adjacent black and white squares. This problem can be interpreted as finding a complete matching on the bipartite graph formed by separating the sets of black and white squares and drawing an edge between contiguous pairs of squares as shown in Figure 20.10. The serious idea behind this is to interpret the dominos as diatomic molecules (common table salt, NaCl, for example) and the game board as a crystalline lattice into which the molecules must fit. This problem has many generalizations.

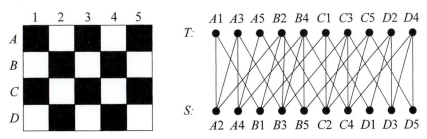

Figure 20.10 A game board and associated bipartite graph

Matroids underlie all these matching problems. To see how requires more terminology:

Definition *Let G be a bipartite graph linking two sets S and T, and let M be a matching on G. The set S_M of those vertices of S that lie on an edge of M is called a **partial transversal**. If $|S_M| = |T|$, that is, the matching is complete, then S_M is simply a **transversal**.*

For example, in Figure 20.11 the sets $\{a, b\}$ and $\{c, d\}$ are partial transversals. The set $\{a, b, d, e\}$ is a transversal. The importance of these ideas is that the partial transversals are the independent sets of a matroid, according to the following theorem.

Theorem *Let G be a bipartite graph linking two sets S and T. Let* **I** *be the collection of all partial transversals of S. Then* **I** *satisfies* (I1), (I2), *and* (I3) *so that S a matroid.*

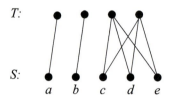

Figure 20.11 Another bipartite graph

Any matroid that arises in this way is called a **transversal matroid**. As an example, observe that partial transversals in Figure 20.11 are the same as the independent sets of the cycle matroid in Figure 20.2. Thus, for example, $\{c, d, e\}$ is a cycle in Figure 20.2, hence is a circuit in the cycle matroid and is also *not* a partial transversal in Figure 20.11 but becomes a partial transversal if any single element is removed.

Proof: Properties (I1) and (I2) are left as an exercise (see Exercise 41). To prove (I3), let I_1 and I_2 be partial transversals, and suppose that $|I_1| < |I_2|$. Since I_1 and I_2 are partial transversals there are matchings M_1 and M_2 that map I_1 and I_2 into subsets of T.

Figure 20.12 gives a simple example. The set I_1 is $\{a, b, c\}$ and I_2 is the set $\{c, d, e, f\}$. The edges of the two matchings have been drawn with different slopes: the edges of M_1 with positive slope, the edges of M_2 with negative slope.

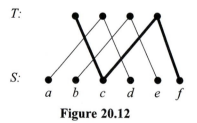

Figure 20.12

Let $M = M_1 \oplus M_2$, that is, let M consist of all the edges in M_1 or M_2 but not the other. It is easy to show that every vertex lies on 0, 1, or 2 edges of M. From this it follows (see Exercise 4) that the components of M are either cycles or paths. In both cycles and paths, M_1 edges and M_2 edges alternate, so the cycles always have an equal number of M_1 edges and M_2 edges. Because M_2 has more edges than M_1, there must therefore be at least one path P in M with an odd number of edges so that it can have one more M_2 edge than it has M_1 edges.

Take the matching M_1 and replace those edges that are in P with the M_2 edges in P. Call this new set of edges M_3. It is easy to show that M_3 is a matching and that it matches the set I_1 plus one more point (from I_2) with a subset of T. ∎

This proof is best understood with the help of a diagram. Figure 20.12 is fairly typical. Here *all* edges are in M because M_1 and M_2 are disjoint. The three components of M are all paths; there are no cycles. Only one path is of odd length; it shown with thick edges. The new matching M_3 consists of four edges, two each from M_1 and M_2. It matches the set $\{a, b, c, f\}$.

This theorem introduces a new kind of matroid, transversal matroids, and shows that finding a complete matching or a transversal is the same as finding a basis in those matroids.

The Marriage Theorem

When do complete matchings exist? The marriage theorem gives a necessary and sufficient condition.   To explain it, we need some more notation.

Definition *Let G be a bipartite graph linking two sets S and T. Let U be a subset of T. The set of all vertices in S lying on edges ending in U, notated G(U), is called the **image** of U under G. In particular, if U consists of only one vertex y of T, G({y}) is the set of all vertices in S whose other endpoint is y.*

An example is given in Figure 20.13. Note that $G(\{x\}) = \{a\}$ and $G(\{x, w\}) = \{a, c, e\}$.

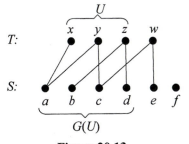

Figure 20.13

L-Does S in Figure 20.13 have a transversal?

M-What is the rank of the set $G(U)$ in the transversal matroid of Figure 20.12?

If G has a complete matching, then $G(T)$ has the same number of elements as T. Even more is true. If G has a complete matching, then $G(U)$ has at least as many elements as U for all subsets U of T. Conversely, it turns out that if $|G(U)| \geq |U|$ for all subsets U of T, then G has a complete matching. This is the marriage theorem.

We will deduce the marriage theorem from a more general result. Suppose that S is not only the set of vertices in the bipartite graph G, and therefore a transversal matroid, but is also a matroid in its own right with separately defined independent sets. The general question is whether S has a transversal that is an independent set in this other matroid. This is answered by Rado's theorem:

Rado's Theorem *Let G be a bipartite graph linking the two sets S and T. In addition, let S be a matroid. Then S has a transversal that is an independent set of S if, and only if,*

$$r(G(U)) \geq |U|$$

for all subsets U of T.

The function r in Rado's Theorem is the **rank** function of the matroid S. For any subset A of S, $r(A)$ is the size of the largest independent subset of A. To prove Rado's Theorem, we need the following property of r.

Lemma *Let A and B be any two subsets of a matroid S. Then*

$$r(A) + r(B) \geq r(A \cup B) + r(A \cap B)$$

Proof: Let I be a complete independent subset of $A \cap B$. Then I is also independent in $A \cup B$. Therefore, I can be enlarged [according to (I3)] to a complete independent subset J of $A \cup B$.

Because $J \cup A$ is an independent subset of A, $|J \cup A| \leq r(A)$. Similarly, $|J \cup B| \leq r(B)$. Therefore

$$r(A) + r(B) \geq |(J \cap A)| + |(J \cap B)|$$
$$= |(J \cap A) \cup (J \cap B)| + |(J \cap A) \cap (J \cap B)|$$
$$= |J \cap (A \cup B)| + |J \cap (A \cap B)|$$
$$= |I| + |J| = r(A \cup B) + r(A \cap B)$$

This is the desired conclusion. ■

Proof of Rado's Theorem: If the set $G(\{x\})$ contains exactly one element for every x in T, then the theorem is evidently true. Otherwise, some subset $G(\{x\})$ contains two or more elements. We will show then that some element of $G(\{x\})$ can be removed from S (and the corresponding edges removed from G) in such a way that the hypothesis of the theorem remains true. Thus the members of S can be removed, one by one, until all sets $G(\{x\})$ have exactly one point. Then the conclusion of the theorem is obvious.

To complete the proof, suppose that there is an element x in T such that $G(\{x\})$ contains more than one element. We will prove that $G(\{x\})$ contains at most *one* element a whose removal invalidates the hypothesis of the theorem.

Proceeding by contradiction, let us suppose that $G(\{x\})$ contains *two* elements, a and b, and that removing either a or b invalidates the hypothesis of the theorem. This means that there exist subsets U and V of T, both containing x, such that

$$r(G(U) - \{a\}) < |U|$$

and

$$r(G(V) - \{b\}) < |V|$$

Since these are whole numbers, this can also be expressed by writing

$$r(G(U) - \{a\}) + 1 \le |U|$$

and

$$r(G(V) - \{b\}) + 1 \le |V|$$

Combining these inequalities,

$$|U| + |V| \ge r(G(U) - \{a\}) + r(G(V) - \{b\}) + 2$$

$$\ge r(G(U) \cup G(V)) + r(G(U) \cap G(V) - \{a, b\}) + 2$$

where the last inequality follows from the lemma. Simple properties of bipartite graphs and rank (see Exercise 42) now imply

$$|U| + |V| \ge r(G(U \cup V)) + r(G(U \cap V - \{x\})) + 2 \qquad (**)$$

and using the hypothesis of the theorem

$$|U| + |V| \ge |U \cup V| + |U \cap V - \{x\}| + 2$$

$$= |U \cup V| + |U \cap V| + 1$$

This contradiction proves the theorem. ∎

From Rado's Theorem, we deduce:

Corollary *(The marriage theorem) Let G be a bipartite graph linking the sets S and T. Then G has a complete matching if, and only if,*

$$|G(U)| \ge |U|$$

for all subsets U of T.

Proof: This follows from Rado's theorem by letting S be a free matroid. Because free matroids have no circuits, every subset A is independent. Hence $r(A) = |A|$ for every subset of S. ∎

Summary

Matroids and graphs are just two among many different types of discrete geometries.   Graphs are used to answer questions about connections among objects. Matroids generalize graph theory and linear algebra. Matroid theory has a variety of cryptomorphic forms. These give matroid theory great flexibility and applicability. Matroids are particularly applicable to extremal problems, including the problem of finding minimal spanning trees and finding complete matchings.

EXERCISES

Graphs

1. Is there a graph with exactly two cycles?   Exactly three? Exactly n, where n is an integer?   What if the graph must be connected?

2. Let K_n be the **complete graph** on n vertices, that is, K_n has n vertices and all possible edges between these vertices. How many edges are there in K_n? How many paths?

3. Let $K_{n,m}$ be the **complete bipartite graph** on $n + m$ vertices, that is, the vertices of $K_{n,m}$ consist of two separate sets of n and m vertices and the edges of $K_{n,m}$ include all edges connecting a vertex of one set with a vertex in the other set.   How many edges are there in $K_{n,m}$? How many paths?

4. Let G be a graph such that every vertex is on 0, 1, or 2 edges. Prove that the components of G are paths, cycles, or isolated points.

Cycles

5. Let C_1 and C_2 be two cycles in a graph G. Prove that $C_1 \oplus C_2$ has the property that every vertex of G is on an even number of edges of $C_1 \oplus C_2$. Show by examples that this is not true for $C_1 \cup C_2$ or for $C_1 \cap C_2$.

6. Prove that the cycles of a graph satisfy the circuit axioms (C1) and (C2).

7. Prove this strengthened lemma on cycles: A nonempty set S of edges from a graph G is a union of disjoint cycles if, and only if, every vertex of G is on an even number of edges of S.

8. Prove the *strong cycle elimination property*: Let C_1 and C_2 be distinct cycles in a graph G. Suppose that x is in $C_1 \cap C_2$, and y is in $C_1 - C_2$. Then there is a circuit C_3 such that $y \in C_3 \subseteq C_1 \cup C_2 - \{x\}$.

9. How many cycles are there in the complete graph K_n?

10. How many cycles are there in the complete bipartite graph $K_{n,m}$?

Cut Sets

11. Prove that the cut sets of a graph satisfy the circuit axioms (C1) and (C2).

12. Let D_1 and D_2 be two cut sets in a graph G. Is $D_1 \oplus D_2$ always a union of cut sets? What about $D_1 \cup D_2$ and $D_1 \cap D_2$?

13. Prove this strengthened lemma on cut sets: A nonempty set S of edges from a graph G is a union of disjoint cut sets if and only if S is of the form $\mathbf{D}(V)$ for some set V of vertices.

14. Prove the *strong cut set elimination property*: Let D_1 and D_2 be distinct cut sets in a graph G. Suppose that x is in $D_1 \cap D_2$, and y is in $D_1 - D_2$. Then there is a circuit D_3 such that $y \in D_3 \subseteq D_1 \cup D_2 - \{x\}$.

15. Prove that a cycle intersects a cut set in an even number of edges.

Examples of Matroids

16. Let F_n be the free matroid with n elements. Show that F_n is a matroid by verifying the circuit axioms.

17. What are the independent sets in F_n? What are the bases? What is the rank of F_n?

18. Let $U_{n,m}$ be the uniform matroid with n elements whose circuits are all subsets of $U_{n,m}$ containing exactly $m + 1$ elements (so that $m < n$). Show that $U_{n,m}$ is a matroid by verifying the circuit axioms.

19. What are the independent sets in $U_{n,m}$? What are the bases? What is the rank of $U_{n,m}$?

20. Find two different graphs with the same cycle matroid.

Hint: Consider graphs whose cycle matroid is a free matroid.

21. Show that $U_{4,2}$ is not a graphic matroid.

22. Show that these graphs have representable cycle matroids:

    (a) K_3 (b) K_4

    (c) C_4 (a cycle of length 4)   (d) C_n

23. Show that these matroids are representable:

    (a) $U_{n,1}$ (b) $U_{n,2}$

Note: It follows from Exercises 21 and 23 that $U_{2,4}$ is representable but not graphic.

Cryptomorphism

24. Let S be a matroid and let **B** be the set of bases of S. Prove that **B** has the following properties:

(B1) **B** is not empty,

(B2) If B_1 and B_2 are bases and $x \in B_1 - B_2$, then there is a y in $B_2 - B_1$ such that $(B_1 - \{x\}) \cup \{y\}$ is a basis.

Note: (B2) is called the **basis exchange axiom***.*

25. Let a set S be given together with a collection of subsets **B** satisfying properties (B1) and (B2). Let **I** be the set of all subsets A of S such that $A \subseteq B$ for some element B in **B**. Prove that **I** satisfies the independence axioms (I1), (I2), and (I3) making S a matroid. Show that the bases of this matroid are exactly the sets in **B**.

26. Let S be a matroid and for every subset A of S define the rank of A, notated $r(A)$, to be the size of the largest independent subset of A. Prove that the function r has these properties:

(R1) For all subsets A, $0 \leq r(A) \leq |A|$,

(R2) If $A \subseteq B$, then $r(A) \leq r(B)$,

(R3) For all subsets A and B,

$$r(A) + r(B) \geq r(A \cup B) + r(A \cap B)$$

27. Let a set S be given together with a function r that assigns a real number to every subset of S and that has the properties (R1), (R2), and (R3). Let **I** be the set of all subsets A of S such that $r(A) = |A|$. Prove that **I** satisfies the independence axioms (I1), (I2), and (I3) making S a matroid. Show that the rank function in that matroid is the given function r.

28. Let S be a matroid and let **S** be the set of all spanning subsets of S. (A set A is **spanning** if every element x of S is either in A or is in a circuit consisting of x plus some elements of A.) Prove that **S** has these properties:

(S1) **S** is not empty,

(S2) If A_1 is in **S**, and $A_2 \supseteq A_1$, then A_2 is in **S**,

(S3) If A_1 and A_2 are in **S**, and $|A_2| > |A_1|$, then there is an element x in A_1 such that $A_1 - \{x\}$ is in **S**.

29. Let a set S be given together with a collection of subsets **S** satisfying properties (S1), (S2), and (S3). Prove that the set **B** of minimal elements of **S** satisfy the basis axioms (B1) and (B2) so that, according to Exercise 24, S is a matroid. Show that the spanning sets of this matroid are exactly the sets in **S**.

Duality

30. Let S be a matroid. Let $\mathbf{I}^*$ be the collection of subsets of S formed by taking the complements of the spanning subsets of S (see Exercise 28). Prove that the sets in $\mathbf{I}^*$ satisfy the independence axioms (I1), (I2), and (I3). Thus the collection $\mathbf{I}^*$ makes S into another matroid, the **dual matroid** of S. The sets in $\mathbf{I}^*$ are called **co-independent**.

31. Describe the dual matroids of free and uniform matroids.

32. Let S be the cycle matroid of a graph G. Prove that the dual matroid S^* is the cut set matroid of G.

33. Let S be a matroid. Let S^* be the dual matroid. Prove that the rank of S^* and the rank of S add up the number of elements in S.

34. Let S be a matroid. Prove that the bases of S^* are the complements of the bases of S.

35. Let S be a matroid. Prove that the dual matroid of the dual matroid of S is the same as S, that is, $(S^*)^* = S$.

The Greedy Algorithm

36. Use the matroid greedy algorithm to find minimum and maximum weight spanning trees for the graph in Figure 20.14.

37. Use the greedy algorithm to find minimum and maximum weight spanning co-forests for the graph in Figure 20.14.

38. Prove equation (*) on page 249.

39. Let S be a matroid with weights assigned to each element and let I be an independent set in S. Develop an algorithm that finds the minimum weight basis for S containing I. Prove that this algorithm is correct.

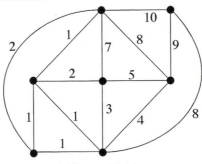

Figure 20.14

Transversal Matroids

40. Show that free matroids and uniform matroids are transversal matroids.

41. Prove that the partial transversals of a bipartite graph satisfy the independence axioms (I1) and (I2).

42. Complete the proof of Rado's theorem by proving equation (**).

43. Find complete matchings in these bipartite graphs:

(a)   (b)

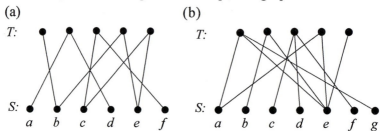

44. Let S be a set of jobs and let T be a set of workers. Let G be the bipartite graph linking S and T containing an edge $\{x, w\}$ if job x can be completed by worker w. Let every job x in S be given a priority $\rho(x)$, a real number with higher values denoting higher priority. A basis $B = \{x_1, x_2, \ldots, x_r\}$ for the transversal matroid S is called a **job assignment**. It assigns jobs so that workers are given jobs that they can complete simultaneously. In such an assignment, we always assume that the jobs are written in order of decreasing priority.

Find an algorithm that outputs a job assignment $B = \{x_1, x_2, \ldots, x_r\}$ that is optimal in the sense that if $C = \{y_1, y_2, \ldots, y_r\}$ is any other job assignment, then $\rho(x_i) \geq \rho(y_i)$ for all i.

45. Let S be a set of jobs to be performed by a single person. Let each job x have a deadline $D(x)$. A **job assignment** is a set $A = \{x_1, x_2, \ldots, x_r\}$ that can be completed by deadline. We always assume that the jobs in a job assignment are listed in order of increasing deadline. Prove that the set I of all job assignments satisfies the independence axioms (I1), (I2) and (I3) making S a matroid. Show that this is a transversal matroid.

46. In the situation described in Exercise 40, suppose that each job is also assigned a penalty $P(x)$ that is paid if x is not completed by its deadline. Find a greedy algorithm that finds the job assignment that minimizes the sum of the penalties paid for late jobs.

21 REFLECTIONS

Reflection in a straight line (also called **mirror reflection**) is a fundamental geometric transformation (see Figure 21.1). It is essentially discrete, in the sense that it cannot be factored into simpler geometric transformations. In this chapter, we will see that reflections can be used to build all other geometric transformations.

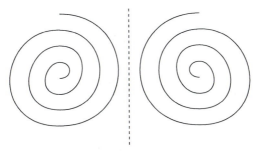

Figure 21.1 Euclidean mirror reflection

Algebraically, the simplest mirror reflection is complex conjugation, $Tz = \bar{z}$. This formula, however, only describes reflection across the x-axis. We need a formula for reflection about a line through the origin making an arbitrary angle θ with the x-axis. To find such a formula, we first rotate the given line to the x-axis, then apply conjugation, then rotate back. This is illustrated in Figure 21.2, where the flag A is reflected in the line I in three stages. At B, the line I has been rotated to the x-axis accompanied by the flag; at C, the rotated flag has been reflected in the x-axis; and at D, the x-axis has been rotated back to I.

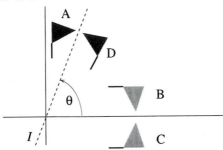

Figure 21.2 Reflection in a line through the origin

The formula that results is

$$Tz = e^{i\theta}\left(\overline{e^{-i\theta}z}\right) = e^{2i\theta}\overline{z}$$

Note: If the line doesn't pass through the origin, a still more complicated formula is needed.

A–What is the formula for reflection in the *y*-axis?

Reflection is atomic. This means that it is impossible to possible to combine other geometric transformations and get a reflection. One practical consequence of this is that it is impossible to transform one of the two spirals in Figure 21.1 into the other *without* using a reflection. Notice that one spiral spirals clockwise, while the other spirals counter clockwise. Rotations and translations always preserve this direction of motion, but reflections reverse it. The difference between the two curves is sometimes described by saying that one of them (it doesn't matter which) is **right handed** while the other is **left handed**.

This raises the question: Should mirror-symmetric figures (such as the spirals in Figure 21.1) be considered congruent in Euclidean geometry? Since reflections preserve distances, a given figure and its reflection have identical **metric** properties (perimeter, arc length, and area, for example). This makes a case for the inclusion of reflections among the group of Euclidean motions. However, reflections do alter the **handedness** of figures. Therefore, if we decide that handedness should be considered a Euclidean geometric property, then reflections must be excluded from the group of Euclidean motions (because they destroy handedness).

Thus, there are actually two groups of motions associated with Euclidean geometry. One is the group **E** (introduced in Chapter 4) consisting of rotations and translations. The group **E** contains no reflections and is, therefore, called the **special Euclidean group**. There is also a larger group defined as follows:

Definition *The **full Euclidean group** E^+ is the group of motions of the Euclidean plane consisting of the transformations of the special Euclidean group **E** (rotations and translations), plus reflections, and all transformations that are combinations of reflections, rotations, and translations.*

Introducing reflections relatively late in this book reflects the author's personal inclination to consider that handedness *is* a Euclidean property and therefore, reflections should not be considered as Euclidean congruence motions. However, in this chapter and the next two, we work with the full transformation groups of Euclidean and non-Euclidean geometries. Paradoxically, handedness now begins to play a role through the introduction of transformations that destroy it.

B–Let us say that a figure has **handedness** if it is *not* congruent to its reflection. Not all figures have handedness. Find examples of triangles with and without handedness. Find other plane figures with and without handedness. Find some three-dimensional figures with and without handedness.

C–Figure 21.3 includes a flag. Draw the reflection of this flag in each of the lines *I, II, III,* and *IV.* Call the original flag **right handed**. Then the flags you have drawn are all **left handed**. Describe in words the difference between the two types of flags.

D–Draw the flag in Figure 21.3 doubly reflected in the *two* lines *II* and *III* (*II,* then *III*). What kind of transformation results? What happens if the order of the two reflections is reversed?

E–Draw the flag in Figure 21.3 doubly reflected in the *two* lines *I* and *II* (*I,* then *II*). What kind of transformation results? What happens if the order of the reflections is reversed?

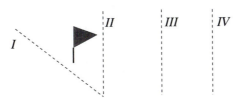

Figure 21.3 Practice with reflections

Combining Two Reflections

To appreciate the value of reflections requires understanding what happens when they are combined. In the first place we have the following theorem:

Theorem 1 *The composition (or **product**) of reflections in two lines is either a rotation or a translation. If the lines intersect, the product is a rotation whose rotation angle is twice the angle between the two lines. If the lines are parallel, the product is a translation in the direction perpendicular to the two lines and the translation distance is twice the distance between the lines.*

Proof: Consider two intersecting lines *I* and *II.* By applying a translation, if necessary, we can assume that they intersect at the origin (as in Figure 21.4).

Let M_I denote reflection in the line *I.* Then as worked out above, M_I is given by the formula $M_I(z) = e^{\angle I\theta}\bar{z}$. Similarly, reflection in line *II* is given by $M_{II}(z) = e^{\angle I\omega}\bar{z}$. The composition of the two reflections (*I* first) is the transformation:

$$M_{II}(M_I(z)) = M_{II}\left(e^{2i\theta}\overline{z}\right) = e^{2i\varpi}\,\overline{e^{2i\theta}\,\overline{z}} = e^{2i(\varpi-\theta)}z$$

This is a rotation about the origin by the angle $2(\omega - \theta)$ (i.e., twice the angle between the lines I and II).

The proof for parallel lines is similar. (See Exercise 4.) ■

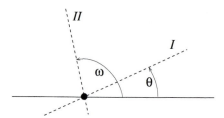

Figure 21.4 Proof of Theorem 1

A Crucial Note

The rotation that results from composing reflections in two intersecting lines depends on the directed *angle* between the lines, not on the *slopes* of the lines. In Figure 21.5, for example, the products $M_{II}M_I$ and $M_{IV}M_{III}$ are the *same rotation* (clockwise by an angle of 2θ), because the angle from line I to line II is the same as the angle from line III to line IV.

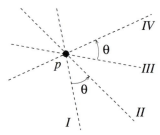

Figure 21.5 $M_{II}M_I$ = rotation about p by $2\theta = M_{IV}M_{III}$!

Similarly, when a translation is generated by combining reflections in two parallel lines, these lines can be chosen anywhere in the plane, as long as they are perpendicular to the direction of translation and half the translation distance apart. (See Figure 21.6.)

F–What kind of transformation results from composing a reflection in the x-axis followed by reflection in the y-axis? Does the order of these reflections matter?

G–What kind of transformation results from composing a reflection in the line $x + y = 1$ followed by reflection in the line $x + y = 2$? Does the order of these reflections matter?

Figure 21.6 $M_{II}M_I$ = **horizontal translation by** $2k = M_{IV}M_{III}$!

Combining Three Reflections

So far, we know what kind of transformation results by combining two reflections: either a rotation or a translation. Considering the composition of three reflections, we encounter a new transformation, the **glide reflection**, defined as the result of combining translation with mirror reflection in a line parallel to the direction of translation. Figure 21.7 illustrates this motion. Flag A is transformed to flag B: combining a translation by k units in the indicated direction with reflection in the dashed line.

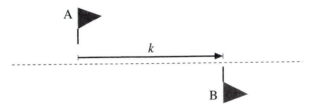

Figure 21.7 A glide reflection

The same result can be obtained by composing three mirror reflections $M_{III}M_{II}M_I$. The "ghost" flags in Figure 21.8 show intermediate stages of the transformation.

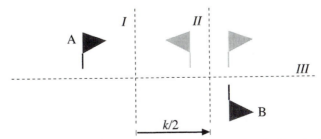

Figure 21.8 Glide reflection as the composition of reflection in the three lines I, II, and III

We now prove

Theorem 2 *The product of three reflections is either a reflection or a glide reflection.*

Proof: **Watch this proof carefully.** The basic technique of "moving the lines" introduced here is used over and over in this and the next two chapters to analyze combinations of transformations.

Let the transformation $S = M_{III}M_{II}M_I$, be the product of mirror reflections in three lines: *I*, *II*, and *III*. There are two cases:

Case 1. Line *II* Intersects Line *I* or Line *III*

In this case, we will prove that S is a glide reflection.

Suppose, for definiteness, that *II* intersects line *I*. (The case where *II* intersects *III* is similar.) Then the three lines may be depicted as in Figure 21.9(a) where lines *I* and *II* intersect each other, but might or might not intersect *III*. Then

$$S = M_{III}M_{II}M_I = M_{III}(M_{II}M_I) = M_{III}R$$

where $R = M_{II}M_I$ (the product of the first two of the reflections) is a rotation by some angle 2θ (because *I* and *II* intersect). Thus, S is the combination of a rotation and a reflection. However, to prove that S is a glide reflection, we must prove that S is the combination of a *translation* (not a rotation) and a reflection *parallel* to this translation. To do this, we will "move the lines"!

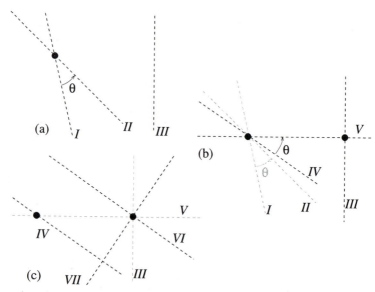

Figure 21.9 Case 1: Intersecting Lines. (b) *I*→*IV* and *II*→*V*, (c) *III*→*VI* and *V*→*VII*

Why can we move the lines? Lines *I* and *II* can be replaced by any other two lines generating the same rotation R; that is, as noted on page 265, any two lines that intersect at the same point and have the same angle between them. Thus, in Figure 21.9(b), lines IV and V intersect at the same point as I and II with the same angle θ.

It follows that $M_V M_{IV} = M_{II} M_I = R$, so that

$$S = M_{III}R = M_{III}M_V M_{IV}$$

Thus, we have moved line I to line IV and line II to line V. So what?

The advantage of "moving the lines" is that in choosing lines IV and V, we can *make* V intersect line III, if we like, and even insist that V be *perpendicular* to III, as in Figure 21.9(b). We now have

$$S = M_{III}M_V M_{IV} = (M_{III}M_V)M_{IV} = R'M_{IV}$$

Note that $R' = M_{III}M_V$ is a rotation (by 180°!).

Next, we move lines V and III. We can replace them with any other pair of perpendicular lines (VI and VII, say) intersecting at the same point. The clever part of this move is that we insist that VI be *parallel* to IV as in Figure 21.9(c). Now we have

$$S = R'M_{IV} = M_{VII}M_{VI}M_{IV} = M_{VII}(M_{VI}M_{IV}) = M_{VII}T$$

where $T = (M_{VI}M_{IV})$ is a translation, and M_{VII} is reflection in a line *parallel* to the direction of this translation. This proves that S is a glide reflection.

Case 2: Parallel Lines

Assuming that line II does not intersect either line I or line III, it follows that I, II, and III are parallel, as in Figure 21.10(a).

We then group the reflections as follows:

$$S = M_{III}M_{II}M_I = M_{III}(M_{II}M_I) = M_{III}T$$

where $T = M_{II}M_I$ is a translation. As in Case 1, we can move the lines I and II. According to the note on page 265, we may replace them by any other parallel lines separated by the same distance. The process is illustrated in Figure 21.10(b). Here $T = M_{II}M_I = M_V M_{IV}$. The clever part of this move is to choose the new lines IV and V so that V *coincides with III*, as in Figure 21.10(c). Then

$$S = M_{III}T = M_{III}M_V M_{IV} = M_{III}M_{III}M_{IV} = M_{IV}$$

(since a reflection is its own inverse). Thus, S is a reflection. ∎

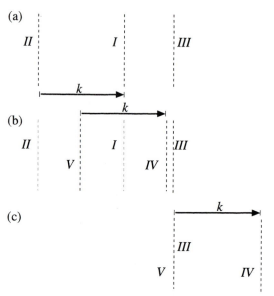

Figure 21.10 Case 2: Parallel lines (b) *I→IV* and *II→V*,
(c) making *V* equal *III*

H–In Case 2, why are the three lines parallel?

I–What cases would have to be considered to prove a theorem analogous to Theorem 2 in hyperbolic geometry?

Combining More Than Three Reflections

It turns out that there are no more new kinds of transformation:

Theorem 3 *Every combination of reflections is either a reflection, a rotation, a translation, or a glide reflection.*

This is a very powerful theorem. It includes information about many combinations of Euclidean transformations that we have not yet considered. For example, what sort of transformation results from combining two rotations? Since rotations are combinations of reflections, a combination of rotations is also a combination of reflections, and Theorem 3 applies. Therefore, according to the theorem, a combination of rotations must be one of the four types of transformation listed above.

The truth is even simpler: Every combination of an *even* number of reflections is either a rotation or a translation (as in Theorem 1); and every combination of an *odd* number of reflections is either a reflection or a glide reflection (as in Theorem 2). To prove this, we first prove that combinations of rotations and translations (i.e., transformations with an even number of reflections) are always either rotations or translations. Then, to handle an odd number of reflections, we have to prove that a reflection combined with a rotation or a translation is either a reflection or a glide reflection.

Outline of the Proof of Theorem 3

Part 1: Even numbers of reflections
  (a) A combination of rotations is a rotation or a translation.
  (b) A combination of translations is a translation.
  (c) A combination of a rotation and a translation is a rotation or a translation.
  (d) Therefore, the combination of any even number of reflections is a rotation or a translation.

Part 2: Odd numbers of reflections
  (a) A combination of a reflection and a rotation is a reflection or a glide reflection.
  (b) A combination of a reflection and a translation is a reflection or a glide reflection.
  (c) Therefore, the combination of any odd number of reflections is a reflection or a glide reflection.

We will prove one part of this, leaving the others as exercises. (See Exercises 6, 7 and 8.) The basic technique is "moving the lines."

Proof of Part 1(a): We will show that the combination of two rotations is either a rotation or a translation. Let R_1 and R_2 be rotations. Let R_1 be the combination of reflections in lines *I* and *II* and let R_2 be the combination of reflections in lines *III* and *IV*. (See Figure 21.11.) Here $R_1 = M_{II}M_I$ and $R_2 = M_{IV}M_{III}$.

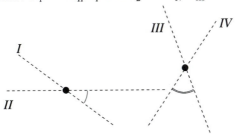

Figure 21.11 Proof of Theorem 3

As in the proof of Theorem 2, we move the lines: *I* to *V*, *II* to *VI*, *III* to *VII*, and *IV* to *VIII*,. The trick is that we make *VI* and *VII* the *same* line, namely, the line joining the two centers of rotation, as in Figure 21.12. Now we have

$$R_2 R_1 = M_{IV}M_{III}M_{II}M_I = M_{VIII}M_{VII}M_{VI}M_V = M_{VIII}M_V$$

(since $M_{VI} = M_{VII}$). This shows that the combination of two rotations is equal to the combination of the two reflections M_V and M_{VIII}, which, by Theorem 1, is either a rotation (if *V* and *VIII* intersect) or a translation (if *V* and *VIII* are parallel). ■

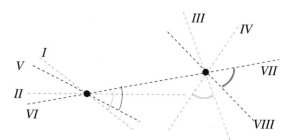

Figure 21.12 Proof of Theorem 3: Moving the lines

J–The proof of Theorem 3, Part 1(a), shows only that the combination of *two* rotations is a rotation or translation. What if there are more than two rotations?

Proper versus Improper Transformations

From Theorem 3, we learn that the extended Euclidean motions we have been studying in this chapter, that is, motions that are combinations of reflections, are of two basic types: those that are the result of an even number of reflections (called **proper**) and those that are the result of an odd number of reflections (called **improper**). The full Euclidean group $\mathbf{E}^+$ includes both types, while the special Euclidean group $\mathbf{E}$ includes only proper motions. Although we have argued that only the proper motions should be used to define Euclidean geometry, improper motions have considerable interest as well. In the next chapter, both types of motion are used to describe the Euclidean symmetry of plane figures.

Note: Our terminology belittles reflections and glide reflections since they are called "improper" while rotations and translations which are easier to manipulate and understand are termed "proper." This is a typical example of **pejorative terminology**, that is, terminology which labels something that is difficult to understand with terms suggesting that it is "defective" or "faulty." Other examples of pejorative terminology in mathematics (and there are many) include "irrational" numbers, "improper" fractions, "imaginary" numbers, and the angular "defect" of a triangle.

K–Is "complex numbers" an example of pejorative terminology?

Point Reflections

Reflections are an example of involutions:

Definition *A transformation T is an* **involution** *if $T \neq I$ and $T^2 = I$, where I is the identity transformation.*

Reflections share these algebraic properties with rotations of 180° about a point (called **half turns**). For this reason, half turns are sometimes called **point reflections**. They have some geometric, as well as algebraic, similarities with **line reflections** (as ordinary reflections are called in this context).   To facilitate working with these two kinds of transformation, we adopt the following systematic notation:

Definition *Let p be a point in the complex plane and λ be a line.*
  (1) The rotation H_p of 180° about p is called the **point reflection** *(or* **half turn***) about p,*
  (2)   The transformation M_λ sending each point to the symmetric point on the other side of λ is called the **line reflection** *about λ.*

Point reflections have simple formulae. The half turn about the origin O is given by the formula $H_O z = -z$. To compute a half turn about a point p that is not the origin, we first translate p to the origin, then apply the half turn about the origin, and finally translate back. The result is the formula

$$H_p z \;=\; -(z - p) + p \;=\; 2p - z$$

Say It with Transformations!

Point reflections and line reflections can be used to express many fundamental concepts of geometry. For example, consider incidence. Let p be a point and λ be a line. How do we determine whether p is on λ? In the usual analytic treatment, we calculate the coordinates of p in some coordinate system, find an equation for λ in this system, and then check whether the coordinates of p satisfy the equation for λ. In other words, the question of whether λ and p are incident is settled by calculation with numbers and equations.
  Alternatively, a computation with H_p and M_λ can accomplish the same thing. Here is how it works:

Theorem *A point p is on a line λ if, and only if, the transformations H_p and M_λ commute, that is, if $H_p M_\lambda = M_\lambda H_p$.*

Proof: By applying a preliminary Euclidean transformation, we may move the line λ to the x-axis. Then

$$M_\lambda(z) = \bar{z}$$

while

$$H_p(z) \;=\; 2p - z$$

Simple computation shows that

$$M_\lambda(H_p(z)) = \overline{2p - z} = 2\overline{p} - \overline{z}$$

and

$$H_p(M_\lambda(z)) = 2p - \overline{z}$$

These are equal if, and only if, $2p = 2\overline{p}$, that is, if p is real, and hence on λ. In other words, $M_\lambda H_p = H_p M_\lambda$ if, and only if, p is on λ. ∎

It follows that the *geometric* question of whether the point p lies on the line λ can be settled *algebraically* by computing the transformations $H_p M_\lambda$ and $M_\lambda H_p$ and then comparing them. In this way, the transformations H_p and M_λ (and the algebra of their composition) can replace the point p and line λ (and their coordinates/equations).   This leads to an alternative algebraic approach to geometry independent of coordinates, an idea explored in more detail in the following exercises, and in Chapters 25 and 26.

EXERCISES

Finding Formulas for Euclidean Transformations

1. Let a, b, and c be three collinear points in the complex plane as shown below:

$$\underset{a}{\bullet} \text{———} \underset{b}{\bullet} \text{————} \underset{c}{\bullet}$$

Assume that the distance from a to b equals the distance from b to c. Then there are four Euclidean transformations that move the line segment ab to the line segment bc. Describe these transformations in geometric terms, and find formulae for them in terms of the complex numbers a, b, and c.

Hint: Two of the four transformations are proper; two are improper. Two of them reverse the line, that is, they map a to c and leave b fixed. There is one rotation, one translation, one reflection, and one glide reflection. It may be useful to place the three points in some special relation to the coordinate axes.

Example: One of these transformations is the half turn about b. Its formula is $Tz = 2b{-}z$.

2. Let a, b, c, and d be four points in the plane as shown below:

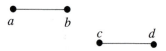

Assuming that the distance from a to b equals the distance from c to d, and that the two line segments are parallel, there are four

Euclidean transformations that move the line segment ab to the line segment cd. Describe these transformations in geometric terms, and find formulae for them in terms of the complex numbers a, b, c, and d.

Hint: Two of the transformations are proper; two are improper. Two of them map a to c; two of them map a to d. Among them are one rotation, one translation, and two glide reflections. It may be useful to place the three points in some special relation to the coordinate axes.

3. Let a, b, c, and d be four points in the plane as shown below:

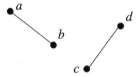

Assuming only that the distance from a to b equals the distance from c to d, what Euclidean transformations, proper or improper, are there that move the line segment ab to the line segment cd?

Moving the Lines

4. Complete the proof of Theorem 1 by proving that the composition of two reflections in parallel lines is a translation.

5. Use the technique of "moving the lines" to prove that the product of a translation and a reflection is either a reflection or a glide reflection.

6. Prove parts 1(b) and 1(c) of Theorem 3.

7. Prove parts 2(a) and 2(b) of Theorem 3.

8. Deduce part 1(d) of Theorem 3 from parts 1(a, b, c). Deduce part 2(c) of Theorem 3 from parts 2(a, b).

Involutions

9. Show that if T is an involution, and S is any other transformation, then STS^{-1} is an involution.

10. Let S and T be involutions. Prove that ST is an involution if, and only if, S and T commute and $S \neq T$.

11. Euclidean transformations include rotations, translations, reflections, and glide reflections. Find all involutions among each of these categories of transformation.

Geometric Ideas and Involutions

12. Let p and q be distinct points. What kind of Euclidean transformation is $H_p H_q$? Prove that $H_p H_q = H_q H_p$ if, and only if, $p = q$.

Hint: To find out what kind of transformation H_pH_q is, move the lines.

13. Let α and β be two lines. Prove that $M_\alpha M_\beta$ is an involution if, and only if, α and β are perpendicular.

14. Let p, q, and r be distinct points. Prove that $H_pH_q = H_qH_r$ if, and only if, q is the midpoint of the line segment from p to r.

15. Let α, β, and ω be three lines. Under what geometric condition does $M_\alpha M_\beta = M_\beta M_\omega$?

16. Let p and q be points and α be a line. Under what geometric condition does H_pM_α equal $M_\alpha H_q$?

17. Let p be a point and α and β be lines. Under what geometric conditions does H_pM_α equal $M_\beta H_p$?

18. Let p, q, and r be distinct points. What kind of Euclidean transformation is $H_pH_qH_r$? When is $H_pH_qH_r$ an involution?

19. Let α, β, and ω be three lines. What kind of Euclidean transformation is $M_\alpha M_\beta M_\omega$? When is $M_\alpha M_\beta M_\omega$ an involution?

22 DISCRETE SYMMETRY

What is Symmetry?

An object is **mirror** (or **bilaterally**) **symmetric** if it is invariant under a mirror reflection. For example, the figure consisting of the two spirals in Figure 20.1 is mirror symmetric. Rectangles are also mirror symmetric, in fact, they are invariant under *two* different mirror reflections. Bubbles, bluebells, butterflies, and the human body, as well as many objects made for human use such as bridges, bricks, beds, books, and bathtubs: All are examples of three-dimensional objects with mirror symmetry.

Figure 22.1 Rotational symmetry

Mirror symmetry is the meaning of the word "symmetry" in everyday use. In mathematics, however, **symmetry** more broadly describes any and all geometric transformations of an object that leave the object invariant. The pinwheel in Figure 22.1, for example, is not mirror symmetric, but does have **rotational symmetry**. To be precise, the pinwheel has *fourfold* rotational symmetry, since the minimal rotation that leaves it invariant (rotation by 90°) has to be repeated four times in order to get the identity transformation.

Figure 22.2 Rotational and reflection symmetry combined

Figures can have several symmetries at once. The design in Figure 22.2 has sevenfold rotational symmetry [the minimal invariant rotation of 51.4° (= 360°/7) must be repeated seven times in order to get the identity]. Figure 22.2 also has mirror symmetry. *Note:* In fact, the figure is invariant under mirror reflection in seven different mirror lines.

A–What rotational symmetry does an equilateral triangle have? A square? An isosceles triangle?

B–What mirror symmetries do the figures in Figure 22.5 have?

The next example is presented in Figure 22.3 but only part of the whole design is there. The whole figure extends infinitely far in both directions. It has a new kind of symmetry: **translational symmetry** (indicated by the arrow below the figure). In addition, Figure 22.3 has twofold rotational symmetry. In fact, there are an infinite number of twofold rotocenters. Furthermore, they are of two different kinds!

Figure 22.3 Translational symmetry

C–Find the centers of twofold rotation in Figure 22.3. (*Hint:* Some are the midpoints of vertical line segments, others of horizontal line segments.)

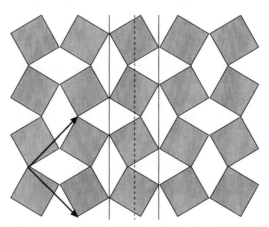

Figure 22.4 A complex combination of symmetries

A final example (in Figure 22.4) exhibits some further possible complication. Like Figure 22.3, this figure extends to infinity, but now in all directions, and there is *two*-dimensional translational symmetry (indicated by the arrows pointing in two different directions). Numerous other symmetry elements are present including both twofold and fourfold **rotocenters** (centers of rotational symmetry), and horizontal and vertical reflection lines. A further new feature is the presence of **glide reflection lines**. (In the figure, two mirror lines are indicated by ordinary lines, and a glide reflection line is indicated by a dashed line.)

D–Find the centers of fourfold rotation in Figure 22.4. Note that they are of two different kinds.

E–Find the centers of twofold rotation in Figure 22.4. Note that these centers are also congruent to each other in two groups.

These examples lead to the following definition:

Definition *Let F be a figure in a geometry. The set of all transformations T such that T(F) = F is called the **symmetry group** of F and is notated σ(F). The elements of σ(F) are called **symmetry elements** of F.*

The geometry in this definition can be any geometry. In this chapter it will be Euclidean plane geometry. The next chapter considers symmetry in other geometries and other dimensions.

F–Identify as many symmetry elements as you can in Figure 22.5.

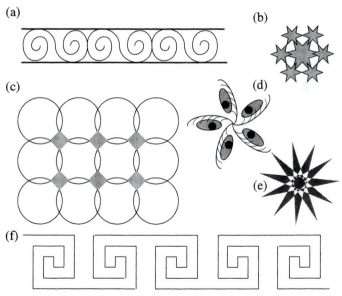

(a) (b) (c) (d) (e) (f)

Figure 22.5 Symmetry practice

Discrete Symmetry Groups

Symmetry groups can be complicated. However, the discrete ones can be completely classified and listed, at least for Euclidean geometry. Here is the technical definition of discreteness.

Definition *Let G be a group of transformations of the complex plane, and let z be a complex number. The set {Tz : T ∈ G} is the **orbit** of z **under the action** of G.*

*A group G is a **discrete** group if every orbit of G has only a finite number of points inside any circle.*

This definition prevents *G* from moving points arbitrarily small distances, and, therefore, means that the transformations of *G* can not be "infinitely divisible" in terms of other transformations of *G*.

Discrete groups include the symmetry groups of all "normal" geometric objects, and, if we restrict our attention to discrete groups, it is possible to list all Euclidean plane symmetry groups. One goal of this chapter is to present an inventory of these groups.

An object *S* and its symmetry group *G* together constitute a geometry (as described in Chapter 4). If the group *G* is discrete, the resulting geometry is a discrete geometry. Such geometries are very different from the other geometries in this book. For example consider the pinwheel in Figure 22.1 and its symmetry group. This is a discrete symmetry group so the resulting geometry is discrete. One aspect of discreteness, in this case, is that not all points are congruent. The tips of the "blades" of the pinwheel, for example, are congruent to each other but are not congruent to any other points. The center of the pinwheel is a unique point, not congruent to any other point of the geometry!

When we search for all the symmetry groups of a certain kind—discrete or continuous, Euclidean or non-Euclidean—we are actually finding all the geometries of this kind, from Klein's point-of-view, although this is not the usual way this search is described.

Classification of Symmetry Groups

Discrete symmetry groups are classified according to their fixed elements, defined as follows:

Definition *Let G be a subgroup of the **full Euclidean group** E⁺. Then, G has a **fixed point** z, if Tz = z for all T in G. G has a **fixed line** λ if T(λ) = λ for all T in G.*

*A figure F in Euclidean geometry (**C, E⁺**) has a **singular point** (or **singular line**) if its symmetry group σ(F) has a fixed point (or fixed line).*

Singular elements of figures are easy to spot. For example, Figures 22.1 and 22.2 have singular points (but no singular lines), Figure 22.3 has singular lines (the central horizontal line), but *no* singular points, and Figure 22.4 has no singular elements at all. Note that a line can be fixed even though no point on the line is fixed (as in Figure 22.3).

Classification of Symmetry Groups *Let F be a figure in the geometry* (**C**, **E**⁺). *There are three cases:*
  *(a) F has a singular point. Then F is called a **rosette**.*
  *(b) F has a singular line, but no singular point. Then F is called a **frieze**.*
  *(c) F has no singular element. Then F is called a **network**.*

Thus, Figures 22.1 and 22.2 are rosettes, Figure 22.3 is a frieze, and Figure 22.4 is a network. Figures with symmetry appear throughout the visual arts. Wallpaper designs, moldings, jewelry, ornaments of all kinds, and patterns in cloth, brickwork, and flooring, all these, and countless other fabricated objects, exhibit symmetry of one sort or another. By classifying and enumerating symmetry groups, we create a catalog of decorative possibilities.

There are also many scientific applications of symmetry. An obvious example is the appearance of networks in the arrangement of atoms and molecules in crystals, and, in fact, much of the pioneering work in symmetry was done by crystallographers. Theoretical physics also makes heavy use of symmetry.

G–Which of the figures in Figure 22.5 are rosettes, which are friezes, and which are networks?

H–Rosette groups contain no translations or glide reflections. Why?

I–What is Figure 22.6: rosette, frieze, or network?

A Symmetry Inventory

In the rest of this chapter, we list *all* possible Euclidean discrete symmetry groups. Each group is described in words and illustrated by a figure. In addition, the crucial symmetry elements of the group are summarized by giving the International Symbol of the group. Many systems have been invented for naming symmetry groups. The International Symbols are those adopted by the International Union of Crystallographers and have achieved nearly universal acceptance.

Figure 22.6 An object with no symmetry

To begin, a figure need not have any symmetry at all. Figure 22.6 is an example. By *no* symmetry, we mean that only the identity transformation leaves this figure invariant. The International Symbol for this (trivial) symmetry group is 1.

Figures with significant symmetry are (usually) built up by applying geometric transformations to so-called "primitive" figures, figures without symmetry such as the flag in Figure 22.6. We will illustrate this by systematically creating figures with more complicated symmetry using copies of this flag.

Symmetry Types of Rosettes

Let F be a rosette, and suppose that $\sigma(F)$ is a *discrete* symmetry group. Consider first just the rotations in $\sigma(F)$. All these rotations are about the fixed point (otherwise the fixed point would not be fixed!). Furthermore, among all these rotations, there is a smallest one, an atomic rotation: If there were no smallest rotation, the existence of an infinite sequence of smaller and smaller rotations would violate the definition of discreteness. It follows that all the rotations can be obtained by repeatedly applying this smallest one. In the case where $\sigma(F)$ consists only of rotations, the whole group is generated by repeatedly applying this one fundamental rotation.

Cyclic Symmetry

Therefore, one type of rosette is generated by applying one rotation repeatedly to a given "primitive" figure until it is brought back to its starting position. Figure 22.7 is an example. It consists of exactly five copies of the flag in Figure 22.6. The symbol ⬟ in Figure 22.7 represents the rotocenter graphically. The International Symbol for this group is 5.

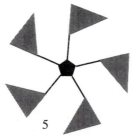

5

Figure 22.7 Cyclic symmetry

Symmetry of this type exhibited by Figure 22.7 is called **cyclic** symmetry. There is a cyclic symmetry group for every positive integer n. This group consists of an n-fold rotation, plus transformations produced by combining this rotation with itself. The corresponding groups (whose primary algebraic feature is that all elements in the group are powers of a single generating element) are called **cyclic** groups.

When copies of a primitive figure combine in this way, *without overlapping*, to create a symmetric figure, the primitive figure is called a **fundamental domain** of the final figure.

Dihedral Symmetry

What other symmetry transformations can a rosette have? There is not much freedom of choice. A rosette group cannot contain translations or glide reflections (since these have no fixed points). Only reflections, it seems, can be added to the rotations with which we have just become acquainted. When a reflection is added to rotational symmetry, the rotations combine with the reflection to create more reflection lines (at multiples of the basic rotation angle to the first reflection line). Therefore, a group with an *n*-fold rotocenter and a reflection line actually has *n* reflection lines meeting at angles of π/n. Figure 22.8 gives two examples.

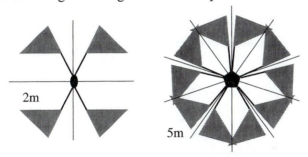

Figure 22.8 Dihedral symmetry

The International Symbol for this symmetry group is *n*m, where *n* is the order of the rotocenter of the group (and the number of reflections). The dashed lines in Figure 22.8 represent the reflection axes. The resulting symmetry is called **dihedral** symmetry, and the group is called a **dihedral** group.

Rosette Group Summary

There are two infinite families of rosette groups:

1. *Cyclic Groups:* pure rotational symmetry: 1, 2, 3, . . .
2. *Dihedral Groups:* *n*-fold rotocenter + reflection: m, 2m, 3m, . . .

Both families contain a subgroup consisting of rotations which is generated by a smallest rotation. The cyclic groups contain only these rotations; the dihedral groups contain reflections as well. The symbols we have introduced for these groups were developed by the International Union of Crystallographers.

J–Identify the symmetry group of the three rosettes in Figure 22.5.

K–Draw a figure of symmetry type 3, and then add more detail so that your figure has symmetry type 3m. How many different mirror reflections are in the symmetry group of your final figure?

Symmetry Types of Friezes

The new feature that appears in friezes is translational symmetry. The simplest frieze group consists of translational symmetry alone. (See Figure 22.9.) The International Symbol for translational symmetry in a single direction is **r**.

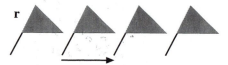

Figure 22.9 A simple frieze

The translations in any symmetry group always form a subgroup. (See Exercise 9.) For a frieze, all translations must be in a single direction. Furthermore, in a discrete group, there is always a smallest translation, an atomic translation (indicated by the arrow in Figure 22.9). Different smallest translations generate different symmetry groups, but these groups differ only in scale.

By adding other symmetry elements to the basic translational symmetry **r**, friezes of different symmetry types are obtained. For example, we may add a reflection line parallel to the direction of translation. (See Figure 22.10.) The International Symbol for this symmetry group is **r11m**. This symbol encodes basic information from which, in principle, the whole symmetry group can be reconstructed. Let us explain how these symbols are created.

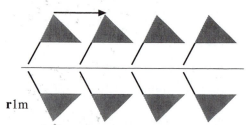

Figure 22.10 A frieze with a single reflection

Creation of the International Symbols

The first letter of the symbol always represents any translational symmetry in the group. For friezes, the translational symmetry is always the same: in one direction only. To construct the International Symbol, the frieze is placed horizontally, with the direction of translation parallel to the *y*-axis. (See Figures 22.11 and 22.12.)

The next three places in the symbol describe the principal symmetry elements (other than translation) in each **direction** in which the three coordinate axes point. In Euclidean geometry, such "extra" symmetry elements, according to Theorem 3 of Chapter 20, can only be rotations, reflections, and glide reflections.   In the International Symbol, rotations are represented by a number (the order of the rotocenter), mirror reflections by the letter "m," and glide reflections by the letter "g."

International Symbol

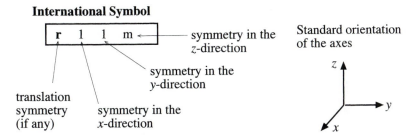

Figure 22.11 Construction of the International Symbols

The whole notational scheme proposed by the Crystallographers Union is based, however, on the assumption that the figure is *three-*dimensional. To apply it to plane figures, one must imagine that they are thickened into solids.   (See Figure 22.12.)   In this process, reflection *lines* of the original figure become reflection *planes*, and rotocenters become rotation axes. (See Figures 22.12 and 22.13.)

Important Note: The *direction* of a plane is taken to be the *direction* of its normal. (See Figure 22.12.)

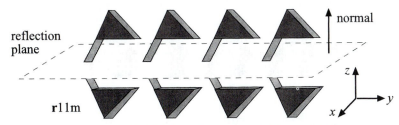

Figure 22.12 Thickened version of Figure 22.10

Example 1: A Frieze

The symbol **r**11m can now be understood in detail as follows:
—"**r**" is the symbol for *one-dimensional translation symmetry* (arranged, by convention, so that the *y*-axis is the translation direction),
—The first "1" indicates that there are *no* extra symmetry

elements in the x-direction,

−The second "1" indicates that there are *no* extra symmetry
elements in the y-direction,

−The "m" indicates that there is a *reflection* plane in the z-direction.

Example 2: A Rosette

As a second example, consider the dihedral group 2mm. (See
Figure 22.13.) The interpretation of this symbol is that

− There is *no* translational symmetry, so the symbol does not
begin with a letter,

− There is a *twofold rotational axis* in the x-direction,

− There is a *reflection plane* in the y-direction,

− There is also a *reflection plane* in the z-direction.

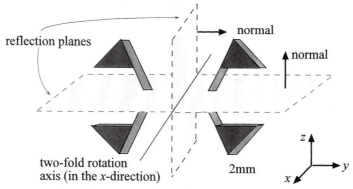

Figure 22.13 Three-dimensional example of the group 2m

This analysis confirms that 2mm should be the symbol of this group.
But, the symbol 2m already contains sufficient information to
distinguish this group from all others (because there is only one
dihedral group whose highest rotation axis is a twofold axis); therefore
the simpler symbol 2m is often used instead. Both 2m and 2mm are
accepted. If we adopt the additional principle that every symbol
should contain only enough information to distinguish its group from
all others, then 2m is preferred.

The Remaining Frieze Groups

So far we have introduced two frieze groups: **r** and **r11m**. The
others are now easily described.

Instead of possessing a reflection line *parallel* to the direction of
translation, a figure can have a reflection line *perpendicular* to the
direction of translation. This is the group **r1m**. (See Figure 22.14.)
The reflections are in planes whose normals point in the y-direction.

Figure 22.14 The frieze group r1m

There is also a group containing both types of reflection: the group **r2mm** (see Figure 22.15). Notice that the perpendicular reflection lines combine to produce centers of twofold rotation. If Figure 22.15 is thickened, these twofold axes point in the *x*-direction.

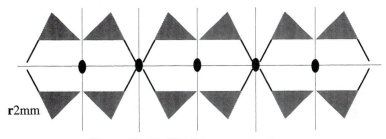

Figure 22.15 The frieze group r2mm

The next group, **r2** (see Figure 22.16), has only twofold axes in addition to translational symmetry. These are the same axes that appeared in the group **r2mm**.

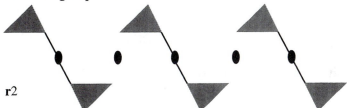

Figure 22.16 The frieze group r2

The last two frieze groups are based on glide reflection. In the case of the group **r11g** this is all it has besides translations. (See Figure 22.17, where the *long* dashed line represents the glide reflection line. Thickened, it is a plane in the *z*-direction.)

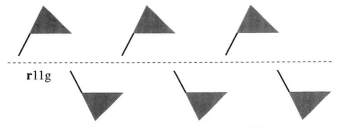

Figure 22.17 The frieze group r11g

The last frieze group, r2mg, has a glide reflection plus perpendicular reflection lines. (See Figure 22.18.) Notice that the twofold axes for the group are now halfway between the reflection lines, not on them, as they were in r2mm. (See Exercise 13.)

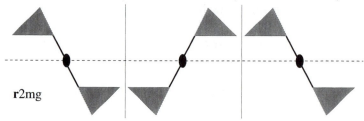

r2mg

Figure 22.18 The frieze group r2mg

Frieze Group Summary

Different frieze groups result from the combination of translational symmetry with added rotations, reflections, and/or glide reflections. Only twofold rotocenters are possible; reflection lines can only be parallel or perpendicular to the translation direction, and glide reflections must be parallel to the translation direction. (See Exercise 11.) These are considerable restrictions, and the following seven groups listed are the only possibilities:

1. **r**   pure translational symmetry in one dimension
2. **r11m**   translational symmetry + a parallel reflection
3. **r1m**   translational symmetry + perpendicular reflection
4. **r2mm**   translational symmetry + perpendicular and parallel reflections
5. **r2**   translational symmetry + perpendicular twofold axes
6. **r11g**   pure glide reflection symmetry in one direction
7. **r2mg**   glide reflection symmetry + perpendicular reflections

L–Give symbols for the symmetry of the friezes in Figure 22.5.
M–Why is frieze group symmetry limited to twofold rotocenters?

Symmetry Groups of Networks

Network groups are characterized by two-dimensional translational symmetry. In the simplest group of this type, translational symmetry is the only symmetry. (See Figure 22.19.)

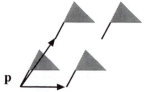

p

Figure 22.19 Pure translational symmetry in two dimensions

The arrows in Figure 22.19 represent two "primitive" or smallest translations. If we take one point (the top of the flag, for example) and follow its orbit under the transformation group, we obtain a pattern of points forming a **parallelogram lattice**. (See Figure 22.20.) The International Symbol for translational symmetry in two dimensions is **p**.

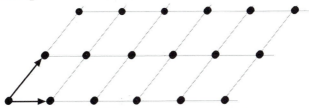

Figure 22.20 A parallelogram lattice

Network symmetry groups with symmetry beyond pure translational symmetry are created by adding more copies of the fundamental pattern (in our examples this is a flag) around the lattice points. The simplest group with such additional symmetry has twofold rotations at each intersection of the network. (See Figure 22.21.)

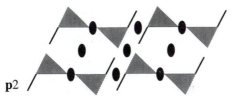

p2

Figure 22.21 The network group p2

This figure is created by placing *two* flags (forming a rosette with symmetry 2) at each lattice point. Figures with the symmetry **p2** actually have *four* essentially different centers of twofold rotation. This is because the spacing of centers of two-fold symmetry must be at *half* the length of the smallest translation. (See Exercise 13.)

N–Find the four *different* kinds of two-fold rotocenter in Figure 22.21.

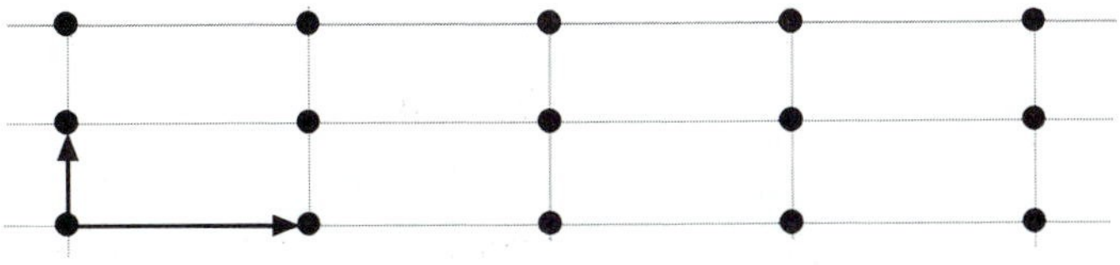

Figure 22.22 A rectangular lattice

The two network groups introduced so far, **p** and **p2**, use the parallelogram lattice of Figure 22.20. However, to introduce symme-

try beyond twofold rotocenters, the translation lattice must itself
have more symmetry.   Therefore, we turn to the **rectangular
lattice**, whose primitive translations are at right angles. (See Figure
22.22.)

To the translations of this rectangular lattice, we can now add
reflections parallel to either one of the translation directions building
a rosette with the symmetry m at each lattice point.   This gives the
group **p**1m (or **p**11m). (See Figure 22.23.)

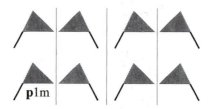

Figure 22.23 The network group p1m

Instead of a reflection, a glide reflection can be created.   This is
the group **p**1g (alternatively **p**11g) illustrated in Figure 22.24.

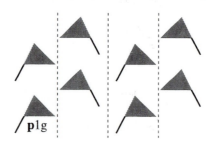

Figure 22.24 The network group p1g

Important Note: In the illustrations, an unbroken line is used for
reflections and a dashed line for glide reflections.)

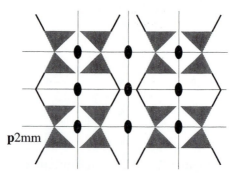

Figure 22.25 The network group p2mm

There are three more groups associated with a rectangular lattice. They have reflections (or glide reflections) in *each* translation direction. These are the groups **p2mm**, **p2mg** and **p2gg**, illustrated in Figures 22.25, 22.26, and 22.27, respectively. In each direction (*x* and *y*) these groups have either reflections or glide reflections *but not both*. Note that these three groups also have twofold rotations.

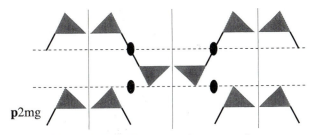

p2mg

Figure 22.26 The network group p2mg

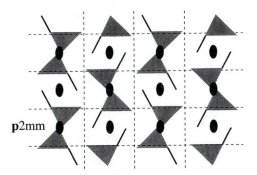

p2mm

Figure 22.27 The network group p2gg

A translation lattice with still higher symmetry is the **centered rectangular lattice**. (See Figure 22.28.) The centered rectangular lattice is based on translation along the sides of a rhombus (an equilateral parallelogram). The lattice points outline a pattern of rectangles, *plus* the centers of those rectangles; hence the name centered rectangular lattice. The higher symmetry of this lattice permits figures with alternating reflections and glide reflections in the *same* direction.

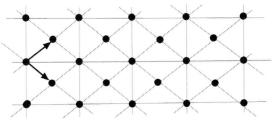

Figure 22.28 A centered rectangular lattice

One of the two groups in this family, **c**1m, has these alternating reflection and glide reflection lines in just one direction. (See Figure 22.29.)

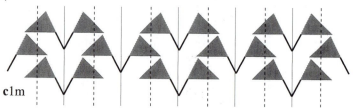

Figure 22.29 The network group c1m

The other group, **c**2mm, has alternating reflections and glide reflections in two directions. (See Figure 10.30.) The letter "**c**" (for "centered rectangular") is used in these symbols to prevent confusion with the groups **p**1m and **p**2mm.

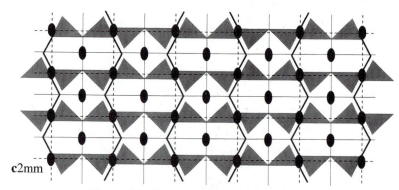

Figure 22.30 The network group c2mm

So far, we have seen only twofold rotation symmetry. Higher rotation symmetry requires a translation lattice with still higher symmetry. An obvious candidate is a **square lattice**. (See Figure 22.31.)

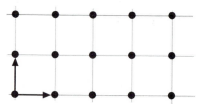

Figure 22.31 A square lattice

There are three groups associated with a square lattice. The group **p**4 (Figure 22.32) has only rotations besides translations.

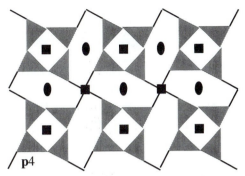

Figure 22.32 The network group p4

The groups **p**4mm and **p**4mg (shown in Figure 22.33) have reflection and glide reflection lines also.

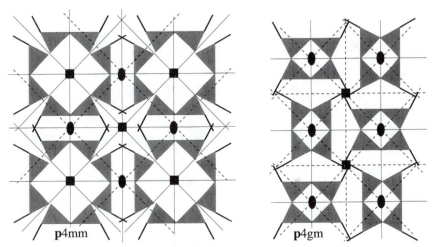

Figure 22.33 The network groups p4mm and p4gm

The lattice with the most symmetry is a **hexagonal lattice** (see Figure 22.34), which has threefold and sixfold symmetry. Three groups exploit the threefold axes of this lattice and two groups exploit the sixfold axes.

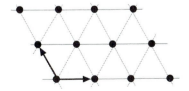

Figure 22.34 A hexagonal lattice

The groups with threefold symmetry are **p**3, **p**3mm and **p**3mg. (See Figures 22.35, 22,36and 22.37.)

Figure 22.35 The network group p3

The groups **p3**mm and **p3**mg are quite complicated. They are the two network groups most difficult to tell apart.

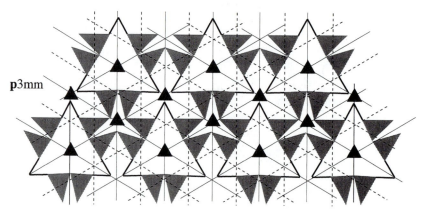

Figure 22.36 The network group p3mm

Figure 22.37 The network group p3mg

The last two network groups, **p6** and **p6**mm, have sixfold rotation axes. (See Figures 22.38 and 22.39.)

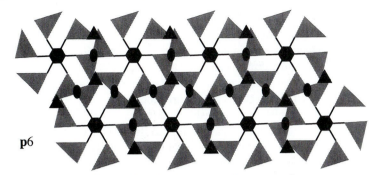

Figure 22.38 The network group p6

Figure 22.39 The network group p6mm

Network Group Summary

Figures with network symmetry are built by stamping a pattern on the plane at the points of a planar lattice. This lattice determines the translational symmetry of the network. There are five different lattices each with different nontranslational symmetry elements. Thus, network symmetry groups divide into five families: parallelogram groups (2), rectangular groups (5), centered rectangular groups (2), square groups (3) and hexagonal groups (5). These 17 groups are the only possible symmetry groups for a network figure:

1. **p**   pure translational symmetry
2. **p**2   parallelogram translational symmetry + twofold axes
3. **p**1m   rectangular translation lattice + reflections in one direction
4. **p**1g   rectangular translation lattice + glide-reflection in one direction
5. **p**2mm   rectangular translation lattice + reflections in two directions

6. **p2mg**   rectangular translation lattice + reflections in one direction and perpendicular glide-reflections

7. **p2gg**   rectangular translation lattice + glide-reflections in two perpendicular directions

8. **c1m**   centered rectangular translation lattice + alternating reflections and glide-reflections in one direction

9. **c2mm**   centered-rectangular translation lattice + alternating reflections and glide-reflections in two perpendicular directions

10. **p4**   square translation lattice + fourfold axes and twofold axes

11. **p4mm**   square translation lattice + fourfold axes + reflection lines through the fourfold axes

12. **p4gm**   square translation lattice + fourfold axes + glide-reflection lines through the fourfold axes

13. **p3**   hexagonal translation lattice + threefold rotation axes

14. **p3mm**   hexagonal translation lattice + threefold axes + reflection lines through the threefold axes

15. **p31m**   hexagonal translation lattice + threefold axes + reflection lines that miss some of the threefold axes

16. **p6**   hexagonal translation symmetry + sixfold axes

17. **p6mm**   hexagonal translation symmetry + sixfold axes + reflection lines

Notice that the only rotational symmetry possible in networks are twofold, threefold, fourfold, and sixfold axes. The absence of fivefold axes and any rotational symmetry higher than sixfold is called the **crystallographic restriction**, and, in fact, the internal symmetry of all substances that form crystals obeys this restriction.

EXERCISES

Identifying discrete symmetry groups

1. Identify the symmetry group of the rosettes in Figure 22.40. Be careful to regard them strictly as plane figures.

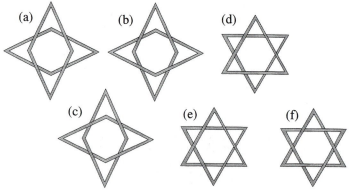

Figure 22.40 Rosettes for Exercise 1

2. Identify the symmetry group of the friezes in Figure 22.41.

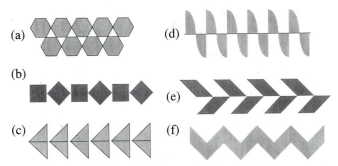

(a)   (d)

(b)   (e)

(c)   (f)

Figure 22.41 Friezes for Exercise 2

3. Here are some infinite tapes. Fill in all the blank cells to create a pattern with the indicated frieze symmetry. An example is included.

r1:

r2:

r11m:

r1m:

r2mm:

r1g:

r2mg:

4. Fill in Table 22.1. Entries should be either Y (yes) or N (no). An example is included. The last column answers the question: Are the highest order rotocenters *always* on a reflection line? The answer to this question distinguishes among the most complicated groups, namely **p4mm** *vs.* **p4mg** and **p3mm** *vs.* **p3mg**.

5. Identify the symmetry groups of the twelve Egyptian designs on page 4.

group	rotations?				reflections?	glide reflections?	all max rotations on reflection line?
	2	3	4	6			
p1							
p2							
p1m							
p1g							
p2mm							
p2mg							
p2gg							
c1m							
c2mm							
p4							
p4mm							
p4gm	Y	N	Y	N	Y	Y	N
p3							
p3mm							
p3mg							
p6							
p6mm							

Table 22.1 Identification of network groups

6. Identify the symmetry group of the facades of the buildings in the *Piazza San Marco, Venice* by Bellotto (page 134).

7. Make a mathematically precise copy of the pattern in the background of Picasso's *Girl before a Mirror* (page 232). What is the symmetry group of this pattern? Make a mathematically precise copy of the pattern on the fabric of the girl's swimsuit in the same painting. Classify the symmetry of this pattern.

Hint: The swimsuit pattern is not a network! (Why?) What is it?

Theoretical Results

8. Let F be the plane figure consisting of all the points z on the unit circle such that $\arg(z)$ is a rational number. Describe the symmetry group of this figure. Explain why this group is not a discrete group. Find another example of a symmetry group that is not discrete.

9. Prove that the subset of translations is a sub*group* of every symmetry group G.

10. Is the smallest rotation in a rosette group unique?

11. Explain why the only rotations possible in a frieze group are twofold rotations. Explain why any reflection line must be parallel or perpendicular to the translation direction. Explain why a glide reflection must be in the same direction as the direction of translation.

12. What transformation is the combination of a glide reflection and a perpendicular reflection. Use this to explain the location of the centers of twofold rotation in **r**2mg.

13. What transformation is a combination of a translation and a twofold rotation? Use transformation to explain the spacing of twofold rotocenters in the group **p**2.

Fundamental Domains

*Discussion: A **fundamental domain** for a symmetry group is a region whose transforms under the group cover the whole plane without significant overlap, meaning that the overlap between any two transforms of the fundamental domain has zero area.*

14. Find a fundamental domain for each of the seven frieze groups.

Example: A fundamental domain for **r**11m *is a half-infinite strip. (See Figure 22.42.) The transforms of this domain are congruent half-strips that overlap only along line segments. Note that the fundamental domain contains the parts of exactly one flag.*

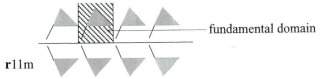

Figure 22.42 Fundamental domain for r11m

15. Find fundamental domains for each network group.

Example: A fundamental domain for **p**4mm *is a 45–45–90 right triangle. (See Figure 22.43.) Transforms of this domain are equilateral right triangles that cover the plane overlapping only along border line segments. Note that the fundamental domain contains the parts of exactly one flag.*

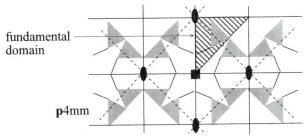

Figure 22.43 Fundamental domain for p4mm

23 NON-EUCLIDEAN SYMMETRY

In principle, the classification and enumeration of discrete symmetry groups, described in Chapter 21, can be carried over to non-Euclidean geometries and extended from the plane to three and higher dimensions. In practice, however, the increased variety and complication are a formidable obstacle to carrying out this program, and, at least in some cases, the problem of finding and describing all discrete symmetry groups is unsolved. In this chapter, we will outline some of the known results, and give examples of discrete symmetry groups in several non-Euclidean geometries.

Non-Euclidean Reflections

The heart of symmetry is the operation of reflection in a straight line. Since, in the models of non-Euclidean geometry we have studied, the role of straight line is often played by a circle, we must define reflection in a circle. Fortunately, the appropriate concept has already been introduced in Möbius geometry. There, we defined two points z and z^* as *symmetric* with respect to a circle, if they lie on the same diameter of the circle, and the product of their distances from the center equals the square of the radius of the circle. (See Figure 23.1.)

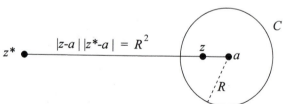

Figure 23.1 Reflection across a circle

*Definition Let C be a cline. The transformation of **reflection in C** is defined by setting Tz $= z^*$, where z^* is the point symmetric to z with respect to C.*

A formula for reflection in any particular circle is easily found using ideas developed in Chapter 5. For example, reflection in the unit circle is $Tz = 1/\bar{z}$. In this case, T combines reflection in a line (i.e., complex conjugation) and inversion in the unit circle.

A–Justify the preceding formula.

Elliptic Symmetry Transformations

Reflections play a role in elliptic geometry very similar to their role in Euclidean geometry. The major difference is that two straight lines always intersect in elliptic geometry. Therefore, the composition of reflections in two elliptic straight lines is always an elliptic rotation.

When it comes to the combination of three reflections, however, a new transformation appears. Called **mirror rotation**, this transformation is the composition of a rotation (which is the composition of reflections in two intersecting lines) and reflection in a third line perpendicular to the first two. Figure 23.2 illustrates mirror rotation in the unit circle model (single elliptic geometry).

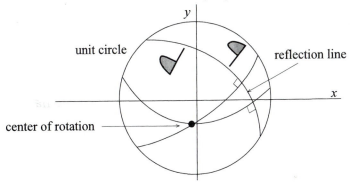

Figure 23.2 Mirror rotation in single elliptic geometry as the composition of three reflections

However, we prefer, in this chapter, to use a spherical model of elliptic geometry because it is easier to visualize transformations there than in the unit disk. On the sphere, the center of rotation becomes an axis of rotation. (See Figure 23.3.)

Mirror rotation is to elliptic geometry what glide reflection is to Euclidean geometry. We now state the fundamental theorem on reflections for elliptic geometry:

Theorem *Every combination of reflections in elliptic geometry is either a reflection, a rotation, or a mirror rotation. Every combination of an even number of reflections is a rotation; every combination of an odd number of reflections is either a reflection or a mirror rotation.*

Proof: This is a "moving lines" proof. (See Exercise 9.) ∎

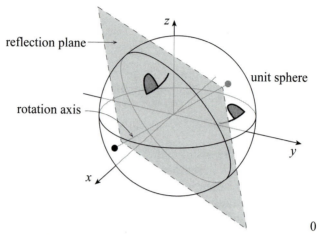

Figure 23.3 Mirror rotation on a sphere

B–The crystal in Figure 23.4 has a mirror *reflection* plane. Find it. This crystal also has mirror *rotation* symmetry.   Where is the rotation axis of the mirror rotation? What is the rotation angle?

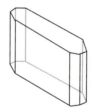

Figure 23.4 A copper sulfate crystal

Elliptic Symmetry Groups

The study of elliptic symmetry uses the *full* elliptic transformation group:

***Definition** The **full elliptic group** S^+ is the group of motions of the elliptic plane consisting of the transformations of the (special) elliptic group S (rotations of the elliptic plane), plus reflections and mirror rotations.*

The discrete subgroups of S^+ are the elliptic symmetry groups, just as the discrete subgroups of E^+ are the Euclidean symmetry groups. Like the Euclidean groups, the elliptic symmetry groups are classified by their fixed (or singular) elements. The theory is simplified by the fact that in elliptic geometry, if a transformation fixes a point, then it also fixes a line: the **polar line** (i.e., the equator) of that point. (See Exercise 11.) Conversely, if an elliptic transformation fixes a line, it also fixes a point. Therefore, there is no distinction in elliptic geometry between rosette and frieze groups. The elliptic symmetry groups with a fixed point (and fixed line) are called **point groups**.

The two families of Euclidean rosette groups reappear in elliptic geometry: the cyclic groups n generated by a single n-fold rotation axis and the dihedral groups nm generated by an n-fold rotation axis plus a reflection line (or plane) passing through the rotation axis.

An example of an object with dihedral symmetry is a pyramid. Figure 23.5 shows two versions of a pyramid, one with conventional, flat, Euclidean planar faces, the other inscribed on the curved surface of a sphere.

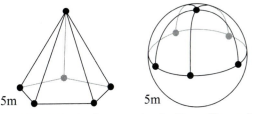

5m   5m

Figure 23.5 Dihedral symmetry in three dimensions

The International Symbols for elliptic symmetry groups are the same as for the corresponding Euclidean symmetry groups. As explained in Chapter 21, the "places" in each symbol indicate the symmetry in the direction of each coordinate axis. By convention, the object whose symmetry is being studied is oriented so that the axis of highest order rotation points in the x-direction. Then, if mirror symmetry is present, the object is rotated so that the mirror is in the y-direction if possible. This is illustrated in Figure 23.6.

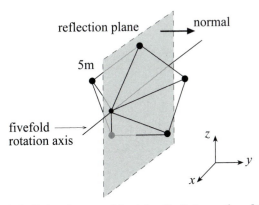

reflection plane   normal

5m

fivefold
rotation axis

Figure 23.6 Orienting an object for its International Symbol

Although simpler than dihedral symmetry, cyclic symmetry is less common in polyhedral form. To find a polyhedral example, requires a little work. Start with a truncated pyramid, and twist the top just a little relative to the base. This destroys the reflection planes. (See

Figure 23.7.)   Propellers are an example of a familiar object with cyclic symmetry.

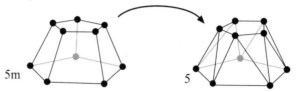

Figure 23.7 Destroying mirror planes creates a figure with cyclic symmetry.

The next family of point groups is formed by combining a rotation axis with a perpendicular reflection plane. Figure 23.8 illustrates this by joining the solid in Figure 23.7 with its mirror image. The twist in this solid is necessary to prevent the appearance of *parallel* (i.e., longitudinal) reflection planes. The symbol 5/m (or more generally, n/m, where $n = 1, 2, 3, \ldots$) is used to indicate that the rotation axis and the reflection plane have the *same* direction; in other words, the "5" and the "m" are in the same "place" in the International Symbol for this group.   Automobile tires are a commonplace object with this symmetry.

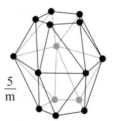

Figure 23.8 The point group 5/m

A fourth family of groups includes both perpendicular and parallel reflection planes. A double pyramid has this symmetry. (See Figure 23.9.)

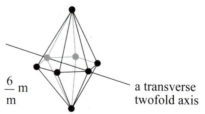

Figure 23.9 The point group (6/m)m

The symbol (6/m)m (more generally, $(n$/m)m where $n = 1, 2, 3, \ldots$) indicates that the group has reflection planes parallel *and* perpendicular to the rotation axis. This solid has still more symmetry than that indicated by the International Symbol, namely, twofold rotation axes perpendicular to the main rotation axis. These result from combining the two types of reflections. In Figure 23.9, there are six of these so-called **transverse** axes, all running through the central hexagon. This type of symmetry is found in many familiar objects both artificial (bricks, gears, nuts, bolts, etc.) and natural (oranges, apples, many seeds, and many crystals).

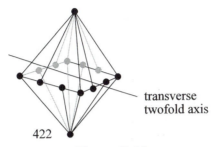

transverse
twofold axis

422

Figure 23.10

A further family of symmetry groups has only the transverse axes described in the previous paragraph, in addition to rotational symmetry. Figure 23.10 contains an example created by drawing extra vertexes and lines on a double pyramid. The extra vertexes and lines destroy the reflection planes (that the unadorned double pyramid has) while preserving the twofold axes. The symbol 423 (more generally, $n23$, where $n = 1, 2, 3, \ldots$) represents the main fourfold rotation symmetry and the traverse twofold axes. This family of groups (and the next two also), while well represented among crystals, does not appear much in manufactured objects.

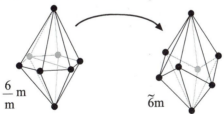

$\dfrac{6}{m}$ m $\tilde{6}$m

Figure 23.11 Creating a figure with mirror rotation symmetry

Two final families of elliptic point groups are built around mirror reflections. To visualize a figure from the first of these families, start with a double pyramid. (See Figure 23.11.) Now, alternately raise and lower the vertexes around the "equator." This creates a solid in which the sixfold axis has become a threefold axis and the perpendicular (equatorial) reflection plane has disappeared. The missing rotation and reflection plane are still there, however, combined in the form of a mirror rotation. That is, the solid is invariant under the motion of a rotation of 60° composed with a perpendicular reflection. The parallel reflection planes of the original double pyramid are still present. This symmetry group is notated $\tilde{6}$m (more generally, $\tilde{n}$m for $n = 2, 4, 6, 8, \ldots$). *Note: n must be even.*

The last family of elliptic point groups has only the mirror rotation from the preceding group without the parallel reflection planes. A solid with this symmetry can be constructed by truncating Figure 23.11 and then twisting the top and bottom (see Figure 23.12).

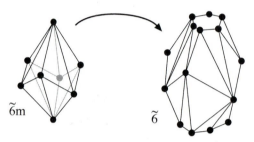

Figure 23.12 The point group $\tilde{6}$

Elliptic Point Group Summary

There are seven families of elliptic point groups:

1. *n*   pure rotational symmetry
2. *n* m   rotation symmetry + parallel reflection plane
3. *n*/m   rotational symmetry + a perpendicular reflection plane
4. *n*/m m rotational symmetry + parallel and perpendicular reflection planes
5. *n*23   rotational symmetry + transverse 2-fold axes
6. $\tilde{n}$ pure mirror-rotation symmetry
7. $\tilde{n}$ m   mirror-rotation symmetry + parallel reflection plane

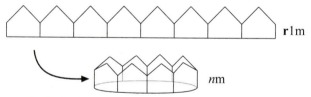

Figure 23.13 The frieze r1m makes a crown with symmetry *n*m

Each family of elliptic point groups is linked with a family of plane Euclidean frieze groups. If a frieze is wrapped around a cylinder (making a crown), it shifts from Euclidean symmetry to elliptic symmetry. (See Figure 23.13 and Exercise 20.)

C–Look for reflections, rotations, transverse twofold axes, and mirror rotations in the objects in Figure 23.14.

D–Identify the major rotation axis of each object in Figure 23.14. What is the symbol for the symmetry of each?

Figure 23.14 A bevel-toothed gear, a ratchet, a crosscut circular-saw blade

Polyhedral Groups

In addition to the point groups, there are seven more elliptic symmetry groups analogous to the network groups of plane Euclidean geometry. As groups of motions of the sphere (i.e., as subgroups of the extended elliptic group S^+), these groups do not have a fixed point. Instead, they have several rotation axes of order 3 or higher. They are called **polyhedral groups** because they include the symmetry groups of the five **Platonic solids**: tetrahedron, cube, octahedron, icosahedron, and dodecahedron. Among these five solids, there are only three distinct symmetry groups (illustrated in Figure 23.15), since the cube and octahedron (likewise, the icosahedron and dodecahedron) have the same symmetry.

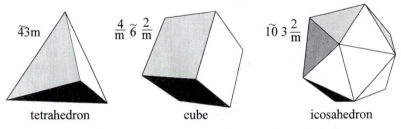

tetrahedron   cube   icosahedron

Figure 23.15 The full polyhedral groups

There are also reduced versions of the polyhedral groups without reflection symmetry. These are illustrated in Figure 23.16, where rosettes are drawn on the faces of the solids to destroy their *reflection* symmetry without altering their *rotational* symmetry.

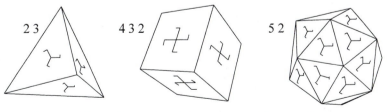

Figure 23.16 The polyhedral groups without reflections

One final group is obtained by drawing on the faces of the cube a pattern that destroys only some of the reflection symmetry. (See Figure 23.17.)

Note: The International Symbols for the polyhedral groups use a different set of axes than that used to create the symbols of the other groups. (See Exercise 21.)

$\tilde{6}\dfrac{2}{m}$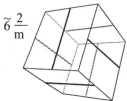

Figure 23.17 The exceptional polyhedral group

E–Choose a polyhedron from one of the Figures 23.15, 23.16, or 23.17. Find the rotation axes, mirror rotation axes, and/or reflection planes indicated by the symbol for the symmetry group of the polyhedron. For example, where are the reflection planes indicated by the "m" in the symbol $\overline{4}$ 3m?

Three-dimensional Euclidean Symmetry

The fundamental symmetry transformation in three-dimensional Euclidean geometry is reflection in a plane. We have already described all transformations that are combinations of reflections, namely, reflections themselves, rotations, translations, glide reflections, mirror rotations, and screw motions. They are called **proper** or **improper**, according to whether they are composed of an even or an odd number of reflections.

Screw motions, which were introduced in Chapter 18, are a combination of a rotation and translation in the same direction as the rotation axis. Screw motions, the most complicated of Euclidean three-dimensional transformations, are composed of four reflections, hence are a proper transformation.

Definition *The **three-dimensional full Euclidean transformation group** $\mathbf{R}^+$ is the group of motions of three-dimensional Euclidean space hyperbolic plane consisting of the transformations of the group $\mathbf{R}$ (rotations, translations, and screw motions), plus reflections and glide reflections.*

The discrete subgroups of $\mathbf{R}^+$ are the three-dimensional symmetry groups. As in our earlier treatment of plane Euclidean and elliptic symmetry, these groups are classified according to their invariant or fixed elements. Now that we are in three dimensions, there are three possible fixed elements: points, lines, and planes. This gives rise to four types of group:

> **point groups**: have a fixed point
> **rod groups**: no fixed point, fixed line

layer groups: no fixed point, no fixed line, fixed plane
space groups: no fixed elements.

The extra freedom of movement in three dimensions means that these groups are much more numerous than the plane symmetry groups. Even counting only groups that satisfy the crystallographic restriction (described later), there are still 32 point groups, 75 rod groups, 80 layer groups, and 230 space groups. We will not enumerate all these groups, but only comment on the two most important types.

F–Can a point group contain a screw motion?

Point Groups

The importance of the point groups is that they are the symmetry groups of ordinary, finite objects in three-dimensional space: the kind of objects that we encounter in everyday life. Every such object F has a **center of mass** or **centroid**, which is a fixed point of all its symmetry transformations. Thus, the symmetry group, $\ast(F)$, is a point group. As it happens, we have already enumerated all of these groups, as indicated by the following theorem:

Theorem *The three-dimensional point groups are the same as the elliptic discrete symmetry groups.*

Proof: Let G be a three-dimensional point group. By applying a Euclidean transformation, we may assume that the fixed point of G is at the origin. Let T be a transformation from G. Since T fixes the origin, T also maps whole lines through the origin to other such lines. These lines are a model for elliptic geometry. (See Chapter 13.) Alternatively, we can point out that T leaves invariant the unit sphere, another model of elliptic geometry. In any event, we see that G is also a group of motions of the elliptic plane. It is easy to see that G is the same group whether viewed as a transformation of three-dimensional Euclidean space or of the elliptic plane. ∎

Therefore, the seven families of discrete elliptic symmetry groups plus the seven polyhedral groups are also the Euclidean point groups and hence the symmetry groups of all ordinary, finite objects.

In particular, the point groups are the symmetry groups of all possible crystals. Crystallography is one of the most important applications of symmetry groups. Because of their *internal* symmetry (explained later), crystal symmetry groups are subject to the **crystallographic restriction**, meaning that they contain no rotations of order 5, nor any of order 7 or greater. If the list of elliptic symmetry groups is purged of the groups that violate this restriction, just 32 are left. (See Exercise 23.)

Therefore, crystals divide into 32 different **crystal classes** according to their point group. This group gives information on the internal structure of the crystal and also physical and chemical properties of the substance out of which the crystal is made. For example, if the point symmetry group of the crystal has no mirror reflections, then the crystal has handedness and exists in two mirror symmetric forms, called **enantiomorphic** forms.

Space Groups

The space groups are the most important of the three-dimensional groups because they are the symmetry groups of all substances that form crystals. The subject is too complicated to present in detail here, but we will describe the basic features of space group symmetry, and the resulting classification of crystals. Given the very small interatomic dimensions in crystals ($\sim 10^{-15}$ m), this classification probably accurately reflects the nature of crystals, even in a non-Euclidean universe.

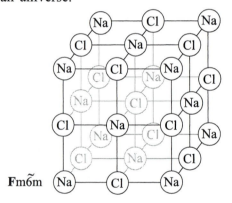

Fm$\tilde{6}$m

Figure 23.18 A typical space group

Consider, for example, rock salt (NaCl), a typical crystalline substance. Figure 23.18 depicts its internal structure. Note that sodium (Na) and chlorine (Cl) atoms both appear in the crystal at regular translation intervals in all three coordinate directions. The locations of each of these elements form a **lattice**: a regular arrangement of points in space with translational symmetry in three independent directions. The lattice appearing in rock salt is called the face-centered cubic lattice. To see why, observe the positions of the sodium atoms in Figure 23.18. They appear at the vertices of a system of cubes with an extra atom at the center of each face of each cube. The sodium and chlorine atoms occupy places in separate but interlocking face-centered cubic lattices.

The International Symbol, **Fm$\tilde{6}$m**, for the space group of rock salt begins with the letter **F** which stands for the face-centered cubic lattice. The remaining three places of the symbol reflect symmetries of the sodium and chlorine molecules, which are also present in the lattice. Alone, each sodium and chlorine molecule has full cubic point symmetry. These symmetries are all also present in the face-centered cubic lattice.

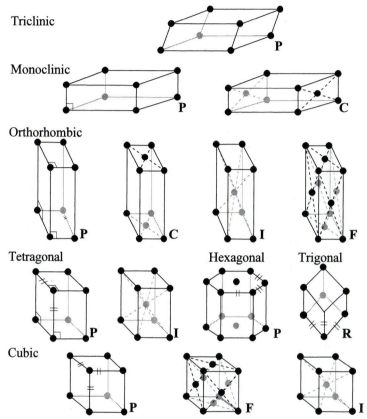

Triclinic

Monoclinic

Orthorhombic

Tetragonal   Hexagonal   Trigonal

Cubic

Figure 23.19 The 14 Bravais lattices and 7 crystal systems

All 230 space groups are fundamentally like the group $Fm\bar{3}m$, built out of a combination of translational and point symmetry. In each case, a lattice describes the translational symmetry. A crystal (or, more generally, any object with space group symmetry) is formed by choosing a translation lattice and placing objects with compatible point symmetry at each point of this lattice. The possible point symmetries are exactly those described in connection with elliptic geometry and Euclidean geometry earlier in the chapter. To be compatible with a lattice, the point symmetry transformations must carry lattice points to lattice points.

In two dimensions, there were only five types of lattice. (See Chapter 21.) In three dimensions, however, there are 14, called the

Bravais lattices. The 14 Bravais lattices divide crystals into 14 different geometric types. Substances that use different lattices will have different chemical and physical behavior.

The 14 Bravais lattices themselves actually have only 7 different symmetry groups. This results in still another classification of space groups and crystals into 7 **crystal systems.** The Bravais lattices and crystal systems are illustrated in Figure 23.19 through their **primitive translation cell**: a polyhedron (usually a parallelepiped) whose edges are formed from the shortest independent translations (called, therefore, the **primitive translation axes**) in the lattice.

G–Draw a body-centered cubic lattice (symbol **I**).

H–The symmetry of rock salt includes vertical fourfold and two-fold axes. Locate these in Figure 23.18.

The Crystallographic Restriction

In all of the 230 space groups there are no fivefold rotation axes, nor any rotation axes of order seven or higher. Inspection of Figure 23.19 explains why: None of the Bravais lattices have any rotation symmetry except of orders 2, 3, 4, or 6. This limited repertoire of rotational symmetry is the **crystallographic restriction**. A real crystal (diamond or other gemstone, for example) obeys this restriction because, although the crystal contains only a finite number of molecules, in principle it could grow to fill three-dimensional space, so its internal structure is dictated by the space groups.

However, it is possible for a single molecule, that is not part of a crystalline structure (i.e., without translational symmetry) to have the symmetries forbidden by the crystallographic restriction. Such molecules of carbon (with iscosahedral symmetry) were discovered only very recently (1989). They are called **fullerenes** (nickname: Buckyballs) after the inventor Buckminster Fuller, whose geodesic domes have approximately the same symmetry.

Hyperbolic Symmetry Transformations

Reflection is also fundamental to hyperbolic symmetry. Thus, the first question is: What transformations in hyperbolic geometry result from the combination of two or more reflections? The first part of the answer is in the next theorem.

Theorem *Every combination of an even number of reflections in hyperbolic straight lines is a transformation from the hyperbolic group* **H** *either a hyperbolic rotation, a parallel displacement, or a hyperbolic translation.*

Proof: Which transformation results from the composition of reflections in two hyperbolic straight lines depends on whether the lines intersect, are parallel, or are hyperparallel. This result is established in Exercise 15. ■

The fact that there are three types of transformation here, rather than two (as in Euclidean geometry) or one (in elliptic geometry) indicates the greater complexity of hyperbolic symmetry.

To further explore combinations of reflections in hyperbolic geometry we import from the Euclidean plane the technique of "moving the lines" exploited in Chapter 20. The following theorem explains how this is done:

Theorem (*Moving the lines in hyperbolic geometry*) *Let λ_1, λ_2, λ_3, and λ_4 be hyperbolic straight lines. Let S be a transformation of hyperbolic geometry that moves λ_1 to λ_3, and λ_2 to λ_4. Let T be the composition of reflections in λ_1 and λ_2, and let T′ be the composition of reflections in λ_3 and λ_4.*

Then T and T′ are the same *transformation under the following circumstances:*

(a) If λ_1 and λ_2 intersect in a point p, and S is a hyperbolic rotation about p,

(b) If λ_1 and λ_2 intersect in an ideal point p, and S is a parallel displacement about p,

(c) If λ_1 and λ_2 are hyperparallel and S is a hyperbolic translation whose axis is their common perpendicular.

Proof: See Exercise 16. ■

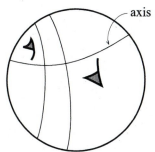

Figure 23.20 A hyperbolic glide reflection

We can now investigate combinations of three reflections. Immediately, we encounter the hyperbolic glide reflection: the composition of a hyperbolic translation, and a reflection in the axis of translation. (See Figure 23.20.) Rather surprisingly, this is the only additional improper transformation in hyperbolic geometry.

Theorem *Every combination of hyperbolic reflections is either a hyperbolic reflection, a hyperbolic glide reflection, a rotation, a parallel displacement, or a hyperbolic translation.*

Proof: This is a "moving the lines" proof. See Exercise 17.

Hyperbolic Symmetry Groups

The enumeration of all discrete symmetry groups of the hyperbolic plane is an extremely complicated subject. There is incredibly greater variety here than in the elliptic or Euclidean case. Problems relating to hyperbolic (and pseudo-Euclidean) symmetry groups have been in the forefront of mathematical research for over 100 years, ever since the subject was initiated by Poincaré. Over the years, interest in it has only increased, because of connections with problems in algebra, number theory, and, of course, relativistic physics. Just recently, the work of Thurston has drawn still more attention to the problem of classifying and enumerating the symmetry groups of *solid* hyperbolic geometry (see [H5]).

To begin we extend the hyperbolic group, as we earlier did the Euclidean and elliptic groups:

Definition The ***full hyperbolic transformation group*** **H**⁺ *is the group of motions of the hyperbolic plane consisting of the transformations of the group* **H** *(hyperbolic rotations, parallel displacements, and hyperbolic translations), plus reflections, and hyperbolic glide reflections.*

As in elliptic and Euclidean geometry, the discrete subgroups of **H**⁺ can be classified into rosette, frieze, and network groups. The rosette and frieze groups are relatively straightforward to enumerate (see Exercises 28 and 29), likewise a new category of symmetry groups: the horocycle groups (see Exercise 30).

These are all essentially one-dimensional groups because they have a fixed curve, cycle, straight line, or horocycle. It is in the network groups, which have truly planar symmetry, where real change appears. For example, there are no network groups based on rectangular or square lattices because there are no rectangles or squares in hyperbolic geometry. On the other hand, there are many, many groups based on triangles.

Definition *A discrete symmetry group G is a **triangle group** if all the transformations of G are combinations of reflections in the three sides of a triangle.*

The group **p4mm** is an example of a triangle group in the Euclidean plane. It is generated by the reflections in the three sides of an isosceles right triangle (see Figure 22.43). A hyperbolic triangle group is illustrated in Figure 23.21.

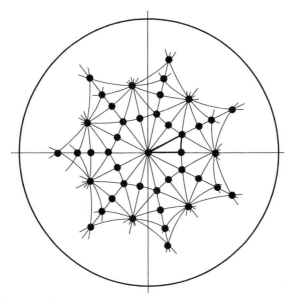

Figure 23.21 Construction of a triangle group network

Among the Euclidean network symmetry groups, there are only three triangle groups. (See Exercise 27.) Why so few? Recall that a discrete group is a group for which the orbits of any point intersect each disk in only a finite number of points. (See Chapter 21.) In order for a triangle group to meet this condition, the angles of the triangle must be of a special form. In a triangle group, rotations are created at every vertex of the defining triangle. If θ is the angle of the triangle, the rotation is by the angle 2θ. (See Figure 23.22.)

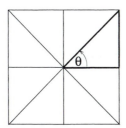

Figure 23.22 Reflection of a triangle around one vertex

In order that the triangle group be discrete, the orbit of a point under these rotations must be finite. Therefore 2θ must evenly divide 2π, or, in other words, θ must be of the form π/k, where k is an integer. (Another way to look at this condition is that it ensures that the triangles reflected around one vertex will cover the region around that vertex without significant overlap.)

For a triangle group in Euclidean geometry, this means that

$$\frac{\pi}{k} + \frac{\pi}{l} + \frac{\pi}{m} = \pi$$

where π/k, π/l, and π/m are the three angles of the triangle. Thus, the integers k, l, and m must satisfy

$$\frac{1}{k} + \frac{1}{l} + \frac{1}{m} = 1$$

There are (essentially) only three solutions of this equation. (See Exercise 27: The group **p4mm** corresponds to the solution $k = 4$, $l = 4$, $m = 2$.) There are also only a finite number of solutions of the corresponding equation in elliptic geometry.

On the other hand, in hyperbolic geometry, the integers k, l, and m need only satisfy

$$\frac{1}{k} + \frac{1}{l} + \frac{1}{m} < 1$$

and this inequality has an infinite number of solutions! One solution is $k = 2$, $l = 3$, and $m = 7$ upon which Figure 23.21 is based. Note how at some vertexes 4 triangles meet, at other vertexes 6, and at others 14, exactly twice the numbers 2, 3, and 7, respectively.

Summary

The enumeration and classification of discrete symmetry groups is carried out by the same process, regardless of the geometry (elliptic, hyperbolic, or Euclidean), and regardless of dimension. In every geometry:

(a) All symmetry transformations are combinations of reflections.

(b) The discrete symmetry groups are subgroups of the full transformation group of the geometry.

(c) Discrete symmetry groups are classified by their fixed elements: points, lines, and/or planes.

(d) The point groups are most numerous, often appearing in infinite families (as in elliptic geometry). As the dimension of the invariant element increases, the number of groups decreases, although the complexity of the individual groups increases.

There are also some significant differences among geometries. As might be expected, elliptic geometry has the simplest theory. Euclidean geometry is the next most complicated, and hyperbolic geometry the most complicated. Thus, for example, there are an infinite number of hyperbolic triangle groups while there are only a finite number of triangle groups in both elliptic and Euclidean geometry. There is also a significant increase in variety and complexity as the dimension of the geometry increases.

EXERCISES

Reflections in Möbius Geometry

1. Show that the composition of reflections in two clines is a Möbius transformation.

Hint: Use the formula for reflection in a cline given in Chapter 5.

2. Find two clines C and C' such that the composition of the reflections in C and C' is inversion.

3. Find two clines C and C' such that the composition of the reflections in C and C' is a homothetic transformation.

4. Prove that the result of composing reflections in two clines C and C' is an elliptic transformation if C and C' intersect, a parabolic transformation if C and C' are mutually tangent, and a hyperbolic transformation if C and C' do not touch.

Hint: Symmetry is a property of Möbius geometry. Use an Erlanger Programm proof.

5. Can every Möbius transformation be obtained as the result of combining reflections in two clines? Why or why not?

Reflections in Elliptic Geometry

6. Let the cline C is be elliptic straight line. Prove that reflection in C preserves pairs of diametrically opposite pairs of points in the elliptic plane.

Hint: Use the Erlanger Programm (in elliptic geometry).

7. Prove that the composition of reflections in two elliptic straight lines is a transformation from the elliptic group **S**.

8. Show that the composition of reflections in three elliptic straight lines is either a reflection or a mirror rotation.

Hint: This is a "moving the lines" proof.

9. Prove that the composition of any number of reflections in elliptic geometry is either a reflection, a rotation, or a mirror reflection.

Hint: Move the lines.

10. Prove that for every point p in elliptic geometry, there is a unique line λ that is perpendicular to every line passing through p. The line λ is called the **polar** of p. Conversely, prove that for every line λ, all the lines perpendicular to λ intersect at a point p. This point is called the **pole** of λ. Draw examples of a point and its polar in both the unit disk model and on the sphere.

11. Prove that if a transformation fixes a point p, then it also fixes the polar line of p and conversely.

12. Let p be a point in the elliptic plane. Let T be the rotation about p of 180 , and let S be the reflection in the polar of p. Show that in double elliptic geometry S and T are different transformations, but that in single elliptic geometry they are the same!

13. In single elliptic geometry, mirror reflection is a proper transformation. What kind of proper transformation is it? Combine this result with that of Exercise 12 to prove that in single elliptic geometry there are no improper transformations.

Reflections in Hyperbolic Geometry

14. Let the cline C be a hyperbolic straight line. Prove that reflection in C maps the unit disk to itself.

Hint: Use the Erlanger Programm *(in Möbius geometry).*

15. Prove that the combination of an even number of reflections in hyperbolic straight lines is a transformation from the hyperbolic group **H**.

16. Prove the theorem on moving the lines in hyperbolic geometry.

Hint: Use the Erlanger Programm *(in hyperbolic geometry). Be clever about which model you use (disk or half plane) for each case of the theorem.*

17. Prove that the combination of three or more hyperbolic reflections is a hyperbolic reflection, hyperbolic glide reflection, rotation, parallel displacement, or a hyperbolic translation.

Hint: Move the lines.

Elliptic Symmetry

18. Draw examples of figures in the unit circle model of elliptic geometry for these symmetry groups: 4, 4m, 4/m, 2/m m, 423, $\tilde{4}$, $\tilde{4}$m.

19. Decorate the faces of a hexagonal prism (see Figure 23.23) to produce a figure with each of these symmetry groups: 6, 3, 2, 6m, 3m, 6/m, (6/m)m, 623, $\tilde{6}$, $\tilde{6}$m.

Figure 23.23 A hexagonal prism

20. Each of the seven families of point groups corresponds (via wrapping, see Figure 23.13) with one of the families of plane frieze groups. Which elliptic point group family corresponds to which Euclidean frieze group family?

21. In Figures 23.15, 23.16, and 23.17 illustrating the polyhedral groups, locate the rotation axes and reflection planes that are indicated by the symbols for the groups. For example, where are the reflection planes indicated by the "m" in the symbol "$\tilde{4}3m$"?

Hint: These groups do not follow exactly the coordinates axes convention of the other International Symbols. For example, the rotation axes indicated by the "4" and the "3" in "$\tilde{4}3m$" are not at right angles! See what you can find!

22. Identify by symbol the symmetry group of these familiar objects: (a) a coffee mug
   - (b) a two-handled sugar bowl
   - (c) a brick
   - (d) an orange
   - (e) a hex nut (disregarding the thread)
   - (f) a hex nut (counting the thread)
   - (g) a snowflake
   - (h) a Phillips-head screw (disregarding the thread)
   - (i) a Phillips-head screw (counting the thread)
   - (j) a copper sulfate crystal (see Figure 23.4)
   - (k) a bicycle wheel (rim, spokes and hub, but no tire)
   - (l) a bicycle tire
   - (m) a bicycle innertube

Hint: You may want to get an example of an object in order to examine it thoroughly. First, find the axis of rotation with the smallest rotation angle. Orient this axis in the x-direction. Then look for other symmetry elements: reflections, mirror reflections, and transverse twofold axes.

23. List the symmetry groups of the 32 crystal classes.

Hint: Only 5 of the 7 polyhedral groups, plus 27 groups from the seven infinite families of point groups, satisfy the crystallographic restriction.

Warning: You may find more than 32 groups. Watch for duplications!

24. Which elliptic symmetry groups are triangle groups?

Hint: Let the angles of the triangle be p/k, p/l, and p/m. The angle sum is greater than p. Solve for k, l, and m.

Euclidean Symmetry

25. Draw and/or make a face-centered cubic lattice. Draw and/or make a centered orthorhombic lattice.

26. Find the point group symmetry of each Bravais lattice.

Hint: There are only seven different point group symmetries among the 14 Bravais lattices. All the cubic lattices, for example, have full cubic symmetry.

27. The symmetry of rock salt includes vertical fourfold and twofold screw motions. Locate these motions in Figure 23.18.

28. Which Euclidean symmetry groups are triangle groups?

29. Examine the objects in Figure 21.38. This time imagine that they are three-dimensional objects. Find the symmetry group of each.

Hyperbolic symmetry

30. Prove that rosette groups in hyperbolic geometry are exactly the same as in Euclidean geometry.

31. Describe the hyperbolic frieze groups.

32. A new type of discrete group appears in hyperbolic geometry: groups with an invariant horocycle. Call these groups **horocycle** groups. Describe all possible horocycle groups.

33. The print *Circle Limit IV*, by Maurits Escher (see page 356), is based on a hyperbolic triangle group. Which one?

PART VI

AXIOM SYSTEMS

GEOMETRY AND ART: PLATE VI

Paul Klee: *Seascape with Heavenly Body* (1920)

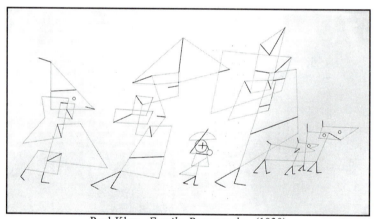

Paul Klee: *Family Promenade* (1930)

Nowhere does geometry so clearly supply the building blocks of art as in Paul Klee's work. Klee (who was well acquainted with mathematics, including parts of the theory of functions, analytic geometry, and spherical trigonometry) systematically uses geometrical motifs to express a playful view of the world. Note how few different shapes (mostly triangles and rectangles) are used in these two drawings. *Seascape with Heavenly Body* portrays a crystalline vision of reality characteristic of much of Klee's work. *Family Promenade* delicately suggests family relationships by the overlapping (or not) of geometric forms. For each of these works, Klee develops a different graphic language, using only a few shapes subject to a few simple rules. This process is similar in spirit to the exploration, in mathematics, of a few undefined terms subject to the postulates of an axiom system.

In contrast with this, one finds frequently now, on the part of persons who are interested only in the logical side of things and not in the side of perception or the general theory of knowledge, the opinion that *the axioms are only arbitrary statements which we set up at pleasure and the fundamental concepts, likewise, are only arbitrary symbols for things with which we wish to operate.* ... For one, I cannot share this point of view. I regard it, rather, as the death of all science. *The axioms of geometry are—according to my way of thinking—not arbitrary, but sensible statements, which are, in general, induced by space perception and are determined as to their precise content by expediency.*

     –Felix Klein (1908) in *Geometry* [B6]

Axiom systems are an alternative to the analytic models, which have hitherto been used to express geometric ideas in this book. Each approach has its own strengths and weaknesses; both are worth understanding. While analytic models make for a swift development of the theory and a certain ease of proof, axiom systems make crystal clear the fundamental ideas upon which each geometry depends.

Furthermore, no single axiom system, *or* model fulfills all purposes. As we have seen from the study of hyperbolic, elliptic, and projective geometry, more than one model is typically used to describe a geometry since different ideas are often more easily expressed in one model than another. Likewise, geometries can be described by more than one axiom system. Alternative axiom systems stress distinct sets of fundamental ideas and suggest different ways to develop a geometry.

Two axiom systems for plane geometry are presented here: Hilbert's and Bachmann's. Each system has its own viewpoint of what is important, in other words, has its own spin on plane geometry.

Hilbert's system (Chapter 24) is important historically. It was deliberately constructed to provide a complete system for Euclidean geometry without the logical gaps left by Euclid. Thus, Hilbert's view of geometry (as expressed by his axiom system) is essentially the same as Euclid's. Nevertheless, the appearance of this system at the beginning of the twentieth century (1899–1903), the work of an internationally renowned mathematician, drew attention to axiom systems and set the stage for much modern work with them.

Bachmann's system (Chapters 25 and 26) is more recent (1959) and embodies a more modern view of geometry. In his system, Bachmann expressed the fundamental ideas of geometry using transformations. Bachmann's system, therefore, fits in well with the geometric philosophy of the rest of this book.

24 HILBERT'S AXIOMS

One of the effects of the discovery of non-Euclidean geometries was the realization that Euclid's axioms are inadequate. Nineteenth-century geometers found numerous ways in which this or that assumption was implicitly used by Euclid without proper foundation. Today, many alternative sets of axioms for Euclidean geometry are known in which these assumptions are made explicit.

One of the first such sets was formulated by David Hilbert (in *Foundations of Geometry* [B4]). In this chapter, we present a version of Hilbert's axioms adapted to plane Euclidean geometry. (Both Hilbert's and Euclid's original systems are for solid geometry.)

Note: Euclid distinguished "axioms" from "postulates," but modern usage treats these terms as synonyms. Both refer to any statement accepted without proof as an underlying assumption of a formal mathematical theory.

Undefined Terms

Before stating a single axiom, it is necessary to list the terms to be used therein. Ideally, of course, every geometric term should be carefully defined. However, in a formal axiomatic system, some terms must be accepted without definition. This is the only way to avoid circular definitions or an infinite regress. Therefore, we need:

***Undefined Terms Points** and **lines** are names for the elements of two (distinct) sets. **Incidence** is a relationship that may (or may not) hold between a particular point and a particular line.*

Axioms of Incidence

Hilbert's first group of axioms, like Euclid's, concerns the determination of lines by points:

***Postulates** (Incidence)*
   (1) For every two points, there exists a line incident with both points.
   (2) For every two points, there is no more than one line incident with both points.
   (3) There exist at least two points incident with each line.

(4) There exist at least three points. Not all points are incident with the same line.

These four axioms replace Euclid's first axiom. Hilbert's language is a bit more precise than Euclid's and includes the uniqueness of the line determined by two points, something not stated by Euclid. Also, the fourth axiom insists that the geometry contain at least three non-collinear points.

On the basis of these few assumptions, not much can be established. Here is the only result Hilbert finds worth mentioning:

Theorem *Two distinct lines have either one or no points in common.*

Proof: See Exercise 1. ■

A–What minimum number of points and lines are needed to satisfy Hilbert's incidence axioms? Draw a picture of this minimal geometry.

Axioms of Order

By order, Hilbert means the arrangement of points on a single line. Euclid says nothing about this topic. Hilbert bases his treatment on a new undefined term:

Undefined Term Betweenness *is a relationship that may (or may not) hold between one point and two further points.*

Postulates *(Order)*

(1) If point p is between points q and r, then p, q, and r are all distinct points on a line, and p is also between r and q.

(2) For any two distinct points p and q, there is at least one point r on the line determined by p and q such that r is between p and q and at least one point s so situated that q lies between p and s.

(3) If p, q, and r are three points on the same line, then exactly one is between the other two.

(4) Any four distinct points on a line can always be labeled p, q, r, and s so that q is between p and r, and r is between q and s.

B–Draw pictures to illustrate these axioms.

These postulates were accepted implicitly by Euclid, along with many of their consequences. One consequence is the fact that there are an infinite number of points on a line. In order to state this result, we introduce the first *defined* term in Hilbert's system:

Definition *Let p and q be distinct points. The **segment** [pq] is the set of points including p, q, and all the points between them. The points* *p* and *q* themselves are called the **extremities** of [*pq*]. Note that the segment [*pq*] is a proper subset of the line *pq*, the straight line determined by the points *p* and *q*.

Theorem   *Every segment contains an infinite number of distinct points.*

Proof: See Exercise 3. ■

The axioms of order only specify the order of points on a single line. Hilbert adds a further axiom concerning the order of points in the plane. We give a simple alternative to Hilbert's original axiom:

Postulate *(Separation)   Every line* λ *separates the points not on* λ *into two sets so that two points p and q are (or are not) in different sets according as the segment* [*pq*] *does (or does not) contain a point of* λ.

The two sets referred to in this axiom are called the **sides** of the line λ. Using this rather subtle postulate, we can prove the following result:

Theorem *(Pasch)   Let p, q, and r be three points that are not on the same line, and let* λ *be a line that does not meet any of these points. Then, if* λ *passes through a point of the segment* [*pq*], *it will also pass through a point of the segment* [*qr*] *or a point of the segment* [*pr*] *(but not both).*

Proof: See Exercise 4. ■

Pasch was the first to point out this gap in Euclid and repair it. Hilbert originally used Pasch's theorem as an axiom, later substituting the separation axiom. To show the power of these results, we will prove that a triangle separates the plane into an inside and an outside. This requires the following definition:

Definition *A system of segments* [*pq*], [*qr*], [*rs*], . . . , [*tu*] *is called a* **broken line** *and abbreviated* [*pqrs* . . . *tu*]. *Given three points p, q, and r, not all on one line, the broken line* [*pqrp*] *is called a* **triangle** *and notated* Δ*pqr*. *The points p, q, and r are called* **vertexes** *of the triangle; the segments* [*pq*], [*qr*], *and* [*rp*] *are called* **sides** *of the triangle.*

Theorem *A triangle* Δ*pqr separates the points not on the triangle into two sets. Two points u and v are in the same set if, and only if, there is a broken line* [*u* . . *v*] *that does not contain a point of the* Δ*pqr. One of these two sets, called the* **exterior** *of the triangle, contains an entire line. The other set, called the* **interior**, *does not contain an entire line.*

Proof: Let Δ*pqr* be a triangle. The first step of our proof is to find a line that does not intersect Δ*pqr*. Figure 24.1 shows how this is accomplished. Formally, the reasoning is as follows. By the second axiom of order, there exist points *u* and *v* such that *q* is between *p* and *u* and *r* is between *p* and *v*. We claim that the line *uv* does not intersect Δ*pqr*. The proof uses a lemma:

Lemma *If a line does not intersect two sides of a triangle, then it does not intersect the third side.*

Proof: See Exercise 8. ∎

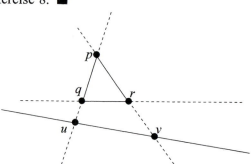

Figure 24.1 A triangle has an interior and an exterior.

Returning to the proof of the theorem, observe that the line *uv* does not intersect the segment [*pq*], since it intersects the line *pq* at the point *u*, which is not between *p* and *q* (by the third axiom of order). Similarly, the line *uv* does *not* intersect the segment [*pr*]. Therefore, by the lemma, *uv* cannot intersect Δ*pqr* at all.

By the separation axiom, *u* and *v* are on the same side of the line *qr*. Call this side 1 of *qr*. Then *p* is on the other side of *qr*, which we call side 2. Note next that *u* cannot be on the line *pr*, for then line *uv* would be the same as *pr*, and we know from the preceding paragraph that they are different. Name the sides of *pr* in such a way that *u* is on side 1 of *pr*. Similarly, name the sides of *pq* so that *v* is on side 1 of *pq*.

We now define the **exterior** of Δ*pqr* as all points that are on side 1 of at least one of the lines *pq*, *pr*, or *qr*. The **interior**, then, consists of the points that are on side #2 of all three of these lines. It follows immediately that Δ*pqr*, its interior, and its exterior are three mutually exclusive sets that together include every point of the plane.

Now, from the separation axiom, it follows that if *s* and *t* are in the interior of Δ*pqr*, then all points on the segment [*st*] are also on side 2 of all three lines *pq*, *pr*, and *qr*, so that [*st*] lies entirely in the interior. This proves that any two points in the interior are connected by a broken line entirely in the interior.

Next, consider a point *w* on the line *uv*. According to the third axiom of order, one of the three points *u*, *v*, *w* is between the other two. In all three cases, we will conclude that w is in the exterior of Δpqr. Thus, if w is between u and v, then, since both u and v are on side 1 of qr, w is also on side #1 of qr, and hence in the exterior. Or, if u is between w and v, then u and w are on the same side of pr; that is, both are on side 1 of pr, and again, w is in the exterior of Δpqr. In the third case, v is between w and u. Then w and v are on the same side of pq, side 1, and again, w is in the exterior of Δpqr. This proves that the whole line uv is exterior to Δpqr.

It remains to prove (a) that any two points in the exterior of Δpqr can be connected by a broken line entirely in the exterior, (b) that any broken line connecting a point in the interior with a point in the exterior must intersect Δpqr, and (c) that the interior of Δpqr does not contain an entire straight line. We will prove (a), leaving (b) and (c) as exercises. (See Exercises 9 and 10.)

Therefore, let w now be any point in the exterior of Δpqr. We will prove that w can be connected to the particular point u by a broken line lying entirely in the exterior of Δpqr. It will then follow that any two points in the exterior can be so connected. Since w is in the exterior of Δpqr, w is on side 1 of one of the three lines pq, pr, or qr. If w is on side 1 of either pr or qr, then the line segment $[wu]$ is entirely contained on the same side, since u is on side 1 of pr and qr. The only interesting case, therefore, is if w is on side 1 of pq. Then the segment $[vw]$ is entirely in side 1 of pq, hence the broken line $[uvw]$ connects u and w and lies entirely in the exterior of Δpqr. ∎

Several features of this proof are characteristic of axiomatic geometry and therefore worth emphasizing. First is the role of figures. Although a figure was drawn to aid in understanding the proof, it was not part of the proof itself. Every step of the proof was justified by appeal to one of the axioms. It is important to realize that a major goal of Hilbert's axiom system is to avoid the casual dependence on figures that, it is believed, caused Euclid and other geometers to overlook crucial assumptions. Figures, of course, are still important. They are a necessary tool for finding proofs and understanding and applying results. But, figures are not part of the logical development of the geometry.

In contrast, proofs in the rest of this book often depend on diagrams and their manipulation. The incorporation of figures in a geometric proof *is* logically justified when the figure is produced by calculation. This use of a figure is the same as using a table or any other device to summarize calculated results. A good example of a calculated figure, of course, is the application of a transformation to a given figure.

Another characteristic feature of the preceding proof is the use of constructed elements, that is, points and lines not mentioned in the statement of the theorem. Auxiliary (or constructed) elements are crucial to most proofs in Euclidean geometry; few can be carried out without them.   Many of Hilbert's postulates clearly prepare for construction by asserting the existence of points and lines with various properties.

The results proved so far using Hilbert's axioms may strike the reader as obvious, as indeed they are from the viewpoint of a convinced Euclidean geometer.   However, no matter how obvious, if they are to be part of an axiomatic development of Euclidean geometry, then they must be made explicit.   This means that they must be assumed outright (as axioms) or proved on the basis of other assumptions.   In further defense of these results, note that they represent ideas that are genuinely difficult to recognize.   Indeed, it took roughly 2,200 years (from 300 BC to 1900 AD) for them to be discovered and written down.

C–How many of the axioms introduced so far assert the existence of lines or points?

D–Verify that every statement in the preceding proof is an application of an axiom or follows from preceding statements by pure logic.

The Parallel Postulate

Hilbert chooses Playfair's axiom, the most popular version of the parallel postulate.   This makes Hilbert's geometry a Euclidean geometry.   Other axioms about parallelism would lead to different geometries.

Postulate *(Playfair's Axiom)   Through a point p not on a line λ, there is exactly one line that does not meet the line λ.*

Theorem *If two lines λ and μ do not meet a third line κ, then they do not meet each other.*

Proof: See Exercise 11. ∎

Axioms of Congruence

Congruence is another major idea not mentioned by Euclid.   At first it is another undefined term in Hilbert's system; later it becomes a defined term as well.

Undefined Term   Congruence *is a relationship that may (or may not) hold between two line segments.   Congruence is later extended to angles (also without definition).*

Postulates *(Congruence of segments)*
(1)   If p and q are two points on a line λ, and if p′ is a point on the same or on another line λ′, then there is exactly one point q′ on a

With this postulate it is possible to establish all the usual Euclidean theorems on the congruence of triangles. In particular:

Theorem *(SAS) If, for two triangles, $\triangle pqr$ and $\triangle p'q'r'$, $[pq]$ is congruent to $[p'q']$, $[pr]$ is congruent to $[p'r']$, and $\angle qpr$ is congruent to $\angle q'p'r'$, then the two triangles are congruent to each other.*

Proof: See Exercise 15. ■

Coordinatizing the Plane

Although no axioms of Euclid or Hilbert mention distance, angle measure, or coordinates, our modern point of view demands the development of these metric concepts because then the convenient techniques of analytic geometry are available for further geometrical developments. It is clear that Hilbert's axioms for congruence of line segments and angles anticipate coordinization by preparing the way for the addition of linear and angular measures. And, indeed, with Hilbert's axioms, coordinates *can* be introduced and the full power of analytic geometry made available. The details of this development are too lengthy, however, to describe here.

Completing Hilbert's Axioms

Hilbert's axiom system *is* complete, in the sense that all the usual theorems of Euclidean geometry can be deduced from his system. But, the axioms do not specify what numbers will appear as distances, angle measures, and coordinates of points. Normally, we think of the real numbers as corresponding to the points on a line. After all, this is why the real axis is called real. But, if no more axioms are added to Hilbert's system, the real axis need *not* be coordinatized by the *real* number system. The *rational* numbers (which exclude, for example, such numbers as π and $\sqrt{2}$, which cannot be expressed as a quotient of whole numbers) are sufficient to coordinatize the Euclidean plane as described by Hilbert's axioms so far. Hilbert, therefore, added two more axioms to ensure that, when coordinatized, the lines in his geometry would contain points corresponding to *all* real numbers. Here are these axioms:

Postulate *(Archimedean property) Given points p and q and a point p_1 between p and q, let points p_2, p_3, . . . be such that p_1 is between p and p_2, p_2 is between p_1 and p_3, and so forth. Suppose also that $[pp_1] \cong [p_1 p_2] \cong [p_2 p_3] \cong \dots$. Then there is a positive integer n so that q is between p and p_n.*

It requires some thought to understand this axiom! In less formal terms, it says that, given two line segments $[pq]$ and $[pp_1]$, where the first is longer than the second, then starting at p and measuring off line segments equal to $[pp_1]$ we will eventually pass q. In other words, any segment can be used to measure off (approximately) a longer segment. Still less formally, the axiom says that no line segment is infinitely small.

G–Draw a picture to illustrate the Archimedean postulate. Explain how this postulate implies that no line segment is infinitely small.

Hilbert's final postulate is very clever:

Postulate *(Hilbert's completeness axiom)   No additional points or lines can be added to this system without violating one of the preceding postulates.*

This axiom says, in effect, that lines in Hilbert's plane have to have the maximum possible number of points consistent with his axioms. Hilbert wrote his completeness axiom knowing that the reals are the largest number system that will coordinatize a line (given some natural assumptions about the arithmetic of the coordinatizing numbers). Thus, (although we present none of the details), Hilbert's completeness axiom demands that the points of a line be coordinatized by the real numbers.

Properties of Axiom Systems

Hilbert's axiom system for Euclidean geometry is concise, descriptive, and elegant. We list here a few of its other important properties.

Consistency

An axiom system is **consistent** if its axioms do not contradict each other, in other words, if it is not possible to deduce a contradiction from the axioms. Consistency is obviously an *essential* property of any mathematical system. An inconsistent system is useless since anything follows from a contradiction.

Unfortunately, it is not possible to prove consistency, except for very simple systems. All that can be proved in most cases is **relative consistency**, that is, that the given axiom system is at least as consistent as some other part of mathematics. Relative consistency is proven by finding a model for the axiom system in another mathematical field. Thus, suppose that **A** is a new axiom system and that we find a model for the system **A** in a more familiar part of mathematics called **B**. It then follows that any inconsistency or contradiction in **A** is also an inconsistency in **B** (since **B** contains a model for **A**). This shows that **A** is at least as consistent as **B**. Usually, **B** is an area of mathematics whose consistency is not proven (no really complicated part of mathematics is provably consistent) but is nonetheless so well known that an inconsistency in **B** is considered very unlikely. In this way, the consistency of **A** is shown to be unlikely. (For more on relative consistency, see Chapter 27.)

Euclidean geometry can be modeled in the Cartesian plane. Such a model was introduced in Chapter 4. This model satisfies all of Hilbert's axioms and it can be developed without any reference to geometry. (See the discussion of this point in Chapter 4.) The existence of such a model demonstrates that Hilbert's axiom system is at least as consistent as Cartesian mathematics.

Independence

Let **x** be an axiom of a system **A**, and suppose, in addition, that a model can be found in which **x** is *false* while the other axioms of **A** are *true*. The existence of such a model shows that the content of the other axioms is not sufficient to deduce the truth of **x**; in other words, **x** supplies information not contained in the other axioms. In this case, **x** is called **independent** of the other axioms of **A**. If all axioms in a system **A** are independent, then the whole system is called **independent.**

Independence is an *elegant* feature of an axiom system. It means that the axioms have been carefully formulated to avoid duplication. A system of independent axioms shows particularly clearly the relationship of each concept to the system as a whole.

Hilbert's system is such a system: Each of Hilbert's axioms is independent of all the others (as Hilbert carefully proved).

Categorical Systems

An axiom system is **categorical** if all models of the system are essentially the same, or, more technically, are isomorphic. Two models are **isomorphic** if there is a one-to-one correspondence between their elements that preserves all significant features of the models. For models of geometries, the concept of isomorphism was formally defined in Chapter 4. It is the same as saying that two models are models of the same abstract geometry. For other mathematical systems, a similar definition of isomorphism can usually be given on the basis of some form of transformation.

Categorical axiom systems are interesting because such systems describe a *unique* mathematical object. Most axiom systems are not categorical; they characterize a *family* of mathematical objects, objects with some common features (the ones described in the axiom system), but otherwise very different.

A noncategorical system is usually more useful than a categorical system because the theorems of a noncategorical system apply to a variety of different models, whereas the theorems of a categorical system apply (essentially) to only one model. In other words, a noncategorical system has a flexibility of application that a categorical system lacks. *That someone has bothered to devise a categorical axiom system for a particular mathematical object is a sign that that object is considered important.*

Hilbert's axiom system is categorical; it models Euclidean geometry alone.

Summary

Hilbert fixed Euclid up by building an axiom system that, as far as we know, is not only consistent, independent, and categorical, but also correct; that is it describes the geometry of Euclid without any logical gaps or missing assumptions.

EXERCISES

Incidence

1. Using Hilbert's axioms of incidence, prove that two distinct lines have either one or no points in common.

2. What is the smallest geometry satisfying Hilbert's axioms of incidence alone?

Order and Separation

3. Using Hilbert's axioms of incidence and order, prove that a line contains an infinite number of points.

4. Derive Pasch's axiom from Hilbert's axioms of separation.

Hint: According to the statement of Pasch's axiom, p and q are on different sides of λ. On which side is r?

5. Let p be a point on a line λ. For any two points q and r on λ we say that q and r are on the **same side** of p if the segment qr does not contain p. Prove that this is an equivalence relation. Prove that there are exactly two sides of a point on a line.

6. Let λ be a line. Define the relation of being on the same side of λ as follows: Two points p and q are on the **same side** of λ if the segment pq does *not* intersect λ. Prove that this relation is an equivalence relation.

7. In Euclidean geometry, lines are not reentrant. Which of Hilbert's axioms forces this conclusion?

8. Prove the lemma used in the proof of the theorem on the interior and exterior of a triangle.

9. Help complete the proof of the theorem on the interior and exterior of a triangle by proving that any broken line connecting a point in the interior with a point in the exterior must intersect the triangle Δpqr.

Hint: Let B be a broken line connecting a point in the interior with a point in the exterior of the triangle. Argue that one segment [st] of B connects a point s in the interior with a point t in the exterior of Δpqr. Then [st] contains a point of one of the lines pq, qr, or pr. By eliminating other possibilities, show that the point of intersection is actually on Δpqr.

10. Help complete the proof of the theorem on the interior and exterior of a triangle by proving that the interior of Δpqr does not contain an entire straight line.

Parallelism

11. Prove the theorem on parallel lines.

12. Prove that parallelism is an equivalence relation for lines.

Note: Assume that a line is parallel to itself.

Congruence

13. Prove that congruence of line segments is an equivalence relation.

14. Prove that congruence of angles is an equivalence relation.

15. Prove the full SAS congruence theorem.

Hint: Use proof by contradiction.

16. Prove the ASA congruence of triangles: For two triangles Δpqr and Δp'q'r', if pq ≅ p'q', ∠pqr ≅ ∠p'q'r', and ∠rpq ≅ ∠r'p'q', then the two triangles are congruent.

Angles

17. Two angles having the same vertex and sharing a ray are called **supplementary** angles if the rays that they do not share form a straight line. Prove: If two angles are congruent, then their supplementary angles are congruent.

Hint: Use congruent triangles. You will have to construct the triangles by laying off line segments on the sides of the given angles. Use the axioms on congruence of segments to justify the construction.

18. Two angles having the same vertex whose rays form two straight lines are called **vertical** angles. Prove: If two angles are congruent, then their vertical angles are congruent.

19. Let pq, pr, ps, and p'q', p'r', p's' be two sets of three half rays, each emanating from a single point. Prove: If ∠qpr ≅ ∠q'p'r' and ∠rps ≅ ∠r'p's', then ∠qps ≅ ∠q'p's'.

20. A **right** angle is an angle that is congruent to its supplementary angle. Hilbert suggests, "deduce the following simple theorem, which Euclid held–although it seems to me wrongly–to be an axiom: All right angles are congruent to one another."

Hint: Use proof by contradiction.

Circles

21. Formulate definitions of "circle," "center," and "radius" appropriate for use with Hilbert's axiom system.

Note: Your definitions should not make any use of distance.

22. Use the definitions from the previous exercise to deduce, from Hilbert's axioms, Euclid's fourth axiom: A circle can be described with any center and any radius.

Properties of axiom systems

23. How many of Hilbert's axioms are existence statements? How many assert the uniqueness of something already the subject of an existence axiom? What is the character of the few remaining axioms that are neither existence nor uniqueness statements?

24. The axiom system for an **equivalence relation** has three axioms: the reflexive, symmetric, and transitive properties (see page 327). Is this axiom system consistent? Are the axioms independent? Is the system categorical?

25. The axiom system for a **metric space** is as follows:

Undefined Terms **Points** *are elements of a set.* **Distance** *is a real-valued function of two points.*

Postulates Let p, q, and r be points. Let d be the distance function.
 (1) $d(p, q) \geq 0$. $d(p, q) = 0$ only if $p = q$.
 (2) **(symmetry)** $d(p, q) = d(q, p)$.
 (3) **(triangle inequality)** $d(p, r) \leq d(p, q) + d(q, r)$.

Is this axiom system consistent? Are the axioms independent? Is the system categorical?

26. The axiom system for a **partial order** is as follows:

Undefined Terms **Points** *are elements of a set.* **Order** *is a relationship that may (or may not) hold between two points.*

Postulates Let p, q, and r be points. Let $<$ be the order relation.
 (1) **(irreflexivity)** It is not true that $p < p$.
 (2) **(antisymmetry)** If $p < q$, then it is not true that $q < p$.
 (3) **(transitivity)** If $p < q$ and $q < r$, then $p < r$.

Is this axiom system consistent? Are the axioms independent? Is the system categorical?

27. Here are Peano's axioms for the natural numbers:

Undefined Terms **Numbers** *are elements of a set.* **Successor** *is a function that transforms a number to another number.* **One** *is a particular number.*

Postulates *Let p and q be numbers. Let S be the successor function. Let 1 be the number one.*
- *(1) 1 is not the successor of any number.*
- *(2) If S(p) = S(q), then p = q.*
- *(3) (mathematical induction) If A is a set of numbers such that*
  - *(a) 1 is in A and*
  - *(b) if p is in A, then S(p) is also in A,*

then A contains all numbers.

Is this axiom system consistent? Are the axioms independent? Is the system categorical?

Where Do Axiom Systems Come From?

28. The quote from Klein (page 319) urges that the axioms of geometry be "sensible statements . . . induced by space perception." Does this description apply to Hilbert's axioms or (on the contrary) are they arbitrary statements? Or is there a middle ground? Read the quote from Poincaré (page 357). What is your view of Hilbert's axioms? Do they come from experience or imagination? Where should axioms come from?

29. What changes in Hilbert's axioms are needed in order to obtain a system that categorically determines elliptic geometry? Of course, the parallel axiom would have to change, but what other axioms require modification?

30. Apply the questions in Exercise 29 to hyperbolic geometry.

Hilbert's Axioms and Dimension

31. Hilbert's axioms are for Euclidean *plane* geometry. Which axiom or axioms require that Hilbert's geometry be at least two dimensional? Which axiom or axioms require that Hilbert's geometry be at most two dimensional?

25 BACHMANN'S AXIOMS

Hilbert's axioms describe Euclidean geometry directly in geometric terms. In contrast, Bachmann's axioms use algebraic relationships and algebraic objects. The existence of such complementary approaches confirms a major theme of modern geometry: Algebra and geometry are two aspects of a single subject.

Bachmann's system (in *Aufbau der Geometrie aus dem Spiegelungsbegriff* [C1]) is based on the wonderful idea, already explored in Chapter 20, that geometric relationships among lines and points can be replaced by algebraic relationships among certain types of transformation, namely, mirror reflections for lines and half turns for points. Here are two examples:

Example 1: Incidence

A point p lies on a line α if, and only if, the transformations M_α (mirror reflection in α) and H_p (half turn about p) commute, that is, $M_\alpha H_p = H_p M_\alpha$. In other words, the algebraic relationship of commutation between M_α and H_p has the same meaning as the geometric relationship of incidence between α and p.

Example 2: Perpendicularity

Two lines α and β are perpendicular if, and only if, their mirror reflections M_α and M_β commute.

These examples (and others in Chapter 20) show that certain geometric ideas can be expressed using the algebraic idea of commutation. Bachmann's system systematically exploits this idea. His success demonstrates that *every* concept and theorem of Euclidean geometry can be encoded this way in algebra.

Abstract Groups

In order to use the algebra of transformations to create an axiom system, we need the concept of an abstract group. This is a transformation group, as defined in Chapter 4, *without* an underlying set. In other words, an abstract group is a transformation group stripped of its geometry. Here is the formal definition:

Definition *A set **G** is an (**abstract**) **group** if, for any* α, β *in **G**, the product* αβ *is defined so that*

(a) *(closure)* αβ *is contained in **G**.*

(b) *(associativity)* σ(αβ) = (σα)β *for* σ, α, β *in **G**.*

(c) *(identity) There is an element* 1 *in **G** such that* 1α = α1 = α *for all* α *in **G**.*

(d) *(inverses) For every* α *in **G**, there is an element* α⁻¹ *in **G** such that*

$$\alpha^{-1}\alpha = \alpha\alpha^{-1} = 1$$

Note that abstract groups have the same *algebraic* properties as a transformation groups. (See Chapter 4.) The difference between them is that the elements of an abstract group are not necessarily transformations; there is, in fact, no underlying set for them to transform. An abstract group simply consists of a set of things that can be multiplied, subject to a few algebraic "laws."

Examples of Abstract Groups

First and foremost, every transformation group is an abstract group. Specifically, the groups we have used to define geometries, the groups **E**, **S**, and **H**, the Euclidean, elliptic, and hyperbolic groups (see Chapters 4, 7, and 11) are abstract groups, as are the full or extended versions of these groups **E⁺**, **S⁺**, and **H⁺** (see Chapters 20 and 22). The latter groups are important in this chapter because they include reflections and, therefore, as we shall see, model Bachmann's axioms.

What about abstract groups that are not transformation groups? How can they be defined? Some arise from subsets of other algebraic systems. For example, let C_4 be the set of the four complex numbers $\{1, -1, i, -i\}$. Then C_4 is closed under ordinary complex multiplication. Its multiplication table is in Figure 25.1.

C_4	1	−1	i	$-i$
1	1	−1	i	$-i$
−1	−1	1	$-i$	i
i	i	$-i$	−1	1
$-i$	$-i$	i	1	−1

Figure 25.1 A cyclic group

The remaining algebraic properties required of a group are also satisfied because the complex numbers as a whole satisfy these laws. The group C_4 is called the **cyclic** group with four elements.

An abstract group can also be presented simply by giving its multiplication table. For example, if K_4 is the set $\{1, a, b, c\}$ with the multiplication table shown in Figure 25.2, then K_4 is an abstract group called the **Klein four-group**.

Note: The **order** of a group is the number of its elements. The groups C_4 and K_4 are of order 4. As it happens, they are the only groups of order 4.

K_4	1	a	b	c
1	1	a	b	c
a	a	1	c	b
b	b	c	1	a
c	c	b	a	1

Figure 25.2 The Klein four-group

A–Find the multiplication tables of all groups of order 2.
B–Find the multiplication tables of all groups of order 3.

Bachmann's Axioms: Algebraic Setup

Developing a formal axiom system requires a conscious act of forgetting. For example, for the purpose of proving theorems in Hilbert's system, one must (in principle) forget everything one knows about Euclidean geometry and use only Hilbert's axioms (and results that have been proved with them). Fortunately, Hilbert's axioms use geometric terminology (lines and points, for instance), and this helps our imagination construct the required proofs.

Bachmann's system requires a more radical act of forgetting. Since the basic setting is an abstract group, algebraic instead of geometric terminology is used. Points and lines are involutions, not physical entities.

Definition *An **involution** is an element α of an abstract group, not the identity element, but such that $\alpha^2 = 1$.*

C–Find the involutions in C_4 and K_4.

Involutions are significant because the all-important mirror reflections and half turns, as discussed previously in Chapter 20, are involutions. With this in mind, we begin the development of Bachmann's system.

There are two undefined terms:

Undefined Terms   Motions *and* **line reflections** *are two sets.*

The first postulates describe the basic algebraic framework:

Postulates *(Group properties)*
    (1) The set of motions is an abstract group.
    (2) Line reflections are motions.
    (3) Line reflections are involutions.

(4) Every motion is a product of line reflections.

Although the word "motion" suggests otherwise, motions are *not* transformations. They are only elements of an abstract group. As such, they can be multiplied, but that is all. Later, as the system develops, it will turn out that motions can be represented as transforming points, but initially we must "forget" this. Indeed, there are no points yet to transform!

A simple example of a system satisfying Bachmann's axioms is the geometry **K** defined as follows: The motions of **K** are the elements of the group K_4; there are just two reflections: a and b. Note that a and b *are* involutions, and, by consulting the multiplication table for K_4, we see that the remaining element, c, of K_4 *can* be expressed as a product of a and b, namely, $c = ab$.

D–Verify Bachmann's group axioms for another "toy" Bachmann geometry: **K**$'$, which has K_4 as the group of motions and $\{a, b, c\}$ as the reflections.

   E–Can C_4 be the group of motions of a Bachmann geometry?

Involutions

Since lines (or rather, line reflections) are involutions, it is crucial to be able to decide when the product of two motions is an involution. Thus, the next result is fundamental.

Lemma Let α *and* β *be involutions. The product* $\alpha\beta$ *is an involution if, and only if,* $\alpha \neq \beta$ *and* $\alpha\beta = \beta\alpha$.

Proof: See Exercise 8. ∎

The following definition sets up basic notation:

Definition Let α *and* β *be involutions. If* $\alpha\beta$ *is an involution, then we write* $\alpha \mid \beta$. *More generally, the notation*

$$\alpha_1, \alpha_2 \mid \beta_1, \beta_2$$

means that $\alpha \mid \beta$ *for* $\alpha = \alpha_1$ *or* α_2, *and* $\beta = \beta_1$ *or* β_2.
If α *and* β *are reflections and* $\alpha \mid \beta$, *then the involution* $p = \alpha\beta$ *is called a **point** or a **point reflection**.*

Thus, point is a *defined* term in this system! We will use small Latin letters for point reflections $(p, q, \ldots)$ and small Greek letters for line reflections $(\alpha, \beta, \omega, \ldots)$.

According to the previous lemma, the relationship $\alpha \mid \beta$ is equivalent to two conditions: $\alpha \neq \beta$ and $\alpha\beta = \beta\alpha$. Another way to phrase this is that $\alpha \mid \beta$ if, and only if, α and β are distinct and commute.

The examples cited at the beginning of the chapter suggest the next definition.

Definition Let α and β be involutions and p be a point. If $\alpha \mid p$, then we say that the line α **contains** the point p or, alternatively, that p **lies** on the line α. If $\alpha \mid \beta$, then we say the lines α and β are **perpendicular**.

In the geometry **K**, with K_4 as the group of motions and just two reflections a and b, we have $a \mid b$, so that $c = ab$ is a point. A picture of this geometry looks something like Figure 25.3.

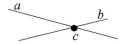

Figure 25.3 The plane of the geometry K

F–Draw the plane of the geometry **K′**. Label all points and lines. Indicate any perpendicular lines. *Note:* A motion can be *both* a point and a line.

Incidence

The next three postulates express the basic incidence relations between points and lines (with the understanding that the relationship $p \mid \alpha$ is interpreted as saying that the point p lies on the line α).

Postulates *(Incidence)*
  (1) Given points p, q, there is a reflection α such that p, $q \mid \alpha$.
  (2) Given p and α, there is a reflection β such that $\beta \mid p$ and $\beta \mid \alpha$.
  (3) From p, $q \mid \alpha$, β, it follows that either $p = q$ or $\alpha = \beta$.

In more usual geometric language, (1) says that two points determine a line.

G–Restate (2) in ordinary geometric language.

The incidence postulates imply some further classical existence and uniqueness results:

Theorem *(Perpendicular lines intersect in a unique point)* Let α and β be distinct reflections such that $\alpha \mid \beta$ and $p \mid \alpha$, β. Then $p = \alpha\beta$.

Proof: By definition, $\alpha\beta$ is a point; call it q. Since, clearly, $q \mid \alpha$, β, we have p, $q \mid \alpha$, β. Then, by the third incidence postulate, either $\alpha = \beta$ or $p = q$. Since $\alpha \neq \beta$, it follows that $p = q$. ∎

Theorem *(Two points determine a unique line)* Given distinct points p and q, there is a unique reflection α such that p, $q \mid \alpha$.

Proof: See Exercise 10. ∎

The next postulate concerns the product of *three* reflections:

Postulate *(Dreispiegelsatz) If either α, β, σ | p or α, β, σ | ω, then αβσ is a reflection.*

In words, the *Dreispiegelsatz* (translation: three-mirror theorem) says that if three reflections have either a common point or a common perpendicular, then their product is a reflection. In Euclidean geometry (and also in elliptic and hyperbolic geometry), the *Dreispiegelsatz* is proven using the technique of "moving the lines." (See Chapter 20.) In Bachmann's system, it is a crucial axiom. Here is one reason why:

Theorem *(Uniqueness of perpendiculars) Suppose that a point p and reflection α satisfy p | α. Then there is a unique reflection β such that β | α and β | p.*

Proof: By definition, $p = \mu\sigma$, where μ and σ are line reflections. Now, we have $\alpha, \mu, \sigma | p$, therefore, by the *Dreispiegelsatz*, $\beta = \mu\sigma\alpha = p\alpha$ is a line reflection. Clearly, $\alpha | \beta$ and $p | \beta$. This proves the existence part of the theorem. The proof of uniqueness is left as an exercise. (See Exercise 11.) ∎

Definition *A pair **G** = (G, R) consisting of an abstract group G and a subset R of G consisting of involutions and satisfying the postulates of this chapter (the group axioms, the incidence axioms, and the three-mirror theorem) is a **Bachmann geometry**. The set of points of a Bachmann geometry is called the **plane** of the geometry.*

H–Show that the geometries **K** and **K′** are Bachmann geometries. *Note:* the three-mirror theorem is *vacuously* true in both **K** and **K′**.

Models of Bachmann's Axioms

Unlike Hilbert's system, Bachmann's axiom system is not categorical, as the examples **K** and **K′** demonstrate. Here are a few more examples of Bachmann geometries:

The Classical Geometries

The three classical plane geometries, parabolic, elliptic and hyperbolic, all satisfy Bachmann's axioms. In each case, the group of motions is the full group $\mathbf{E}^+$, $\mathbf{S}^+$, or $\mathbf{H}^+$ of the classical geometry, and the line reflections are the mirror reflections appropriate to each geometry (as set forth in Chapters 21, 22, and 23). In each case, every motion is a product of reflections and the incidence axioms are easily proven. (See Exercises 12, 13, and 14.)

A Finite Geometry

Bachmann's geometries include geometries with only a finite number of points and lines. A more elaborate example than the geometries **K** and **K'** is presented in Figure 25.4. It consists of a lattice of points and lines confined to a square. The geometry **T**, based on this figure, includes *only* the indicated points and lines. To make this scheme work, the boundary of the square is identified, as indicated by the labels in the figure.

To understand the effect of these identifications, imagine yourself a two-dimensional creature moving around inside the square. If you walk off the square at a side, the identifications immediately cause you to appear at the opposite side. For example, the line starting at the top of the square at *b* and continuing diagonally to *f* leaves the square at *g* (on the right side) but then immediately appears at the *g* on the left side and goes on, returning to *b* on the bottom of the square.

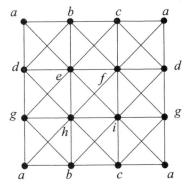

Figure 25.4 A geometry on an unglued torus

To put it another way, if you cut Figure 25.2 out of this book and actually glue the points labeled *a*, *b*, *c*, and so on., together, you get a **torus**, a doughnut-shaped surface in space. (See Figure 25.5.) The paper has to stretch!

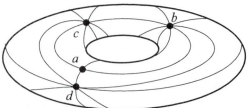

Figure 25.5 A geometry on the glued torus

The *reflections* of the geometry **T** are the reflections in the indicated lines, with the understanding that if a point is reflected outside the square in Figure 25.4, it is brought back in by a horizontal or vertical translation, as suggested by the identifications. Thus, reflection in the line *abc* fixes this line (of course) and interchanges the other two horizontal lines, for example, mapping the point *e* to *h*.   And reflection in the line *dhc* maps *e* to *g* and *b* to *i*!   The *motions* of **T** are all the transformations of Figure 25.5 that are compositions of the reflections.

I–To what point is *a* mapped by reflection in the line *dhc*?

Summary

Bachmann's vision of geometry is Klein's *Erlanger Programm* taken to the algebraic limit.   Bachmann's axiom system is entirely algebraic in character, and all results are proved by carrying out computations within an abstract group.   The whole theory is based on the geometric idea of mirror reflection, or rather, its algebraic counterpart, involution.

Although based on only a few algebraic ideas, Bachmann's system produces a geometric theory rich enough to encompass all the classical plane geometries, as well as some exotic geometries.

EXERCISES

Abstract Groups

1. Let C be the set of complex numbers z such that $|z| = 1$ with the usual complex multiplication.   Verify that C is a group and find all involutions in C.

2. Let D be the set of two-by-two real matrices with nonzero determinant with the usual multiplication of matrices.   Verify that D is a group and find all involutions in D.

3. Let $\Re$ be the set of real numbers.   Verify that $\Re$ is a group with multiplication operation $\oplus$ defined by $a \oplus b = (a + b)/(1 + ab)$. Find all involutions in $\Re$.

4. Let α be the reflection of three-dimensional Euclidean space in the plane $x = 0$: $\alpha(x, y, z) = (-x, y, z)$.   Let β and σ be the reflections in the planes $y = 0$ and $z = 0$, respectively. Additionally, let X, Y and Z be 180° rotations about the x-, y-, and z-axes, respectively.   And let D be the transformation of reflection in the origin: $D(x, y, z) = (-x, -y, -z)$, and let 1 be the identity transformation.   Finally, let H be the set $\{\alpha, \beta, \sigma, X, Y, Z, D, 1\}$ with the operation of composition of transformations. Find the multiplication table of H.   Is H a group?

5. Let Q be the particular set of the quaternions $\{ \pm 1, \pm \mathbf{i}, \pm \mathbf{j}, \pm \mathbf{k} \}$ with the usual quaternion multiplication.   (See Chapter 17). Verify that Q is a group and find all involutions in Q.

6. Let S_n be the set of all one-to-one transformations of the set $\{1, 2, 3, ..., n\}$. These transformations are called the **permutations** of this set. Verify that S_n is a group with composition of transformations as multiplication. Describe all involutions in S_n.

Hint: Work a few examples first, say, S_2 and S_3.

7. Find multiplication tables for two different groups of order 6.

8. Prove that the product ab of two involutions a and b, from any group G, is an involution if, and only if, $a \neq b$ and $ab = ba$.

9. Are the axioms for an abstract group categorical? Are they consistent? Are they independent?

Incidence Axioms

10. Let **G** be a Bachmann geometry. Prove that two distinct points p and q in **G** lie on a unique line α.

11. Let **G** be a Bachmann geometry. Prove that there is a unique line σ perpendicular to a given line β at a given point p on β.

Examples of Bachmann Geometries

12. Verify that the full elliptic group $\mathbf{S}^+$ satisfies Bachmann's axioms, if line reflections are understood to be the reflections in elliptic straight lines.

13. Repeat Exercise 12 for the full Euclidean group $\mathbf{E}^+$.

14. Repeat Exercise 12 for the full hyperbolic group $\mathbf{H}^+$.

15. How many lines are there in the finite geometry **T**? How many points? Besides the reflections described in the text, what kinds of motions are in this geometry? Verify that **T** satisfies Bachmann's axioms.

16. Find an example of a Bachmann geometry other than those discussed in the text.

Hint: Consider some of the groups in exercises 1-7.

Bachmann's axioms and dimensionality

17. Consider the three-dimensional geometry whose motions are the group H of Exercise 4 and whose reflections are the elements α, β, σ and D of H. Which of Bachmann's axioms are satisfied and which are not?

18. Bachmann geometries are plane geometries. Which axiom or axioms require that a Bachmann geometry be at least two dimensional? Which axiom or axioms require that a Bachmann geometry be at most two dimensional?

26 Metric Absolute Geometry

This chapter continues the development of Bachmann geometry. We focus on connecting Bachmann's algebraic vision with other approaches to geometry used in the book. For example, given one further axiom, we show that the motions of a Bachmann geometry can be interpreted as transformations of the plane of the geometry.

Existence Axioms

So far, Bachmann's axiom system does not require that anything exist. In order to develop the theory further, it is necessary to add at least one axiom of existence. There are several possibilities. The next definition gives the one Bachmann explored in greatest detail. (Another possibility is explored in Exercises 15–21.)

Definition *A Bachmann geometry* $\mathbf{G} = (G, R)$ *is a **metric absolute geometry** if there are three reflections* α, β, σ *in R, such that* $\alpha \mid \beta$, *but neither* $\beta \mid \sigma$, *nor* $\alpha \mid \sigma$, *nor* $\sigma \mid \alpha\beta$.

In other words, a metric absolute geometry is a Bachmann geometry that contains a right triangle.

A–Which is the hypotenuse of the triangle α, β, or σ?

B–What is the significance of the condition that $\sigma \mid \alpha\beta$ not hold?

For metric absolute geometries we can prove such mundane theorems as the following:

Theorem *The plane of a metric absolute geometry is nonempty.*

And a stronger result:

Theorem *(Every line contains a point)* *For every* α *in a metric absolute geometry, there is a p such that* $\alpha \mid p$.

The proofs of these theorems are left to the reader in Exercises 1 and 2.

C–Is either **K** or **K′** a metric absolute geometry?

Transformations

We now show that the abstract motions of a metric absolute geometry really are transformations of the plane of the geometry. This connects Bachmann's system with transformation groups and the *Erlanger Programm*.

Let **G** be a metric absolute geometry, and let g be a motion. The first of the following two theorems explains how g defines a transformation T_g of the set of motions itself and gives some fundamental properties of T_g. The second theorem shows that T_g not only transforms motions to motions but also transforms points to points, and lines to lines.

Theorem *(Motions transform motions) Let g be a motion of an metric absolute geometry* **G**. *Let w be a second motion, and set $T_g w = gwg^{-1}$. Then*

(1) $T_g w$ is also a motion, so that T_g is a transformation of the motions of **G**.

(2) If $w = uv$, then $T_g w = T_g(uv) = (T_g u)\,(T_g v)$.

(3) If $g = uv$, then $T_g w = T_{uv} w = T_u T_v w$.

(4) If $u \mid v$, then $T_g u \mid T_g v$.

Proof: (1) follows immediately because the motions form a group and so are closed under multiplication. The remaining parts of the proof are relatively routine. (See Exercises 5 and 6.) ∎

Theorem *(Motions transform lines and points) Let g be a motion of a metric absolute geometry* **G**. *Then T_g maps point reflections to point reflections and line reflections to line reflections.*

Proof: According to Bachmann's group postulate, every motion is a product of reflections. Therefore,

$$g = \alpha\beta\gamma\ldots$$

for suitable reflections α, β, γ, and so forth. Using part (3) of the previous theorem, we can conclude that $T_g = T_\alpha\,T_\beta\,T_\gamma\ldots$. In other words, T_g is a composition of transformations of the special form T_μ, where μ is a line reflection. If we prove that a transformation of the form T_μ maps point reflections to point reflections and line reflections to line reflections, it will follow that T_g has the same properties. Therefore, for the remainder of the proof we consider T_μ instead of T_g.

Let p be a point. By the incidence postulates, there is a line reflection β such that $p\mid\beta$ and $\mu\mid\beta$. Furthermore, $\sigma = p\beta$ is a line reflection such that $\sigma|\beta$. (See Exercise 11 of Chapter 25.) So far the elements we have described can be visualized as in Figure 26.1.

By the *Dreispiegelsatz*, we conclude that

$$\omega = T_\mu\sigma = \mu\sigma\mu$$

is a line reflection (since σ, $\mu \mid \beta$). From $\sigma \mid \beta$, we conclude that $T_\mu\sigma \mid T_\mu\beta$ [by part (4) of the preceding theorem] and then that $\omega \mid \beta$, since $T_\mu\beta = \beta$. (See Exercise 7.) Thus, using part (2) of the preceding theorem, we finally conclude that

$$T_\mu p = T_\mu(\sigma\beta) = (T_\mu\sigma)(T_\mu\beta) = \omega\beta$$

is a point reflection. This proves that T_μ maps point reflections to point reflections.

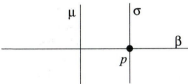

Figure 26.1 Proof that motions transform points

D–Add $T_\mu\sigma$ and $T_\mu p$ to Figure 26.1.

Now, consider a line reflection α. As we know, μ contains at least one point p. By the second incidence postulate, there is a line β such that $p \mid \beta$ and $\alpha \mid \beta$. Therefore, $q = \alpha\beta$ is a point. Furthermore, $q \mid \beta$ and $\alpha = q\beta$. Now, $T_\mu q$ is a point by the previous paragraph, and $T_\mu\beta$ is a line by the *Dreispiegelsatz* (since μ, $\beta \mid p$). Therefore,

$$T_\mu\alpha = (T_\mu q)(T_\mu\beta)$$

Also, from $q \mid \beta$, it follows that $T_\mu q \mid T_\mu\beta$. Hence, $T_\mu\alpha$, being the product of a commuting point and line, is a line reflection. This proves that T_μ maps line reflections to line reflections. ∎

E–Draw a figure to illustrate the preceding paragraph.

The Fundamental Theorem

Even with an existence postulate, Bachmann's system is still far from categorical. If our aim is to obtain an axiom system characterizing Euclidean geometry (as Hilbert's system does), then we must add more postulates. A categorical system, however, is not Bachmann's goal. What is interesting about his system is that so few algebraic axioms describe such a variety of significant geometric spaces.

Nonetheless, it would be nice to know just how far from the classical geometries one can get and still satisfy Bachmann's axioms. A precise answer to this question is given by Bachmann's fundamental theorem, which says that every metric absolute geometry is a subgeometry of a projective geometry.

This result is fundamental in several ways. In the first place, it shows precisely how close metric absolute geometries are, individually and collectively, to the classical plane geometries. Collectively, it is clear that metric absolute geometries do not stray very far from the classical geometries, since all are subgeometries of projective geometry. On the other hand the particular projective geometry in which the fundamental theorem places a given metric absolute geometry can be one of the unusual projective geometries studied in Chapter 16: a complex geometry, for example, or a finite geometry.

In the second place, the fundamental theorem provides a means for systematically studying and classifying all metric absolute geometries. The fundamental theorem says: To find all metric absolute geometries look at subgeometries of projective geometry.

In the third place, the fundamental theorem tells us that a given metric absolute geometry, once we identify the particular projective geometry containing it, can be studied using projective techniques. The fundamental theorem says that projective geometry is the right tool for the study of metric absolute geometry.

Thus, Bachmann's fundamental theorem provides further evidence of the universality of projective geometry (the theme of Chapter 16) by showing, once again, that projective geometry encompasses other geometric theories.

Bundles

Although the proof of the fundamental theorem is too lengthy to be included here, some of the main ideas will be outlined.

So far, metric absolute geometry has been based entirely on one relationship: |. Now Bachmann introduces a second powerful relationship.

Defintion *Let α, β, and σ be line reflections. If $\alpha\beta\sigma$ is also a line reflection, we say that α, β, and σ are **bundled**, written $[[\alpha\,\beta\,\sigma]]$.*

Bundling is rather like an equivalence relation, except that it involves three items rather than two! The next theorem explains how.

Theorem *(Bundling laws)   The bundle relation has the following properties:*

(1) (reflexivity) $[[\alpha\,\beta\,\sigma]]$ *holds (i.e.,* α, β, *and* σ *are bundled) whenever two of* α, β, *and* σ *are equal,*

(2) (symmetry) If $[[\alpha\,\beta\,\sigma]]$, *then the same is true with* α, β, *and* σ *in any order,*

(3) (transitivity) If $\alpha \neq \beta$, *then* $[[\alpha\,\beta\,\sigma]]$ *and* $[[\alpha\,\beta\,\omega]]$ *imply that* $[[\alpha\,\sigma\,\omega]]$.

Proof: Parts (1) and (2) are fairly easy to prove using the previous theorem. (See Exercise 8). The proof of part (3), on the other hand is too lengthy to include here. We assume it for the purpose of exploring bundles further. ∎

Corollary *Two further properties of bundling are:*

(4) If $\alpha \neq \beta$, *then* $[[\alpha\,\beta\,\sigma]]$, $[[\alpha\,\beta\,\mu]]$, *and* $[[\alpha\,\beta\,\omega]]$ *together imply that* $[[\sigma\,\mu\,\omega]]$.

(5) If $\sigma \neq \mu$, *then* $[[\alpha\,\beta\,\sigma]]$, $[[\alpha\,\beta\,\mu]]$, *and* $[[\sigma\,\mu\,\omega]]$ *together imply that* $[[\alpha\,\beta\,\omega]]$.

Proof: See Exercise 9. ∎

F–Suppose that three lines α, β, and σ share a common point? Which of Bachmann's axioms implies $[[\alpha\,\beta\,\sigma]]$? Under what other condition does this axiom imply $[[\alpha\,\beta\,\sigma]]$?

We can now give a key definition leading to the embedding of the plane of a metric absolute geometry in a projective plane:

Definition *Let* α *and* β *be lines such that* $\alpha \neq \beta$. *The set of all lines bundled with* α *and* β *is called the* **bundle** *determined by* α *and* β *and is written* $B[\alpha\beta]$.

The most important properties of bundles are summarized as follows:

Theorem *(Properties of bundles) Let* α *and* β *be lines such that* $\alpha \neq \beta$.

(1) $B[\alpha\beta]$ *is determined by the product* $\alpha\beta$, *not by* α *and* β *individually; thus, if* α' *and* β' *are two other lines such that* $\alpha'\beta' = \alpha\beta$, *then* $B[\alpha'\beta'] = B[\alpha\beta]$.

(2) Any three lines σ, μ, *and* ω *in a bundle* $B[\alpha\beta]$ *are bundled, that is,* $[[\sigma\,\mu\,\omega]]$.

(3) If two distinct lines σ *and* μ *in a bundle* $B[\alpha\beta]$ *and a third line* ω *are bundled, then* ω *is also in* $B[\alpha\beta]$.

(4) If σ *and* μ *are two distinct lines in the bundle* $B[\alpha\beta]$ *then* $B[\sigma\mu] = B[\alpha\beta]$.

(5) Two distinct bundles share at most one line.

Proof: See Exercise 10. ∎

Transformation of Bundles

There are three types of bundles. First, the set of all lines *containing* a given point p is a bundle $B[p]$ called a **point bundle**. Second, the set of all lines *perpendicular* to a given line α is a bundle, called a **line bundle** and denoted $B(\alpha) = B[\sigma\mu]$, where σ and μ are two distinct lines perpendicular to α. Now, let α and β be two distinct lines. If α and β have a common point p, then $B[\alpha\beta] = B[p]$ is a point bundle. On the other hand, if α and β have a common perpendicular ω, then $B[\alpha\beta] = B(\omega)$ is a line bundle. Otherwise α and β have neither a common point nor a common perpendicular, and $B[\alpha\beta]$ is called an **end**.

Next, we consider the action of motions on bundles:

Theorem *(Motions transform bundles) Let g be a motion of a metric absolute geometry* **G**. *Then T_g maps bundles to bundles. That is, if B is a bundle, then $T_g(B)$ is also a bundle. Furthermore:*

(1) If $B = B[p]$ is a point bundle, then $T_g(B)$ is a point bundle and $T_g(B) = B[T_g p]$.

(2) If $B = B(\alpha)$ is a line bundle, then $T_g(B)$ is a line bundle and $T_g(B) = B(T_g\alpha)$.

(3) If B is an end, then $T_g(B)$ is an end.

Proof: This is a straightforward argument based on previous results and the following lemma. (See Exercise 11.) ∎

Lemma $[[\alpha\ \beta\ \sigma]]$ *if, and only if,* $[[T_g(\alpha)\ T_g(\beta)\ T_g(\sigma)]]$.

Definition *The set of bundles $B[\mathbf{G}]$ of a metric absolute geometry* **G** *is called the **ideal plane** of* **G**.

Here is Bachmann's fundamental theorem:

Fundamental Theorem *Let* **G** *be a metric absolute geometry. Then* **G** *is a subgeometry of a projective geometry whose underlying set is the ideal plane of* **G**.

Proof: The goal of the theorem is to embed **G** in a projective geometry. Half that goal is accomplished by the previous theorems, which have shown that the motions of **G** act as transformations not only of the plane of **G**, but also of the ideal plane $B[\mathbf{G}]$. The remaining steps of the proof, which we do not have space to include, show that these motions are a subgroup of a projective group on $B[\mathbf{G}]$. Details are in Bachmann's book [C1]. ∎

Examples

Although we have not proved Bachmann's fundamental theorem, we can make it more plausible by examining some examples. Here we consider each of the metric plane geometries; elliptic, parabolic, and hyperbolic, as metric absolute geometries and interpret the fundamental theorem in each case.

Elliptic Geometry

In elliptic geometry, any two lines intersect (Chapter 11). It follows that all elliptic bundles are point bundles, so that the elliptic plane is the same as its ideal plane.

Therefore, according to the fundamental theorem, the elliptic plane should already be a projective plane, without the addition of any more points (although more motions may be needed). This is exactly what we proved in Chapter 13, where, from a completely different point of view, we discovered that elliptic geometry *is* a model for the real projective plane. This confirms the fundamental theorem for elliptic geometry.

Euclidean Geometry

In Euclidean geometry, two lines either intersect or are parallel. Furthermore, parallel lines always have a common perpendicular. It follows that all Euclidean bundles are either point bundles or line bundles. Now, the line bundles are families of parallel lines, that is, lines with a given slope. In this way the line bundles represent the set of directions in the Euclidean plane.

Therefore, according to the fundamental theorem, the Euclidean plane can be made into a projective plane by adding to the usual Euclidean plane (i.e., the point bundles) extra points: one for each direction (i.e., the line bundles). It is natural to regard these extra points as constituting, as a whole, an extra line surrounding the Euclidean plane. This is exactly what we proved in Chapter 13, where, from a completely different point of view, we discovered that the Euclidean plane plus a circular line "at infinity" *is* a model for the real projective plane. This confirms the fundamental theorem for Euclidean geometry.

Hyperbolic Geometry

In hyperbolic geometry, two lines either intersect, are hyperparallel (in which case they have a common perpendicular), or are parallel. It follows that hyperbolic geometry has all three types of bundles: point bundles (for intersecting lines), line bundles (for hyperparallel lines), and ends (for parallel lines).

In the circle model, these bundles may be further described as follows:

(1) **Point bundles** There is one for every point of the geometry itself, every point of the complex plane *inside* the unit disk.

(2) **Line bundles** There is one for every hyperbolic straight line.   Line bundles consists of all hyperbolic straight lines perpendicular to a given line. There are two subtypes:

(a) *Line bundles for hyperbolic straight lines that are arcs of circles perpendicular to the unit circle.* The centers of these circles are outside the unit disk and every point outside the unit disk is the center of exactly one hyperbolic straight line.   Thus, these hyperbolic straight lines (and their associated line bundles) are parametrized by the points *outside* the unit disk.

(b) *Line bundles for hyperbolic lines that are Euclidean straight lines through the origin.* There is one bundle for each direction through the origin.

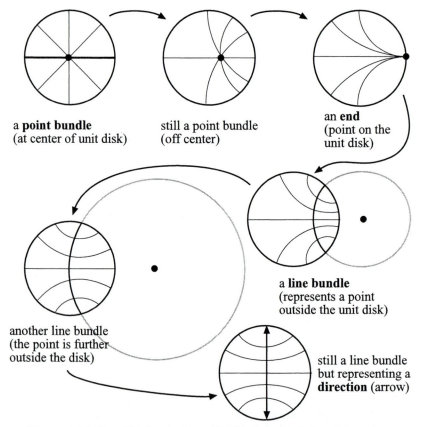

a **point bundle**
(at center of unit disk)

still a point bundle
(off center)

an **end**
(point on the
unit disk)

a **line bundle**
(represents a point
outside the unit disk)

another line bundle
(the point is further
outside the disk)

still a line bundle
but representing a
direction (arrow)

Figure 26.2 Bundles in the hyperbolic plane and the points they determine in the Euclidean plane

(3) **Ends** Finally, there is a bundle of mutually parallel lines for each point on the unit circle.

Figure 26.2 displays all three types of bundles by showing how a single bundle changes as the point determining it moves through the plane.

Therefore, the ideal plane of hyperbolic geometry consists of the points inside the unit circle (point bundles), the points outside the unit circle (line bundles), the points on the unit circle (ends), and a further set of points, one for each direction from the origin (more line bundles). In other words, the ideal plane of hyperbolic geometry corresponds to the whole Euclidean plane plus a line at infinity. which, as we just mentioned, is a model for the real projective plane. This confirms the fundamental theorem for hyperbolic geometry.

Summary

Adding an existence axiom to Bachmann's system brings Bachmann geometries, now called metric absolute geometries, closer to the classical geometries: elliptic, parabolic, hyperbolic, and especially projective. The fundamental theorem extends the plane of a metric absolute geometry to a projective plane by taking families of lines, called bundles, as the points of the ideal plane of the geometry.

EXERCISES

Existence Results

1. Prove that every metric absolute geometry **G** has at least one point.

2. Prove that every line in a metric absolute geometry contains at least one point. Is there a metric absolute geometry in which each line contains *exactly* one point? Two points? Three points?

3. Let **G** be a metric absolute geometry. Prove that every point p in **G** is on a line.

4. Let **G** be a metric absolute geometry. Prove that no line is perpendicular to all other lines.

Motions as Transformations

5. Let g be a motion of a metric absolute geometry **G**. Let $T_g w = gwg^{-1}$, where w is another motion. Prove that if $w = uv$, then $T_g w = (T_g u)(T_g v)$, and that if $g = uv$, then $T_g w = T_u T_v w$.

6. Let g be a motion of a metric absolute geometry **G**. Prove that if $u|v$, then $T_g u \mid T_g v$. Conclude that if α and β are perpendicular then so are $T_g w$ and $T_g w$. Also, deduce that if a point p lies on a line α, then $T_g p$ lies on $T_g \alpha$.

7. Show that Ta, where α is a line reflection, fixes a point p if, and only if, $\alpha \mid p$.

Bundles

8. Prove that the bundle relation is reflexive and symmetric.

Hint: Use the fact that motions transform lines to lines.

9. Prove the two additional bundle laws (4) and (5).

10. Prove the five properties of bundles.

11. Prove the theorem on the transformation of bundles.

12. In classical elliptic geometry, two distinct bundles intersect in a common line. Prove this. What happens in Euclidean geometry?

13. When do bundles in hyperbolic geometry have a common line?

14. Find all bundles in the geometry **T**.

Elliptic Geometries

This set of exercises is based on the following alternative existence axiom for Bachmann geometries:

Definition A Bachmann geometry is a **metric elliptic geometry** if there are three lines α, β, and σ such that $\alpha\beta\sigma = 1$.

15. Let **G** be a metric elliptic geometry. Show that **G** has a triangle with three right angles.

16. Let p be a point reflection and α be a line reflection in a Bachmann geometry **G**. It can happen that $p = \alpha$, in which case p and α are called **polar**. Let p and α be polar. Prove that **G** is a metric elliptic geometry.

17. Prove the converse of Exercise 16: If **G** is a metric elliptic geometry, then there exists a polar pair in **G**.

18. Prove that every line in a metric elliptic geometry contains at least one point. Is there a metric elliptic geometry in which each line contains *exactly* one point? Two points? Three points?

19. Let **G** be a metric elliptic geometry. Let p be a point and α be a line. Suppose that α does not contain p. Prove that there is more than one perpendicular from p to α. Does this contradict the theorem on the uniqueness of perpendiculars?

20. Prove the converse of Exercise 19: If a Bachmann geometry **G** contains a point p and line α such that there is more than one perpendicular from p to α, then **G** is a metric elliptic geometry.

21. Let **G** = (G, R) be any Bachmann geometry. Prove that the motions g in G that are the product of an even number of reflections form a subgroup of all motions. Prove that, for a metric elliptic geometry, this subgroup contains all motions.

PART VII

CONCLUSION

GEOMETRY AND ART: PLATE VII

M.C. Escher: *Circle Limit IV*

As art has become less representational and more conceptual, mathematics, particularly geometry, has appeared in artistic works more and more, not simply as an adjunct, but as the primary subject matter. The Dutch artist Maurits Escher is an obvious example. His designs using repeating patterns involved substantial preliminary mathematical work of classification and construction. Escher was introduced to non-Euclidean geometry by the mathematician H. S. M. Coxeter. The angel-and-devil design (shown here in hyperbolic geometry) is the only pattern Escher used in all three plane metric geometries: Euclidean, elliptic, and hyperbolic.

. . . The geometrical axioms are therefore, neither synthetic a priori intuitions nor experimental facts. They are conventions. Our choice among all possible conventions is *guided* by experimental facts; but it remains *free*, and is only limited by the necessity of avoiding every contradiction, and thus, it is that postulates may remain rigorously true even when the experimental laws which have determined their adoption are only approximate. In other words *the axioms of geometry* (I do not speak of those of arithmetic) *are only definitions in disguise.* What then are we to think of the question: Is Euclidean geometry true? It has no meaning. We might as well ask if the metric system is true, and if the old weights and measures are false; if Cartesian coordinates are true and polar coordinates are false. One geometry cannot be more true than another; it can only be more convenient.

–from *Science and Hypothesis* by Henri Poincaré [F4]

This part treats the scientific and philosophical impact of both non-Euclidean geometry and the other geometrical ideas presented in the book. Chapter 27 surveys cultural issues influenced by the discovery of non-Euclidean geometry. Chapter 28 is devoted to the most prominent of these issues, the question with which the book began: What is the geometric nature of physical space?

27 THE CULTURAL IMPACT OF NON-EUCLIDEAN GEOMETRY

From the golden age of Greek geometry to the beginning of the golden age of nineteenth century geometry, it was accepted without question that Euclidean geometry is concerned with the real world. Euclidean geometry was inextricably linked to properties of our universe: the physical space in which we live.

The discovery of non-Euclidean geometry severed that link. Suddenly, physical space might be described by other geometries. For the early investigators of non-Euclidean geometry, this was the most disturbing consequence of their work: Our own space may not be Euclidean. It is on account of this conclusion that Gauss refused to publish his work.

Geometry and Curved Space

How could we have lived so long without realizing that space might be curved? Our blindness stems from the fact that all absolute geometries appear approximately Euclidean to inhabitants whose knowledge is restricted to a small portion of their space. Since we personally know only a small portion of three-dimensional space, we can form no idea on the basis of our senses that our space is curved.

The fact that plane absolute geometry is locally Euclidean is part of our experience living on the two-dimensional surface of a sphere. The situation regarding our perception of the three-dimensional space in which we move is similar. The analogy with the plane led Gauss and others to advance the revolutionary idea that the three-dimensional geometry we live in may be curved.

Of course someone with an off-planet view of the surface of the Earth easily sees that it is curved. And, by now, everyone has seen photographs of such a view. But, who has seen an out-of-space or off-universe view of three-dimensional space?

One of Gauss' fundamental discoveries is a way to measure the curvature of a space intrinsically, that is, by measurements taken within it. His method applies to three-dimensional as well as two-dimensional spaces. Since this discovery many attempts have been made to determine the curvature of our universe, for example, using light rays as straight lines and observing distant stars. (Gauss is reputed to have used the peaks of the mountains Hohenhagen, Brocken, and Inselberg.)

Unfortunately, all attempts to measure the curvature of (three-dimensional) space have yielded values so close to zero that, taking experimental error into account, it is impossible to decide whether space is elliptic, parabolic (Euclidean), or hyperbolic.

The truth is that physical measurements can never prove that the curvature of space is zero, and hence that the geometry of physical space is Euclidean. For, even if some experimental measurement of the curvature of space came out dead on zero, there would always be some doubt, since every measurement has some margin of error. Ironically, if space is curved, there is some hope that we can know it. Future experiments may yield values for the curvature of space that are unambiguously positive or negative. But if space is Euclidean, we will never know.

The Question of Meaning

By detaching geometry from direct connection with the physical world, non-Euclidean geometry threw into doubt the fundamental meaning of mathematics, which for centuries depended on the idea that mathematics, generally, and geometry, in particular, were unquestionably concerned with the real world. As a result, the feeling grew among mathematicians that mathematics is really just an abstract game that happens at times to be applicable to the real world, but whose own existence is quite separate from physical reality. For some mathematicians, the nature of reality became abstract and divorced of *any* objective physical nature. Thus, Poincaré wrote (see [E3], page 270):

> Does the harmony the human intelligence thinks it discovers in nature exist outside of this intelligence? No, beyond doubt a reality completely independent of the mind which conceives it, sees it, feels it, is an impossibility. A world as exterior as that, even if it existed, would for us be forever inaccessible. But what we call objective reality is, in the last analysis, what is common to many thinking beings, and could be common to all, this common part, we shall see, can only be the harmony expressed by mathematical laws. It is this harmony then which is the sole objective reality, the only truth we can attain.

Here, Poincaré denies the existence of objective reality, or at least our ability to know it, yet at the same time maintains that the nature of objective reality must be mathematical!

In contrast, Einstein, a scientist who used non-Euclidean geometry to deal with physical reality, wrote (see [E3], page 314):

> Of all sciences mathematics is held in especially high esteem, for its theorems are absolutely true and indisputable while the areas of other sciences are to some extent debatable and there is always the danger of their results being overturned by new discoveries. But it is not fitting for a researcher working in some other area of science to envy mathematics, for the propositions of mathematics rest not on real objects but exclusively on objects of our imagination. . . .
>
> In this connection there arises a riddle that has worried researchers of all times.   Whence the remarkable correspondence between mathematics and real things if mathematics is just a product of human thought unrelated to any experience? Can the human mind understand the properties of real things without any experience, just by way of reflection?
>
> To my mind, a concise answer to this question is this: to the extent to which the theorems of mathematics can be applied to reflect the real world they are not exact; they are exact to the extent to which they do not refer to reality.

Thus, Einstein believes that mathematical ideas have no necessary physical meaning.

Whether one doubts objective reality (like Poincaré) or believes in reality, but feels that mathematics has no necessary connection with it (like Einstein), the point is that the discovery of non-Euclidean geometry created a problem: What is the relationship between mathematics and reality?

From the perspective of the many generations of mathematicians who lived before the discovery of non-Euclidean geometry, the existence of this problem would be revolutionary, an overturning of the nature of mathematical truth.

Among the effects of this revolution is the twentieth century emphasis on mathematics for its own sake. The past 80 years have seen an extraordinary growth in mathematics, most of it quite abstract and divorced from contact with what most people would call objective or physical reality.

In this way, the discovery of non-Euclidean geometries contributed to a loss of meaning for mathematics.

The Question of Rigor

While attacking Euclid, geometers of the nineteenth century evolved new standards of proof. The efforts in this direction culminated in the development of the new fields of **set theory** and **mathematical logic**. To be honest, geometry was by no means the only cause of these developments. Equally, or perhaps even more important, was the effort to place the calculus on a rigorous foundation, a movement that, like the development of non-Euclidean geometry, extended over the entire nineteenth century. Nonetheless, the discovery of non-Euclidean geometries certainly contributed to a critical examination of the work of past geometers which in turn raised important logical issues.

A good example is the problem of consistency. A mathematical system is **consistent** if it is not possible to derive a contradiction in the system. Consistency is important; without it, a system is meaningless, since (in principle) anything follows from a contradiction.

Consistency is also the foundation of proof by contradiction. For example, Saccheri *thought* he had derived contradictory results from the *negation* of the fifth postulate. He also assumed that mathematics was consistent, so that contradictory results were impossible. Since the impossible had happened, he deduced that the cause was his assumption of the negation of the fifth postulate. On this basis, he decided that he could deny the negation and deduce the fifth postulate itself.

Most of the time, like Saccheri, we *assume* the consistency of the mathematical systems within which we work. Why don't we *prove* this consistency? Unfortunately, theorems of set theory, discovered this past century, indicate that a proof of consistency for all but the simplest mathematical systems is impossible.

Although no one has ever proven that Euclidean or hyperbolic geometry is consistent, we do know that Euclidean and hyperbolic geometry are **relatively consistent**; that is, if a contradiction exists in one of them, then a contradiction exists in the other. In particular, if a contradiction exists in hyperbolic geometry, as Saccheri thought (although he did not realize that he was working in hyperbolic geometry), then a contradiction must exist in Euclidean geometry as well.

We know this because of **relative consistency arguments** based on models of both geometries. Recall that our development of hyperbolic geometry occurred in the Euclidean plane (inside the unit disk to be precise). It follows that, if there is a contradiction in hyperbolic geometry, then this contradiction is also part of Euclidean geometry (i.e., the Euclidean geometry of the unit disk)! Thus, hyperbolic geometry is at least as consistent as Euclidean geometry, since, if the former is inconsistent, then so is the latter. A similar argument shows that Euclidean geometry is at least as consistent as hyperbolic geometry, since Euclidean geometry can be developed within solid hyperbolic geometry as the natural geometry of a horosphere. (See Chapter 19.)

Relative consistency arguments have led to some of the most spectacular results of mathematical logic in this century. Their roots go back to the discovery of non-Euclidean geometry.

The Synthetic a Priori

The discovery of non-Euclidean geometries also had an impact on a philosophical controversy: the question of the existence of synthetic a priori statements.

A statement is **synthetic** if it adds to knowledge of its subject. Synthetic statements are to be contrasted with **analytic** statements, in which the information conveyed by the statement is already contained in the very definition of the subject of the statement. For example, the statement "An orange is a fruit" is *analytic*, since, by definition, an orange is the fruit of a certain tree. The statement "An orange is a good source of vitamin C," on the other hand, is *synthetic*, as it expands our knowledge of oranges beyond the information contained in the definition of an orange.

A second distinction is that between a priori and a posteriori statements. A statement is **a priori** if it can be derived and tested without reference to sense experience (that is, experience of the real world). In contrast, a statement is **a posteriori** if its derivation and testing requires reference to sense data.

Thus, a synthetic a priori statement is a statement that reveals something about its subject (going beyond definition of the subject) and yet can be derived or verified without reference to the real world.

These distinctions were formulated by the influential philosopher Immanuel Kant in the *Critique of Pure Reason*, published in 1781. In this work, Kant aimed at a "Copernican revolution" in philosophy, meaning that just as Copernicus reversed the way scientists viewed the relationship between the earth and the sun, so Kant would reverse the way philosophers would view the world of experience and the world of the mind. In the *Critique*, Kant defended the existence of synthetic, a priori statements. His main examples of such statements were in mathematics, principally geometry.

For example, consider the statement "To enclose a shape requires at least three straight lines." According to Kant, a geometric statement such as this one is a priori, since it is derived by "pure reason" (that is, deductively), but is also synthetic in that it is not part of the definition of "straight line" or "shape" that three of the former are needed in order to enclose the latter.

The discovery of non-Euclidean geometry cast doubt on the existence of synthetic a priori statements. For instance, the statement "To enclose a shape requires at least three straight lines" is not true in elliptic geometry! (Recall the 2-gon introduced in Chapter 11.) More generally, since the discovery of non-Euclidean geometry appears to refute the idea that Euclidean geometric statements are necessarily about the real world, it undermines Kant's judgment that such statements are synthetic. In fact, the modern point of view is that all mathematical statements are analytic.

In Gauss's lifetime, the followers of Kant had such a grip on German philosophy, that Gauss feared their reaction to his discoveries. Thus, in a letter written in 1818 (quoted in [E3], p. 215), he says

> I am glad that you have the courage to express yourself as if you acknowledged the falsity of our theory of parallels and with it all of our geometry [meaning Euclidean geometry]. But the wasps whose nest you stir up will fly at your head.

In retrospect, Gauss' fear of an attack of Kantian wasps seems exaggerated. In any event, although he erred concerning the synthetic a priori, Kant made so many other important contributions to philosophy, that his reputation and influence have in no way diminished over time.

Geometrical Ideas and Physics

While mathematicians and philosophers, in response to the discovery of non-Euclidean geometry, had to revise their ideas concerning the relationship of their subjects to reality, physicists simply continued to use geometry as a mathematical tool with which to model physical reality. Here are some examples:

The Special Theory of Relativity

Einstein postulated Lorentz invariance for the laws of physics. More precisely he stated that, as viewed by an observer in an inertial frame of reference (i.e., not under the influence of gravity), the laws of physics would be invariant under Lorentz transformations. From the point of view of the *Erlanger Programm*, in effect, his assumption is that the laws of physics are theorems of a certain four-dimensional geometry. (See Chapter 18.) The mathematical groundwork for the special theory of relativity is to be found not only in Klein's work, but also in the work of Poincaré, Lorentz, and Minkowski, all of whom noted special cases of Lorentz invariance.

The General Theory of Relativity

Einstein incorporated gravitation into his theory by supposing that the presence of mass changes the curvature of space-time from point to point. In this view, a particle of matter seems to attract other particles because, by changing the curvature of space-time in its immediate neighborhood, matter bends the paths of particles passing near it.

The mathematical groundwork for the general theory of relativity was the work of Riemann, who first noticed that elliptic geometry was non-Euclidean and who first formulated the idea of a geometry in which curvature could vary from point to point. Also noteworthy is the work of numerous differential geometers (including Ricci and Levi-Città) who studied differential invariants including the integrals for length and area in hyperbolic, elliptic, and absolute geometry given in Chapters 9, 10, 11, and 12.

The Importance of Invariants

Invariants are now recognized as a central theme in physics, from classical mechanics to quantum mechanics. Invariants are an expression of the underlying symmetry of a physical situation and are closely tied to the most famous laws of physics: the conservation laws. The law of conservation of momentum, for example, can be shown to be a consequence of the invariance of physics under the operation of translation in space. In other words, because the laws of physics are believed to be the same at every point in the universe (translation in space), momentum must be conserved. Similarly, the conservation of energy is connected to the invariance of physical law under translation in time; that is, if the laws of physics do not change over time, then energy must be conserved.

The connection between symmetries in physical law and the conservation laws was discovered by Emmy Noether in 1930. This work was a direct outgrowth of nineteenth century work on geometrical invariants. Ever since, the search for symmetry and invariants has been a guiding principle of theoretical physical research.

Non-Euclidean Geometry and the Arts

Non-Euclidean geometry has had a variable influence on artistic and popular culture. For a brief period in the nineteenth century, non-Euclidean geometry was an intellectual fad (as chaos theory is today). The novel *The Brothers Karamazov*, by Dostoyevsky (see page 133), dates from this period. Later, multidimensional geometry and (still later) the geometrical ideas of relativity had a similar vogue.

Conclusion

This chapter has described some of the ways that the discovery of non-Euclidean geometry proved unsettling to cosmologists, philosophers, physicists, and mathematicians. But let us not forget the positive side of that discovery. All the concern about rigor, mathematical meaning, and the concept of space has led to a genuinely improved understanding of all these areas. Furthermore, in non-Euclidean geometry, mathematics acquired an important and powerful tool.

For example consider the proof of Fermat's Last Theorem, which asserts that the equation,

$$x^n + y^n = z^n \quad (*)$$

has no integer solutions, except when $n = 2$ when there are many solutions called **Pythagorean triples**. The proof of this statement has eluded mathematicians ever since it was formulated by Pierre Fermat (~1650). For centuries it was the best known unsolved problem in mathematics. Then, in June 1993, Andrew Wiles announced a proof that, while not yet completely confirmed, is definitely a major contribution to solving the problem.

The modern approach to Fermat's last theorem, encompassing Wiles' work and the work of numerous other mathematicians of this and the previous century, involves the geometries studied in this book in several ways. In the first place, equation (*) naturally defines a curve in the projective plane. (See Chapters 13, 14, and 15.) However, since integer solutions are sought, the relevant planes are discrete projective planes, defined over finite fields. (See Chapter 16.) In the second place, the projective curves associated with (*) are closely connected with *modular forms*, which, in turn, are intimately related to groups of symmetries of the hyperbolic plane. (See Chapter 22.)

Incidentally, the dependence of Wiles' proof on non-Euclidean geometry was denounced by the popular journalist Marilyn vos Savant, who maintained that hyperbolic geometry is inconsistent. Indeed, the consistency of hyperbolic geometry (and Euclidean geometry also) has truly not been proved. However, neither has it been disproved, nor, has the consistency of *any* significant part of mathematics been proved. Furthermore, as we learned in this chapter, hyperbolic geometry is relatively just as consistent as Euclidean geometry. It is remarkable that nearly 200 years after its discovery, non-Euclidean geometry remains a source of controversy.

SUGGESTIONS FOR ESSAYS

Philosophical Issues

1. **Objective Reality**. What does this term mean? Do you believe that objective reality exists? What kind of evidence can be presented to justify belief in objective reality? If objective reality exists, what is it like? Must its nature, as many philosophers have suggested, be essentially abstract and mathematical?

28 THE GEOMETRIC IDEA OF SPACE

What *is* the geometry of physical space? Or, to paraphrase Poincaré: What is the most *convenient* geometry with which to model physical space? This open question is the subject of intense contemporary investigation and speculation.

The Constant Ω

So far as is known at present, from physical observation, it appears that the universe has a geometry that is **homogeneous** (meaning that it appears the same to observers at different points in space), and **isotropic** (meaning that it looks the same in different directions). Based on these two properties, there are only three three-dimensional geometries possible for the universe: elliptic, Euclidean, and hyperbolic. These solid geometries were studied in Chapters 18 and 19. They have properties analogous to the corresponding plane geometries studied in Chapters 6 through 11.

The key quantity that determines the nature of physical space is Ω, a constant defined by

$$\Omega = \frac{8\pi G \rho}{3H^2}$$

where G, H, and ρ are other physical constants. As a consequence of the general theory of relativity, $\Omega > 1$ implies that the universe is elliptic, $\Omega = 1$ implies a Euclidean universe, and $\Omega < 1$ implies a hyperbolic universe.

Of the constants making up Ω, the best known (apart from 3, 8, and π) is G, the **universal constant of gravitation**. The value of G is known to great precision:

$$G = 6.67 \, 10^{-11} \, \frac{\text{m}^3}{\text{kg sec}^2}$$

H, the **Hubble constant**, is determined by the rate of expansion of the universe. It is presently estimated as

$$1.58 \, 10^{-18} \, \frac{1}{\text{sec}} \leq H \leq 3.17 \, 10^{-18} \, \frac{1}{\text{sec}}$$

The third constant, ρ, the **density of the universe**, is estimated to be

$$4.46\,10^{-27}\,\frac{kg}{m^3} \leq \rho \leq 17.87\,10^{-27}\,\frac{kg}{m^3}$$

Apparently, then,

$$.02 \leq \Omega \leq 4$$

It is impossible to determine on this basis what the geometry of the universe is. Note that all three constants are *very small numbers*!

There is a remarkable connection between the geometry of the universe and its long-term fate. It turns out, that if the geometry of the universe is Euclidean or hyperbolic, then the process of expansion (which began with the "big bang") will continue forever (although in the case of a Euclidean universe the expansion will gradually slow down, approaching a velocity of zero). On the other hand, if the geometry of the universe is elliptic, then the universe is doomed eventually to recollapse upon itself. This collapse will be the result of gravitational forces: The larger force of gravitation in an elliptic universe slows, and eventually reverses, its expansion.

Figure 28.1 (based on a table in *The Shape of Space*, by Jeffrey R. Weeks [F6]) summarizes this discussion:

Geometry of the Universe	Fate of the Universe	Gravitational Density	Volume of the Universe
Elliptic	Eventual Recollapse	Dense $\Omega > 1$	Finite
Euclidean	Eternal Expansion	Borderline $\Omega = 1$	Finite or Infinite
Hyperbolic	Eternal Expansion	Sparse $\Omega < 1$	Finite or Infinite

Figure 28.1 Possible universes

Finite or Infinite?

Noteworthy, too, is the question of the finiteness of the universe, which also depends, in part at least, on the geometry of space. If the universe is elliptic, for example, then it turns out that it must be finite in volume. This accords with the example of plane elliptic geometry, which is finite in area. On the other hand, if the universe is Euclidean or hyperbolic, it may be either finite or infinite in volume. This may be rather surprising since both the Euclidean and hyperbolic planes are infinite. However, models of possible Euclidean and hyperbolic universes exist in both infinite and *finite* forms.

An example will demonstrate how space might be simultaneously Euclidean and finite. Take the Euclidean plane, cover it with a square lattice. Then identify all points that occupy congruent positions in the squares. (See Figure 28.2.)

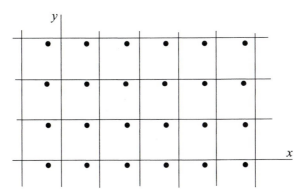

Figure 28.2 A square lattice in the Euclidean plane (the indicated points are all identified)

The resulting space is equivalent to a single square with opposite sides sewn together. The effect of the identification (or sewing) is that a path traveling off one edge of the square immediately reenters it at the opposite edge (as shown in Figure 28.3).

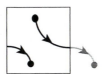

Figure 28.3 A path on a torus

A–Find two different straight lines connecting the two points in Figure 28.3.

The space created by these identifications is called a **torus**. To describe its geometry, we must specify its congruence transformations. For this purpose, we will use Euclidean transformations that preserve the identification of points, in other words, Euclidean transformations T such that if two points p and q (in the plane) are identified, then so are Tp and Tq. Such transformations include all Euclidean transformations that map the lattice of squares in Figure 28.2 onto itself. These include all translations plus a few reflections and rotations but, obviously, not all rotations or reflections.

The geometry we have just defined on the torus is *not* Euclidean overall, that is **globally**. For example, several straight lines connect the two points in Figure 28.3. However, each point on the torus has a (circular) "neighborhood" identical to a circle in the Euclidean plane.

The geometry of this neighborhood *is* Euclidean (not approximately, but exactly). We therefore, say that the torus is **locally Euclidean**.

To be locally Euclidean (or elliptic or hyperbolic) is all we can honestly insist of the geometry of the universe, because our own observations of the universe are limited to a neighborhood of the earth. Note also that the geometry of the torus is homogeneous but *not* isotropic, since not all Euclidean rotations are congruence transformations of the torus. Finally, observe that, as promised, the total area of the torus is finite. Here is an example of a space that is locally Euclidean (in agreement with observation) but finite in extent.

B–Verify that Euclidean translations preserve pairs of identified points and hence are congruence transformations of the torus.

C–What rotations are in the congruence group of the torus?

Quotient Geometries

The construction of the geometry of the torus can be generalized. Other geometries can be substituted for Euclidean geometry and other identification schemes for the square lattice. Here is a formal definition:

Definition *Let* **G** *= (S, G) be a geometry with underlying set S and transformation group G. Let D be a discrete subgroup of G. The* **quotient geometry**, **G/D**, *is defined as follows: The underlying set of* **G/D** *is the set S with points p and q of S identified if there is a transformation T of D such that Tp = q. The transformations of* **G/D** *are the transformations S of G with the property that if p and q are identified, then Sp and Sq are also identified.*

The term "quotient" is used because the foregoing process is the inverse of the process of forming product geometries (defined in Chapter 12). A quotient geometry is also called an **orbifold**. Although not introduced formally until now, several important geometries in this book have already been defined as quotient geometries.

D–Name some.

Now imagine the **three-torus**: the space constructed from a cube by identifying opposite faces. The geometry of the three-torus is the quotient geometry of three-dimensional Euclidean geometry by the discrete group of integer translations in three dimensions. Figure 28.4 shows the lattice of points upon which this geometry is based.

The three-torus has the seemingly paradoxical combination of properties that (a) every point has a (spherical) neighborhood in which the geometry is Euclidean and (b) the total volume of the figure is finite. And our universe is conceivably arranged in just such a way:

every point with a neighborhood (an enormous neighborhood, much larger than our galaxy perhaps) in which the geometry is precisely Euclidean, yet with the whole universe finite in volume.

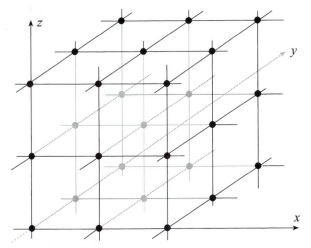

Figure 28.4 A lattice in three dimensions

There are many examples of quotient geometries based on hyperbolic geometry. Figure 28.5 shows the lattice used to define the surface known as a **double torus** (a two-holed torus). Each point of the double torus consists of an infinite number of points identified from congruent positions in a lattice of *octagons* covering the hyperbolic plane. The result is a space wherein each point has a neighborhood in which the geometry is hyperbolic but also wherein the total area is finite.

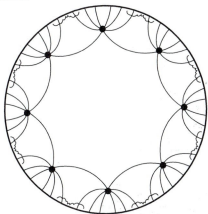

Figure 28.5 A lattice of octagons in the hyperbolic plane

Quotient geometries of three-dimensional hyperbolic geometry represent possible models for the universe in which each point has a neighborhood in which the geometry is hyperbolic, yet the universe as a whole is finite.

Conclusion

Is there such a thing as objective reality?

If so, is there such a thing as physical space?

If so, does physical space obey mathematical laws?

If so, is physical space by nature geometric?

If so, is the geometry of the universe Euclidean?

Some of these questions are likely to remain forever unanswered; others may eventually receive definitive answers. In this unsettled state of things, readers are free to frame answers to suit themselves.

By suggesting that the answer to the last question might be no, the discovery of non-Euclidean geometry in the early nineteenth century opened (or reopened) these questions and thereby caused fundamental change in many areas. Mathematics underwent a crisis of meaning and grew more abstract, less worldly and less applied, a trend that became especially pronounced in the twentieth century. Kant's theory of knowledge was attacked, and new problems concerning the nature of the universe opened up in the philosophy of science and mathematics. As a result of the ferment caused by these developments, major new geometric tools were prepared for use in the radically new physical theories that appeared in the early twentieth century. Above all, the concept of the nature of physical space and its relationship to geometry was forever changed.

Non-Euclidean geometry (particularly hyperbolic geometry) has continued to play an important role in mathematics and physics throughout the twentieth century and is a prominent feature of contemporary mathematics. Topology, complex function theory, analytic number theory, abstract group theory, quantum mechanics, the theory of relativity, and cosmology are some subjects that have made major use either of non-Euclidean geometric ideas or of geometric ideas stemming from the *Erlanger Programm*.

The question of the nature of space is a physical problem: Its resolution will depend on observations and experiments. But the conceptual framework for those observations and experiments has historically been geometry and is likely to remain so for the foreseeable future. For 2,000 years, geometry has remained an impressively flexible tool with which to investigate and consider physical problems, keeping pace and even, on occasion, anticipating experimental developments.

BIBLIOGRAPHY

A. COMPLEX NUMBERS

[A1] Ahlfors, Lars. *Complex Analysis*. McGraw-Hill Book Company, New York, 1966.

[A2] Greenleaf, Frederick P. *Introduction to Complex Variables.* W.B. Saunders Company, Philadelphia 1972.

[A3] Sansone, Giovanni, and Johan Gerretsen. *Lectures on the Theory of Functions of a Complex Variable. I. Holomorphic Functions.* P. Noordhoff, Groningen, 1960.

[A4] Sansone, Giovanni, and Johan Gerretsen. *Lectures on the Theory of Functions of a Complex Variable II. Geometric Theory.* P. Noordhoff, Groningen, 1969.

B. GEOMETRY IN GENERAL

[B1] Abbott, Edwin A. *Flatland*. Dover Publications, New York, 1952.

[B2] Coxeter, H. S. M. *Introduction to Geometry.* John Wiley and Sons, Inc., New York, 1969.

[B3] Eves, Howard. *A Survey of Geometry.* Allyn and Bacon, Boston, 1972.

[B4] Hilbert, D. *Foundations of Geometry.* The Open Court Publishing Company, Chicago, 1902.

[B5] Hilbert, D., and S. Cohn-Vossen. *Geometry and the Imagination.* Chelsea Publishing Company, New York, 1952.

[B6] Klein, Felix. *Geometry.* Dover Publications, Inc., New York, 1939.

[B7] Mader, Adolf. "A Euclidean Model for Euclidean Geometry," *The American Mathematical Monthly,* January 1989.

[B8] Malkevitch, Joseph [ed]. *Geometry's Future.* COMAP, Lexington, MA, 1991.

[B9] Rucker, Rudy. *The Fourth Dimension.* Houghton-Mifflin, 1984.

C. NON-EUCLIDEAN GEOMETRY

[C1] Bachmann, F. *Aufbau der Geometrie aus dem Spiegelungsbegriff.* Springer-Verlag, Berlin, 1959.

[C2] Fenchel, Werner. *Elementary Geometry in Hyperbolic Space.* Walter de Gruyter, Berlin, 1989.

[C3] Martin, George E. *The Foundations of Geometry and the Non-Euclidean Plane.* Intext Educational Publishers, New York, 1975.

[C4] Milnor, John. "Hyperbolic Geometry: The First 150 Years," *Bulletin of the American Mathematical Society* (New Series) 6, pp. 9-24, January 1982.

[C5] Nikulin, V. V., and I. R. Shafarevich. *Geometries and Groups.* Springer-Verlag, Berlin, 1987.

D. PROJECTIVE GEOMETRY

[D1] Dorwart, Harold L. *The Geometry of Incidence.* Prentice Hall, Inc., Englewood Cliffs, NJ, 1966.

[D2] Fishback, W. T. *Projective and Euclidean Geometry.* John Wiley and Sons, Inc., New York, 1962.

E. HISTORY

[E1] Kline, Morris. *Mathematical Thought from Ancient to Modern Times.* Oxford University Press, New York, 1972.

[E2] Richards, Joan L. *Mathematical Visions: The Pursuit of Geometry in Victorian England.* Academic Press, Inc., Boston, 1988.

[E3] Rosenfeld, B. A. *A History of Non-Euclidean Geometry: Evolution of the Concept of a Geometric Space.* Springer-Verlag, New York, 1988.

[E4] Stillwell, John. *Mathematics and Its History*. Springer-Verlag, New York, 1989.

F. THE IDEA OF SPACE

[F1] Gray, Jeremy. *Ideas of Space: Euclidean, Non-Euclidean, and Relativistic*. Clarendon Press, Oxford, 1989.

[F2] Lanczos, Cornelius. *Space through the Ages*. Academic Press, London, 1970.

[F3] Misner, Charles, Kip S. Thorne, and John Archibald Wheeler. *GRAVITATION*. W. S. Freeman and Co., San Fransisco, 1973.

[F4] Poincaré, Henri, *Science and Hypothesis*. Dover Publications, Inc. 1952.

[F5] Taylor, Edwin, and John Archibald Wheeler. *Spacetime Physics*. W.H. Freeman and Co., San Francisco, 1966.

[F6] Weeks, Jeffrey, R. The Shape of Space: How to Visualize Surfaces and Three-Dimensional Manifolds. Marcel Dekker, Inc., New York, 1985.

G. SYMMETRY

[G1] Gardner, Martin. *The Ambidextrous Universe: Mirror Asymmetry and Time-Reversed Worlds*. New York, Scribner, 1979.

[G2] Hargitai, I. [Ed]. *Symmetry: Unifying Human Understanding*. Pergamon Press, New York, 1986.

[G3] Jones, Owen. *The Grammar of Ornament*. Dover Publications, New York, 1987.

[G4] Lockwood, E. H., and R. H. Macmillan. *Geometric Symmetry*. Cambridge University Press, Cambridge, 1978.

[G5] Schattschneider, Doris. *Visions of Symmetry*. W. H. Freeman and Co., New York, 1990.

[G6] Schubnikov, A. V., and V. A. Koptsik. *Symmetry in Science and Art*. Plenum, New York, 1974.

[G7] Weyl, Hermann. *Symmetry*. Princeton University Press, Princeton NJ, 1952.

[G8] Zee, A. *Fearful Symmetry*. Macmillan Publishing Company, New York, 1989.

H. MATROIDS

[H1] Holton, D. and J. Sheehan. *The Peterson Graph*. Cambridge University Press, 1993.

[H2] Lee, Jon and Jennifer Ryan. "Matroid Applications and Algorithms," *ORSA Journal of Computing*, Vol. 4, No. 1 (1982).

[H3] Oxley, James. *Matroid Theory*. Oxford University Press, Oxford, 1992.

[H4] Welsh, D. J. A. *Matroid Theory*. Academic Press, London, 1976.

[H5] Wilson, Robin. *Introduction to Graph Theory* (4th Edition). Longman, Harlow, England, 1996.

I. MISCELLANEOUS APPLICATIONS

[I1] Grünbaum, Branko, and G. C. Shepard. *Tilings and Patterns*. W. H. Freeman and Co., New York, 1987.

[I2] Kappraff, Jay. *Connections: The Geometric Bridge between Art and Science*. McGraw-Hill, New York, 1991.

[I3] Scott, Peter. "The Geometries of 3-Manifolds," *Bulletin of the London Mathematical Society*, 15 (1983), pp. 401–487.

[I4] Stevens, Garry. *The Reasoning Architect: Mathematics and Science in Design*. McGraw-Hill Publishing Co., New York, 1990.

[I5] Thurston, William. "Three Dimensional Manifolds, Kleinian Groups, and Hyperbolic Geometry" *Bulletin of the American Mathematical Society* (New Series), 6 pp. 357–381, May 1982.

[I6] Washburn, D. K., and D. W. Crowe. *Symmetries of Culture: Handbook of Plane Pattern Analysis*. Washington Press, Seattle, 1988.

INDEX

Pages in bold type contain a formal definition of the term cited.